C000320042

ATM NETWORKS

Concepts, Protocols, Applications

THIRD EDITION

ATM NETWORKS

Concepts, Protocols, Applications

Rainer Händel, Manfred N. Huber,
Stefan Schröder

Siemens AG, Munich

Addison-Wesley

Harlow, England • Reading, Massachusetts • Menlo Park, California
New York • Don Mills, Ontario • Amsterdam • Bonn • Sydney • Singapore
Tokyo • Madrid • San Juan • Milan • Mexico City • Seoul • Taipei

Addison Wesley Longman Limited
Edinburgh Gate
Harlow
Essex CM20 2JE
England

and Associated Companies throughout the World.

Cover designed by The Senate
and printed by The Riverside Printing Co. (Reading) Ltd.
Typeset by 32
Printed and bound in Great Britain by Biddles Ltd., Guildford and King's Lynn

First published 1991 with the title Integrated Broadband Networks: An introduction to ATM-based networks
2nd edition published 1994 with the title ATM Networks

3rd edition
First printed 1998

ISBN 0-201-17817-6
(ISBN 0-201-42274-3 2nd edition)

British Library Cataloguing-in-Publication Data
A catalogue record for this book is available from the British Library

Foreword

It has been nearly four years since the second edition of *ATM Networks* was published. Over the past four years, much has changed.

ATM's role in the future of networking has become more clear. Once thought to be a revolutionary protocol that would unify LAN, MAN and WAN technology, ATM has settled down to take its rightful role as the major high-speed WAN technology of the present and future.

Clarifying ATM's role in networks makes explaining it somewhat easier. But my 1994 observation that ATM was hard to explain remains true. ATM combines aspects of TDM and packet switching. The fixed sized cells can at various times be thought of as either large bits or small packets. Thus the author is challenged to explain ATM from two perspectives at once. One of the great strengths of this book is its ability to clearly present ATM from both sides. I believe this third edition justly reflects that strength.

Craig Partridge
September 1997

Preface

ATM is commonplace now, isn't it? Even in ordinary newspapers the term *ATM* can be found now and then, mostly in the context of the 'information society' and its need for powerful and efficient data highways.

So all's well that ends well? Despite the tremendous developments in terms of ATM standardization, product roll-outs, network implementations and, last but not least, public service offerings, the big commercial success of ATM is still missing. Many analysts agree that ATM has a big future, but it is a future that still has to be realized.

The ATM story so far has not been straightforward: originally invented by researchers as a hybrid switching technique combining the merits of channel and packet-switching and (at least in Europe) optimized for real-time applications such as telephony, its success started only after the data transmission experts had had a closer look at the new network transfer (that is, switching and transmission) method and decided that after some modifications and adaptations to the data world's needs it would be a concept promising enough to follow. This approach first led to the well-known 53-byte ATM cell size compromise and then to a lot of other measures to define a standard which reconciles the data community with the telecommunications community.

'Most service providers, equipment vendors, and industry analysts agree that ATM is the technology of the future, unfortunately it is a future that is yet to arrive' (*Broadband Networking News*, 21 January 1997). In 1996, ATM equipment sales represented more than 90% of total ATM market revenue, and this tendency will last for some years to come according to market forecasts. This clearly indicates that ATM is still in its implementation phase, and the time for ATM service revenue will only come after 2000.

ATM networks are true multi-service networks with the potential to offer broadband services; they support network consolidation and ATM has the capability to meet different quality of service requirements. ATM network infrastructure is appealing for both private and public operators even if the market for native ATM services may be limited for a while:

- Corporate networks will be increasingly based on ATM as it is a 'key technology to reduce the overall cost of ownership of an enterprise ... ATM's most important strengths are QoS and ABR services' (Gartner Group presentation at ATM-F TC Meeting, Vancouver, December 1996)

- Public ATM network implementations will mainly be used as broadband backbone networks to support:
  - IP traffic
  - transit telephony traffic
  - leased lines
  - ATM native traffic
  - others (for example, frame relay, SMDS).

A common network infrastructure for the provision of multiple services is a very cost-effective solution, it especially assists in traffic consolidation by allowing optimum usage of transmission facilities.

These ATM backbones will be based on SDH/SONET transport systems, ATM nodes and IP routers. Optimization of connectionless service support and, at the same time, provision of native (connection-oriented) ATM may lead to two distinct subnetworks based on a unique ATM technique.

Currently ATM technology succeeds mainly in backbone networks which require more and more bandwidth; however, genuine ATM to the desktop is still a rare bird. Nevertheless the ATM Forum is pushing ahead towards specification of multimedia applications including voice to the desktop, assuming that in the near or medium term bit rates of some 10 Mbit/s provided via ATM interfaces will be used by personal computers or workstations. In wide area networks, ATM-based backbone infrastructure is being deployed in corporate networks and in public networks where customers can either lease a (virtual) permanent line or demand a switched (virtual) connection – the latter deemed the most cost-effective solution in the future and therefore favored especially by network operators.

ATM and the Internet: the two may well liaise to their mutual benefit. ATM can provide the reliable and efficient network infrastructure that the Internet badly needs to become a commercially usable network which really can satisfy the needs of its users in terms of availability, throughput and other performance parameters, security and privacy, and accountability. ATM network providers may draw from the Internet's globally established user base and know-how concerning 'real-life' applications. Internet applications would then generate huge amounts of traffic which large ATM networks can handle, and which *per se* justify large-scale ATM implementations. So one can expect that a next-generation Internet offering advanced multimedia capabilities and better supporting real-time services may boost ATM deployment.

This third edition of *ATM Networks – Concepts, Protocols, Applications* is a completely revised and enlarged version of our book on ATM networking. The following new items have been included in the third edition:

- Recently defined ATM interfaces (such as UTOPIA, WIRE, FUNI, IMA) and protocols (P-NNI, B-ICI, AINI)
- New AAL type 2
- LAN emulation and Multi-Protocol Over ATM (MPOA)
- Internet supported by ATM networks
- Service classes, quality of service definitions, and traffic management for variable bit rate services (statistical multiplexing)
- Voice and telephony over ATM
- Wireless ATM and mobile ATM
- Residential broadband solutions
- Security in ATM networks
- ATM Application Programming Interface (API).

The host of recently defined new standards and specifications and the current focus of ATM developments forced us to concentrate on key issues, to give sometimes short overviews rather than detailed technical descriptions, and to restructure the book in some parts. We think this will assist the reader in finding his/her path through the ATM jungle. Information on where to find the complete up-to-date version of a standard specification is provided, as was the case with former editions. A brief glossary illustrating the key terms of ATM or ATM-related topics has been added, and we also provide the reader with checking questions relating to the ATM networking techniques described.

R. Händel
M. N. Huber
S. Schröder
October 1997

Contents

1

Brief history of B-ISDN and ATM

1.1 The pre-ISDN situation

At present, most networks are still dedicated to specific purposes such as telephony, TV distribution and circuit-switched or packetized data transfer.

Some applications, such as facsimile, make use of the widespread telephone network. Using pre-existing networks for new applications may lead to characteristic shortcomings, however, as such networks are not usually tailored to the needs of services that were unknown when the networks were implemented. So data transfer over the telephone network is confined by a lack of bandwidth, flexibility and quality of analog voice transmission equipment. Telephone networks were engineered for a constant bandwidth service, and using them for variable bit rate data traffic therefore requires costly adaptation.

Since in general the public telephone network could not effectively support non-voice services to the extent that was required by the customer, other dedicated networks arose, such as public data networks or private data networks connecting, say, a big company's plants or several research institutes. *The* example of a large data network is the Internet.

Private networks often deploy equipment, interfaces and protocols which are unable to offer access to other networks and users. If in such an environment gateways are required to the outside world, their implementation may be tedious and costly.

Table 1.1 illustrates the variety of existing data transmission schemes. Note that this table contains only standardized user classes according to ITU-T Recommendation X.1 (ITU-T, 1996m). ITU-T is a sector of the International Telecommunication Union (ITU), which is in charge of setting network standards for public telecommunication. ITU-T is the new name for the well-known CCITT, which is the acronym for *Comité Consultatif International Télégraphique et Téléphonique.*

Table 1.1 User classes within public data networks.

Class	*Bit rate*	*Characteristics*
1–2	300/50 ... 200 bit/s	Start/Stop mode
3–7	0.6/2.4/4.8/9.6/48 kbit/s	Synchronous operation mode, using X.21 (ITU-T, 1992d) or X.21 bis interfaces (ITU-T, 1988h)
8–12	2.4/4.8/9.6/48/1.2 kbit/s	Synchronous operation mode, using X.25 (ITU-T, 1996n) or X.32 (ITU-T, 1996o) interfaces
14–18	0.6/1.2/2.4/4.8/9.6 kbit/s	Start/Stop mode
19	64 kbit/s	Synchronous operation mode, using X.21 or X.21 bis interfaces
20–25	50 ... 300/75 ... 1200/ 1200/2400/4800/9600 kbit/s	Start/Stop mode, X.28 (ITU-T, 1993x) interface
26	14.4 kbit/s	Synchronous operation mode, using X.25 or X.32 interfaces, or Start/Stop mode, X.28 interface
29	0.3 ... 14.4 kbit/s	Fax terminals, X.38 (ITU-T, 1996p) interface
30–33, 35, 37, 45, 53, 59	64/128/192/256/384/512/ 1024/1536/1920 kbit/s	Synchronous operation mode, using X.21 or X.21 bis or X.25 or X.32 interfaces

1.2 The idea of the integrated services digital network

In 1984, the Plenary Assembly of the CCITT adopted the I series recommendations dealing with the integrated services digital network (ISDN) matters. The CCITT stated that 'an ISDN is a network ... that provides end-to-end digital connectivity to support a wide range of services, including voice and non-voice services, to which users have access by a limited set of standard multipurpose user–network interfaces' (ITU-T, 1984). Such an ISDN standard interface was defined and called **basic access**, comprising two 64 kbit/s B channels and a 16 kbit/s signaling D channel. Another type of interface, the **primary rate access**, with a gross bit rate of about 1.5 Mbit/s or 2 Mbit/s, offers the flexibility to allocate high-speed H channels or mixtures of B and H channels and a 64 kbit/s signaling channel (see Table 1.2).

The ISDN concept laid down in the 1984 recommendations has since been further elaborated; its evolution is documented in the 1988 CCITT Blue Books, and later CCITT/ITU-T recommendations.

This original ISDN is based on the digitized telephone network which is characterized by the 64 kbit/s channel. The channel bit rate of 64 kbit/s is derived from 3.4 kHz voice transmission requirements (8-bit sampling with a frequency of 8 kHz).

Table 1.2 ISDN channels and interface structures.

	ISDN channels	
Channel	*Bit rate*	*Interface*
B	64 kbit/s	Basic access
H0	384 kbit/s	Primary rate access
H11	1536 kbit/s	Primary rate access
H12	1920 kbit/s	Primary rate access
D16	16 kbit/s	Basic access
D64	64 kbit/s	Primary rate access

	Interface structures	
Interface	*Gross bit rate*	*Structure*
Basic access	192 kbit/s	2B + D16
Primary rate access	1544 kbit/s	23B + D64 3H0 + D64 H11 etc.
Primary rate access	2048 kbit/s	30B + D64 5H0 + D64 H12 + D64 etc.

The 64 kbit/s ISDN is basically a circuit-switched network but it can offer access to packet-switched services (Bocker, 1992).

ISDNs are being implemented in this decade. Their benefits for the user and network provider include (Bocker, 1992):

- common user–network interface for access to a variety of services
- enhanced (out-of-band) signaling capabilities
- service integration
- provision of new and improved services.

1.3 B-ISDN and ATM

The highest bit rate a 64 kbit/s based ISDN can offer to the user is about 1.5 Mbit/s or 2 Mbit/s, that is, the H1 channel bit rates (see Table 1.2). Connection of local area networks (LANs), however, or transmission of moving images with good resolution, may, in many cases, require considerably higher bit rates (see Chapter 2). Consequently, the conception and realization of a broadband ISDN (B-ISDN) was desirable.

ITU-T Recommendation I.113 (ITU-T, 1993e) ('Vocabulary of Terms for Broadband Aspects of ISDN') defines **broadband** as:

'... a service or system requiring transmission channels capable of supporting rates greater than the primary rate.'

B-ISDN thus includes 64 kbit/s ISDN capabilities but in addition opens the door to applications utilizing bit rates above 1.5 Mbit/s or 2 Mbit/s. The bit rate available to a broadband user is typically from several Mbit/s up to hundreds of Mbit/s (see Section 5.2.2).

This definition of broadband does not indicate anything about its technical concept. Whereas this definition was settled from the beginning, the final technical concept for B-ISDN, as described in the following chapters, only emerged after long and controversial discussions within the standardization bodies, reflecting the differing backgrounds and intentions of the participants.

The first concrete idea of B-ISDN was simply to:

- add new high-speed channels to the existing channel spectrum
- define new broadband user–network interfaces
- rely on existing 64 kbit/s ISDN protocols and only to modify or enhance them when absolutely unavoidable.

So in the dawn of B-ISDN, channel bit rates of 32–34 Mbit/s and around 45 Mbit/s, 70 Mbit/s and 135–139 Mbit/s were foreseen. The corresponding channels were denominated H2, H3, H4.

These bit rates (and also the interface bit rate of about 140 Mbit/s) were oriented towards the bit rates of the plesiochronous hierarchy (ITU-T Recommendation G.702 (ITU-T, 1988a)) so that the H channels could be transmitted within the signals of the corresponding hierarchical level. (The plesiochronous hierarchy is defined by a set of bit rates and multiplexing schemes for the multiplexing of several, not necessarily synchronous, 64 kbit/s ISDN channels into higher bit rate signals.)

These broadband channels would have provided a rigid bit rate scheme to be applied to all future broadband services which then were not yet fully described. This led to some concern about the suitability of the H channel concept.

Moreover, a decision regarding which and how many H and B channels should be incorporated into the broadband interface could not be achieved at all. There were proposals such as:

$$\text{H4} + 4\text{H1} + n \times \text{B} + \text{signaling channel} \qquad (\text{e.g. } n = 30)$$

to which a lot of critical questions were immediately raised:

- Is this the only interface option or will other channel combinations be allowed?
- Can the H4 channel be subdivided into smaller pieces (for example, into 4 × H2 or into B/H1/H2 combinations) and can the

four H1 channels be combined to yield a 6 Mbit/s or 8 Mbit/s entity?

- If several channel structures can be used as options at the interface, do they have to be fixed at subscription time or can they be dynamically changed?

These issues were never completely resolved, as other ideas entered the discussion.

As another intermediate step, so-called hybrid interface structures were proposed comprising channels for both circuit-oriented (or stream) traffic and capacity to be used for burst-type traffic. Obviously, such structures would have been more flexible than merely channel-oriented ones.

Again, there were never-ending talks about where to draw the border between the stream and burst parts of the interface, whether to allow it to change dynamically, and the steps (in terms of bit rate) by which changes could be made.

This deadlock was finally overcome by adopting an interface model based on a complete breakdown of its payload capacity into small pieces called *cells*, each of which can serve any purpose. Each cell may be employed to carry information relating to any type of connection.

In the following chapters, this principle, embodied by the **asynchronous transfer mode** (ATM), and its impact on B-ISDN will be considered in more detail.

ATM is the technique used in B-ISDNs. B-ISDN is a typical ITU term denoting that ATM-based networks should be embedded into the ISDN environment. Therefore B-ISDNs should be characterized by their compliance to the ITU networking principles established for ISDNs.

However, ATM network implementations – especially private ones – do not always strictly follow ITU rules mainly due to cost considerations, lack of standards-conforming equipment, or because the ATM Forum, a major player on the ATM scene, has put forward alternative solutions.

Review questions

1. What is an ISDN?
2. What does *B-ISDN* denote and what networking technique does it employ?

2

ATM-based services and applications

2.1 B-ISDN services according to ITU

B-ISDN development can be justified and will be successful only if it meets the needs of potential customers. Therefore, a brief outline of foreseeable broadband applications will be given before entering into a discussion of network aspects.

In principle, B-ISDN should be suitable for both business and residential customers. Thus as well as data communication, the provision of TV program distribution and other entertainment facilities has to be considered.

B-ISDN will support services with both constant and variable bit rates, data, voice (sound), still and moving picture transmission and, of particular note, multimedia applications which may combine, say, data, voice and picture service components.

Some examples may be used to illustrate the capabilities of B-ISDN. In the business area, videoconferencing is already a well-established, but still not commonly used, method which facilitates the rapid exchange of information between people. As traveling can be avoided, videoconferencing helps to save time and costs. B-ISDN may considerably improve the current situation and allows for high picture quality (at least today's TV quality or even better), which is crucial for its acceptance, and can provide connections between all potential users via standard interfaces.

Another salient feature of B-ISDN is the (cost-effective) provision of high-speed data links with flexible bit rate allocation for the interconnection of customer networks.

The residential B-ISDN user may appreciate the combined offer of text, graphics, sound, still images and films giving information about such things as holiday resorts, shops or cultural events, as well as interactive video services and video on demand.

Tables 2.1, 2.2, 2.3, 2.4 and 2.5 (based on ITU-T (1993g)) give an overview of possible broadband services and applications as presented by ITU-T.

According to ITU-T Recommendation I.211 (ITU-T, 1993g), services are classified into **interactive** and **distribution** services. Interactive services comprise conversational, messaging and retrieval services; distribution services can be split into services with or without user-individual presentation control.

Table 2.1 Messaging services.

Type of information	*Examples of broadband services*	*Applications*
Moving pictures (video) and sound	Video mail service	Electronic mailbox service for the transfer of moving pictures and accompanying sound
Document	Document mail service	Electronic mailbox service for mixed documents[1]

[1] Mixed document means that a document may contain text, graphics, still and moving picture information as well as voice annotations.

Table 2.2 Retrieval services.

Type of information	*Examples of broadband services*	*Applications*
Text, data, graphics, sound, still images, moving pictures	Broadband videotex	• Videotex including moving pictures • Remote education and training • Tele-software • Tele-shopping • Tele-advertising • News retrieval
	Video retrieval service	• Entertainment purposes • Remote education and training
	High-resolution image retrieval service	• Entertainment purposes • Remote education and training • Professional image communications • Medical image communications
	Document retrieval service	Mixed documents[1] retrieval from information centers, archives, etc.
	Data retrieval service	Tele-software

[1] Mixed document means that a document may contain text, graphics, still and moving picture information as well as voice annotations.

Table 2.3 Conversational services.

Type of information	*Examples of broadband services*	*Applications*
Moving pictures and sound	Broadband videotelephony	Communication for the transfer of voice (sound), moving pictures, and video-scanned still images and documents between two locations (person-to-person) • Tele-education • Tele-shopping • Tele-advertising
	Broadband videoconference	Multipoint communication for the transfer of voice (sound), moving pictures, and video-scanned still images and documents between two or more locations (person-to-group, group-to-group) • Tele-education • Business conference • Tele-advertising
	Video surveillance	• Building security • Traffic monitoring
	Video/audio information transmission service	• TV signal transfer • Video/audio dialog • Contribution of information
Sound	Multiple sound-program signals	• Multilingual commentary channels • Multiple program transfers
Data	High-speed unrestricted digital information transmission service	• High-speed data transfer – LAN interconnection – MAN interconnection – Computer–computer interconnection • Transfer of video information • Transfer of other information types • Still image transfer • Multi-site interactive computer aided design • Multi-site interactive computer aided manufacturing
	High-volume file transfer service	• Data file transfer
	High-speed tele-action	• Real-time control • Telemetry • Alarms
Document	High-speed telefax	User-to-user transfer of text, images, drawings, etc.
	High-resolution image communication service	• Professional images • Medical images • Remote games
	Document communication service	User-to-user transfer of mixed documents[1]

[1] Mixed document means that a document may contain text, graphics, still and moving picture information as well as voice annotations.

Table 2.4 Distribution services without user-individual presentation control.

Type of information	*Examples of broadband services*	*Applications*
Data	High-speed unrestricted digital information distribution service	• Distribution of unrestricted data
Text, graphics, still images	Document distribution service	• Electronic newspaper • Electronic publishing
Moving pictures and sound	Video information distribution service	• Distribution of video/audio signals
Video	Existing quality TV distribution service (NTSC, PAL, SECAM)	TV program distribution
	Extended quality TV distribution service • Enhanced definition TV distribution service • High-quality TV	TV program distribution
	High-definition TV distribution service	TV program distribution
	PayTV (pay-per-view, pay-per-channel)	TV program distribution

Table 2.5 Distribution services with user-individual presentation control.

Type of information	*Examples of broadband services*	*Applications*
Text, graphics, sound, still images	Full channel broadcast videography	• Remote education and training • Tele-advertising • News retrieval • Tele-software

B-ISDN messaging services include mailbox services for the transfer of sound, pictures and/or documents (Table 2.1). Retrieval services (Table 2.2) can be used, for example, to obtain video films at any time or to access a remote software library. Conversational services allow the mutual exchange of data, whole documents, pictures and sound. Examples are given in Table 2.3. Finally, examples of distribution services (Tables 2.4 and 2.5) are electronic publishing and TV program distribution with existing and, in the future, enhanced picture quality – for example, high-definition TV (HDTV).

To be able to derive the network requirements to be met by B-ISDN from potential broadband services, technical characteristics for major B-ISDN applications must be investigated.

Table 2.6 Characteristics of broadband services.

Service	*Peak bit rate (Mbit/s)*	*Burstiness*[1]
Data transmission	1–100	1–50
Document transfer/retrieval	1–50	1–20
Videoconference	1–10	1–5
Broadband videotex/video retrieval	1–10	1–20
TV distribution	1–5	1
HDTV distribution	30 or more	1
TV contribution	~ 100	1
3D moving images	some 100s	1

[1] Burstiness = peak bit rate/average bit rate

Table 2.6 illustrates the following remarkable properties of broadband applications:

- Not all services require very high bit rates, although some do, especially moving picture services with high resolution. The resulting bit rates are in some cases above those employed for conventional ISDN services.
- Several types of communication are highly bursty in nature. If this feature were adequately reflected in network design, considerable economizing on network resources might be achieved (statistical multiplexing gain). In the case of TV and HDTV distribution (and contribution between studios), the statistical multiplexing gain is hard to realize because of the nature of the source signals, so in Table 2.6 the burstiness is set to 1.

The variety of possible B-ISDN services and applications shown in the table obviously requires a network with universal transfer capabilities to:

- cater for services which may employ quite different bit rates
- support burst-type traffic
- take into account both delay and loss-sensitive applications.

The network concept which is assumed to meet all these requirements will be presented in the following chapters.

2.2 Possible implementation scenario for B-ISDN services

Although B-ISDN is intended to support all types of services, at least in the long run, the customer's main interest in the new B-ISDN will be concentrated on services that cannot be offered (or only at greater expense) by existing networks.

B-ISDN will therefore at least initially coexist with other networks like public data networks, analog telephone networks, 64 kbit/s ISDN, TV distribution networks and so on. The services offered by B-ISDN might be restricted to 'typical' broadband services, such as interactive services with bit rates above 1.5 Mbit/s or 2 Mbit/s. Access to other services, like 64 kbit/s ISDN bearer services, would still be provided by existing interfaces like the basic access (ITU-T, 1995f).

Interworking facilities between B-ISDN and the other networks have to be provided. For example, a customer using a dual function videotelephony/telephone set should be able to communicate with 64 kbit/s ISDN telephone users. This simple B-ISDN overlay model is shown in Figure 2.1.

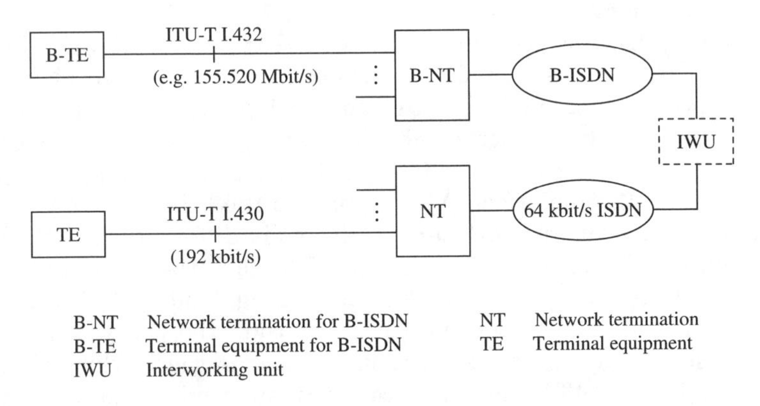

Figure 2.1 Pure B-ISDN overlay network.

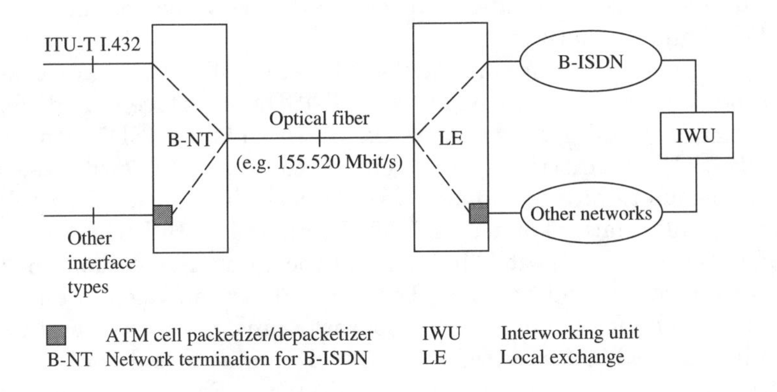

Figure 2.2 Integrated access configuration.

Another step might be to introduce integrated access to different networks via a single, optical fiber-based B-ISDN interface on the network side of B-NT1. This configuration is depicted in Figure 2.2.

On the customer side, this configuration requires non-ATM interfaces to be adapted to the ATM format ('ATM-izing'). This may comprise analog/digital conversion in the case of analog telephone interfaces or analog TV interfaces. These functions could, for example, be implemented in the B-NT as shown in the figure. Note that on the right-hand side of the B-NT, all user data is conveyed by ATM systems.

The integrated access configuration of Figure 2.2 reduces operation and maintenance (OAM) expenditure as only a single customer access has to be maintained. On the other hand, it requires conversion and adaptation equipment at the customer termination and the exchange termination as long as non-ATM interfaces, terminals and networks exist. Interworking functions are still necessary to connect B-ISDN with other networks.

Some specific problems arise with configurations like those of Figure 2.2. As signals from different networks may be multiplexed onto a single access line, OAM activities on this line must not interfere with connections that have been established by any network. Therefore, an entity is required to coordinate OAM activities on the customer's access.

The ATM-izing of non-ATM signals in the B-NT and in the corresponding unit in the network can be effected in different ways. Consider, for example, the basic access signal which comprises two 64 kbit/s B channels, a 16 kbit/s D channel for signaling, and transmission and OAM overhead. Either the entire signal could be ATM-ized or each channel separately. In the latter case mapping of the 16 kbit/s signaling D channel onto ATM-signaling connections would also have to be defined. Information contained in the overhead of the basic access signal need not be transmitted completely to the network. For example, the framing bits of ITU-T Recommendation I.430 (ITU-T, 1995f) are no longer required. However, OAM information like the activation/deactivation indication may still have to be changed between the B-NT and the network.

The choice of mapping of the 64 kbit/s ISDN basic access signal onto the cell stream of the ATM-based B-ISDN interface may depend on the signal processing in the network. When 64 kbit/s ISDN connections and B-ISDN connections are routed to separate switches, it is advantageous to have a compact basic access signal (including relevant signaling information) on a single ATM connection. But when a common ATM-based B-ISDN switch handles all incoming connections, 64 kbit/s ISDN connections and broadband connections are only separated behind the local exchange (that is, onto separate trunk networks), and the use of individual ATM connections for the B channels and the D channel information seems to be more suitable. (For more details on interworking see Chapter 10.)

2.3 Existing ATM network services

The full set of possible B-ISDN services listed in Section 2.1 cannot be offered from the very beginning of ATM networking for several reasons. First, most of the envisaged services are not completely defined. Second, an implementable subset of network services that will attract customers has to be identified. These services must include existing applications (and similar services evolving in parallel). However, new ATM-specific features should be visible for marketing reasons.

The basic service ('native ATM') of an ATM network is the transport and routing (that is, multiplexing, transmission and switching) of ATM cells. This ATM (bearer) service is also named *cell relaying*. The network does not need to know anything about the end-to-end application which is running on an ATM connection. Users can employ this ATM service to exchange data, voice, pictures, or a combination thereof, across the network. This will stimulate users to experiment with new applications, such as multimedia, via ATM. The practical experience gained from such trials will assist in eventually defining appropriate service characteristics.

ATM implementations serve as backbone networks mainly for data communications. Therefore data services such as X.25, frame relay, **switched multi-megabit data service** (SMDS) (Bellcore, 1989), LAN interconnection, and access to the Internet have to be supported. (This support, in terms of adaptation/interworking equipment and interworking protocols, will be specified in more detail in Chapter 10.)

Frame relay is an enhanced packet-type service. Higher throughputs and less delay are achieved by reducing error control and forgoing end-to-end flow control (in contrast to X.25 (ITU-T, 1996n). Frame relay is a connection-oriented service offering bit rates from some kbit/s up to 2 Mbit/s or possibly higher.

SMDS was introduced by Bellcore as a high-speed, connectionless packet-type data service at bit rates up to 45 Mbit/s and, subsequently, 155 Mbit/s. SMDS uses the ITU-T defined address scheme of Recommendation E.164 (ITU-T, 1991a) to support global addressing. SMDS was first run on top of a distributed queue dual bus (DQDB) metropolitan area network, but can also use an ATM network. (The European version of SMDS is called **connectionless broadband data service** (CBDS), see Section 10.3.4.)

Another service to be offered initially to customers is the constant bit rate leased line service operating, for example, at 1.5/2 Mbit/s or 34/45 Mbit/s (*circuit emulation*). As in the case of frame relay and SMDS, such a leased line service can, of course, also be provided by conventional networks. The merit of an ATM backbone network for the network operator is that a common, unique network infrastructure can be deployed flexibly to support all the existing and future services.

Circuit emulation is also used for narrowband (for example, voice) trunking via ATM networks (see Sections 10.2 and 14.1).

The first ATM implementations only offered *permanent virtual connections* (PVCs) established/released by network management, whereas now *switched virtual connections* established/released via signaling procedures are available.

Review questions

1. Break B-ISDN services down into ITU-defined categories.
2. B-ISDN applications may differ in terms of what?
3. What does *ATM-izing* mean? Where is it performed?
4. What is circuit emulation via ATM?

3

Principles and building blocks of B-ISDN

3.1 B-ISDN principles

The motivation behind incorporating broadband features into ISDN is neatly documented in ITU-T Recommendation I.121 ('Broadband Aspects of ISDN') (ITU-T, 1991c):

The B-ISDN recommendations were written taking into account the following:

- The emerging demand for broadband services (candidate services are listed in Chapter 2).
- The availability of high-speed transmission, switching and signal processing technologies (bit rates of hundreds of Mbit/s are being offered).
- The improved data and image processing capabilities available to the user.
- The advances in software application processing in the computer and telecommunication industries.
- The need to integrate interactive and distribution services and circuit and packet transfer modes into a universal broadband network. In comparison to several dedicated networks, service and network integration has major advantages in economic planning, development, implementation, operation and maintenance. While dedicated networks require several distinct and costly customer access lines, the B-ISDN access can be based on a single optical fiber for each customer. The large-scale production of highly integrated system components of a unique B-ISDN will lead to cost-effective solutions.
- The need to provide flexibility in satisfying the requirements of both user and operator (in terms of bit rate, quality of service, and so on).

ISDN is intended to support 'a wide range of audio, video and data applications in the same network' (ITU-T, 1991c). B-ISDN thus follows the same principles at 64 kbit/s based ISDN (cf. ITU-T Recommendation I.120 (ITU-T, 1993f)) and is a natural extension of the latter (ITU-T, 1991c):

> 'A key element of service integration is the provision of a wide range of services to a broad variety of users utilizing a limited set of connection types and multipurpose user–network interfaces.'

Whereas most pre-ISDN telecommunication networks have been specialized networks (for telephony or data) with rather limited bandwidths or throughput and processing capabilities, the future B-ISDN is conceived as a universal (standardized) network supporting different kinds of applications and customer categories. ITU-T Recommendation I.121 (ITU-T, 1991c) presents an overview of B-ISDN capabilities:

> 'B-ISDN supports switched, semi-permanent and permanent, point-to-point and point-to-multipoint connections and provides on-demand, reserved and permanent services. Connections in B-ISDN support both circuit mode and packet mode services of a mono- and/or multimedia type and of a connectionless or connection-oriented nature and in a bidirectional or unidirectional configuration.
>
> A B-ISDN will contain intelligent capabilities for the purpose of providing advanced service characteristics, supporting powerful operation and maintenance (OAM) tools, network control and management.'

We believe the reader of this list of intended B-ISDN capabilities must be deeply impressed; B-ISDN is designed to become *the* universal future network!

B-ISDN implementations will, according to the ITU-T, be based on the asynchronous transfer mode. This transfer mode will be briefly introduced in the next section (and its technical details will be discussed later, see Chapters 4 and 5).

3.2 Asynchronous transfer mode

The asynchronous transfer mode is considered to be the ground on which B-ISDN is to be built (ITU-T, 1991c):

> 'Asynchronous transfer mode (ATM) is the transfer mode for implementing B-ISDN ...'

The term *transfer* comprises both transmission and switching aspects, so a *transfer mode* is a specific way of transmitting and switching information in a network.

In ATM, all information to be transferred is packed into fixed-size slots called **cells**. These cells have a 48-octet information field and a 5-octet header. Whereas the information field is available for the user, the header field carries information that pertains to the ATM layer functionality itself, mainly the identification of cells by means of a label (see Figure 3.1).

A detailed description of the ATM layer functions and ATM header structure and coding will be given in Section 5.6. The protocol reference model for ATM-based networks will be addressed in Section 5.1; the boundaries between the ATM layer and other layers are also discussed in Section 5.1.

ATM uses a label field inside each cell header to define and recognize individual communications. In this respect, ATM resembles conventional packet transfer modes. Like packet switching techniques, ATM can provide a communication with a bit rate that is individually tailored to the actual need, including time-variant bit rates.

The term **asynchronous** in the name of the new transfer mode refers to the fact that, in the context of multiplexed transmission, cells allocated to the same connection may exhibit an irregular recurrence pattern as they are filled according to the actual demand. This is shown in Figure 3.2(b).

In the **synchronous transfer mode** (STM) (see Figure 3.2(a)), a data unit associated with a given channel is identified by its position in the transmission frame, while in ATM (Figure 3.2(b)) a data unit or cell associated with a specific virtual channel may occur at essentially any position. The flexibility of a bit rate allocation to a connection in STM is restricted as it uses predefined channel bit rates (for example, B, H2, see Chapter 1) and the conventional transmission frames have a rigid structure. These normally will not permit individual structuring of the payload or will only permit a quite limited selection of channel mixes at the corresponding interface at subscription time. Otherwise the network provider would have to manage a host of different interface types, a situation that the designer would try to avoid for obvious reasons (for example, STM switching of varying B and H channel mixes per interface requires switching equipment that can simultaneously handle all sorts of channels potentially used by customers at any time).

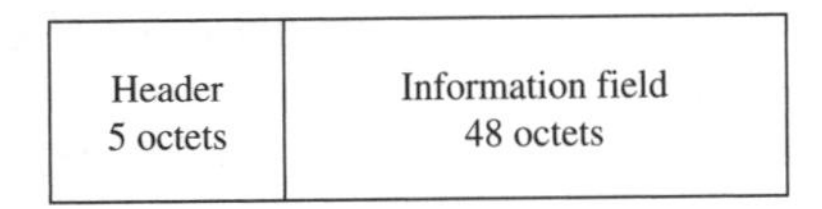

Figure 3.1 ATM cell structure.

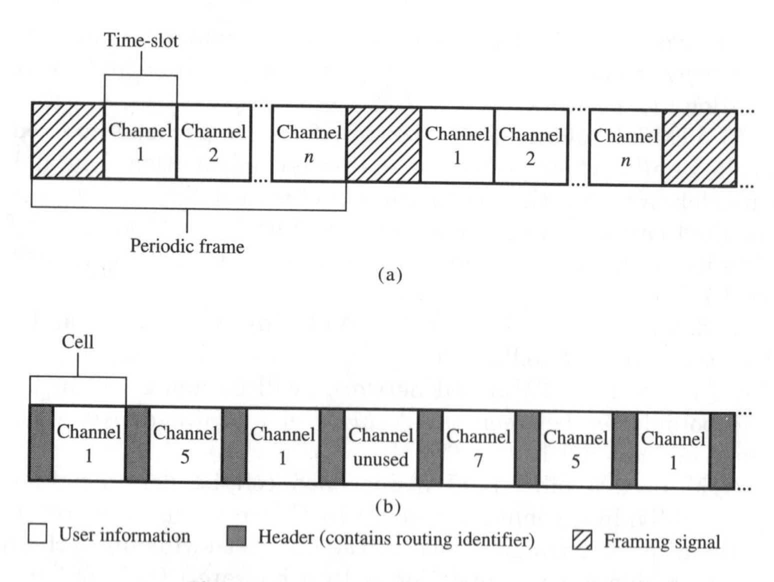

Figure 3.2 STM and ATM principles.
(a) Synchronous transfer mode; (b) asynchronous transfer mode.

In ATM-based networks the multiplexing and switching of cells is independent of the actual application. Thus the same piece of equipment can, in principle, handle a low bit rate connection as well as a high bit rate connection, be it of stream or burst nature. Dynamic bandwidth allocation on demand with a fine degree of granularity is provided. Consequently the definition of high-speed channel bit rates is now, in contrast to the situation in an STM environment, a second-rank task.

The flexibility of the ATM-based B-ISDN network access resulting from the cell transport concept strongly supports the idea of a unique interface which can be employed by a variety of customers with quite different service needs. However, the ATM concept requires many new problems to be solved. For example, the impacts of possible cell loss, cell transfer delay and cell delay variation on service quality need to be determined (see Section 4.4). Other ATM-inherent difficulties – voice echo and tariffs – will be addressed in Sections 14.1 and 14.5.

To sum up, whereas today's networks are characterized by the coexistence of circuit switching and packet switching, B-ISDN will rely on a single new method called ATM which combines advantageous features of both circuit- and packet-oriented techniques. The former requires only low overhead and processing, and, once a circuit-switched connection is established, the transfer delay of the information being carried is low and constant. The latter is much more flexible in terms of the bit rate assigned to individual (virtual) connections. ATM is a circuit-oriented, hardware-controlled, low-overhead concept of virtual channels which (in contrast to X.25 access (ITU-T, 1996n)) have no flow control or error recovery. The implementation of these virtual channels is done by fixed-size (relatively short) cells and provides the basis for both switching and multiplexed transmission. The use of short cells in ATM and the high transfer rates involved (for example, 150 Mbit/s, see Section 3.3) result in transfer delays and delay variations which are sufficiently small to enable it to be applied to a wide range of services, including real-time services such as voice and video. The ability of ATM to multiplex and switch on the cell level supports flexible bit rate allocation, as is known from packet networks.

The overall protocol architecture of ATM networks comprises:

- A single link-by-link cell transfer capability common to all services.
- Service-specific adaptation functions for mapping higher layer information into ATM cells on an end-to-end basis. Examples are packetization/depacketization of continuous bit streams into/from ATM cells, and segmentation/reassembly of larger blocks of user information into/from ATM cells (core-and-edge concept).

Another important feature of ATM networks is the possibility of grouping several virtual channels into one so-called virtual path. The impact of this technique on the B-ISDN structure will be addressed in the following chapter.

3.3 Optical transmission

The development of powerful and economic optical transmission equipment was a major driving force behind B-ISDN. Optical transmission is characterized by:

- low fiber attenuation (allowing for large repeater distances)
- high transmission bandwidths (up to several hundred Mbit/s)
- small diameter (low weight and volume)
- high mechanical flexibility of the fiber
- immunity against electromagnetic fields
- low transmission error probability
- no crosstalk between fibers
- tapping much more difficult.

The high bandwidth of optical transmission systems – currently up to several Gbit/s can be transported on one optical link – has led to implementations in public networks to support existing services like telephony. Fiber-based local area networks are also widely in use nowadays, providing bit rates of the order of some hundred Mbit/s to the users. Thus for B-ISDN the use of optical fiber-based transmission systems is straightforward from a technical viewpoint, at least in the trunk network and in the local access part of the network where considerable distances have to be bridged. (Technical details on optical transmission to be deployed in B-ISDNs will be discussed in Chapter 13.)

In B-ISDN a bit rate of about 150 Mbit/s can typically be offered to the user across the broadband user–network interface (see Section 5.2). Although much higher bit rates could comfortably be transmitted on optical fiber links, the cost of the electronics involved in the transmission equipment (sender/receiver in network terminations, terminals, and so on) together with considerations on expected service needs – that is, the bit rates simultaneously required at the interface – led to the conclusion that a B-ISDN 'basic' interface at about 150 Mbit/s would be sufficient in many cases.

In addition, a second interface type operating at about 600 Mbit/s, at least in the direction from the network to the user, is also foreseen (see Section 5.2), and ATM cell transport on even higher bit rate systems has been defined. Handling of such ATM signals is technically feasible; their implementation may, however, be an economic challenge.

Driven by such economic considerations, additional (not necessarily optical) lower bit rate ATM interfaces – in the range of 1.5 or 2 Mbit/s up to 150 Mbit/s – are also quite important for ATM access (see Section 5.4).

The deployment of highly reliable (mainly optical) transmission systems with low bit error probabilities benefits a simplified network concept with, for example, potentially reducible data link layer functionality.

Review questions

1. What does the ATM cell structure look like?
2. What does *asynchronous* mean in the term *ATM*?
3. Why is ATM a *core-edge network* concept?
4. What is the range of bit rates currently provided by ATM interfaces?

4

B-ISDN network concept

4.1 General architecture of B-ISDN

The architectural model of B-ISDN is described in ITU-T Recommendation I.327 (ITU-T, 1993i). According to this recommendation, the information transfer and signaling capabilities of B-ISDN comprise:

- broadband capabilities
- 64 kbit/s ISDN capabilities
- user-to-network signaling
- inter-exchange signaling
- user-to-user signaling.

This is depicted in Figure 4.1.

Broadband information transfer is provided by ATM. The ATM data unit is the cell, a fixed-size block of 53 octets (see Section 5.6). The 5-octet cell header carries the necessary information to identify cells belonging to the same virtual channel. Cells are assigned on demand, depending on source activity and the available resources.

ATM guarantees (under normal fault-free conditions) **cell sequence integrity**. This means that nowhere in the network can a cell belonging to a specific virtual channel connection overtake another cell of the same virtual channel connection that has been sent earlier.

ATM is a **connection-oriented technique**. A connection within the ATM layer consists of one or more links, each of which is assigned an identifier. These identifiers remain unchanged for the duration of the connection.

Signaling information for a given connection is conveyed using a separate identifier (out-of-band signaling).

Although ATM is a connection-oriented technique, it offers a flexible transfer capability common to all services, including connectionless data services. The proposed provision of connectionless data services via the ATM-based B-ISDN will be discussed in Section 10.3.

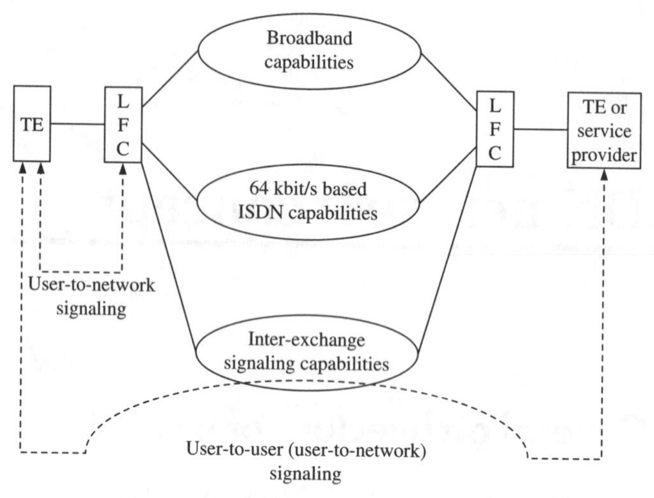

Figure 4.1 Information transfer and signaling capabilities.

4.2 Networking techniques

4.2.1 Network layering

ITU-T Recommendation I.311 (ITU-T, 1996f) presents the layered structure of the B-ISDN depicted in Figure 4.2 (see also Section 5.1).

In this section, we address only the ATM transport network for which the functions are split into two parts, namely physical layer transport functions and ATM layer transport functions. Both the physical layer and the ATM layer are hierarchically structured. The physical layer consists of:

- transmission path level
- digital section level
- regenerator section level.

These are defined as follows:

Transmission path: The transmission path extends between network elements that assemble and disassemble the payload of a transmission system (the payload will be used to carry user information; together with the necessary transmission overhead it forms the complete signal).

Digital section: The digital section extends between network elements which assemble and disassemble continuous bit or byte streams.

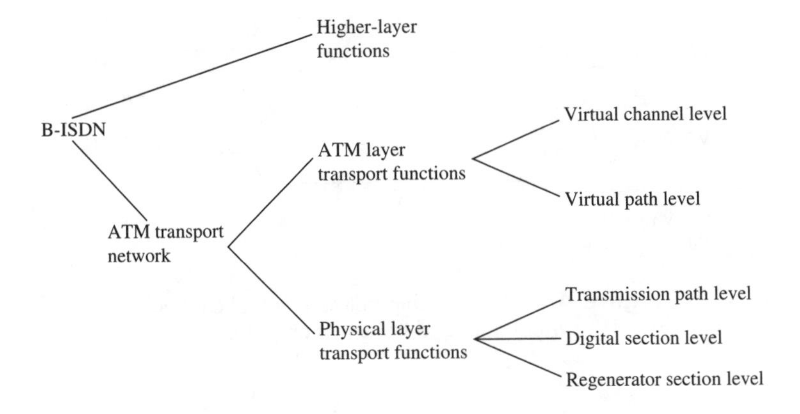

Figure 4.2 B-ISDN layered structure.

> **Regenerator section:** The regenerator section is a portion of a digital section extending between two adjacent regenerators.

The ATM layer has two hierarchical levels, namely:

- virtual channel level
- virtual path level.

Both are defined in ITU-T Recommendation I.113 ('Vocabulary of Terms for Broadband Aspects of ISDN') (ITU-T, 1993e):

> **Virtual channel** (VC): 'A concept used to describe unidirectional transport of ATM cells associated by a common unique identifier value.' This identifier is called the **virtual channel identifier** (VCI) and is part of the cell header (see Section 5.6.2).
>
> **Virtual path** (VP): 'A concept used to describe unidirectional transport of cells belonging to virtual channels that are associated by a common identifier value.' This identifier is called the **virtual path identifier** (VPI) and is also part of the cell header (see Section 5.6.2).

Figure 4.3 demonstrates the relationship between virtual channel, virtual path and transmission path. A transmission path may comprise several virtual paths and each virtual path may carry several virtual channels. The virtual path concept allows the grouping of several virtual channels. Its purpose and application will be explained in Section 4.2.3.

Concerning the levels of the ATM layer (virtual channel and virtual path), it is helpful to distinguish between links and connections (ITU-T, 1993e):

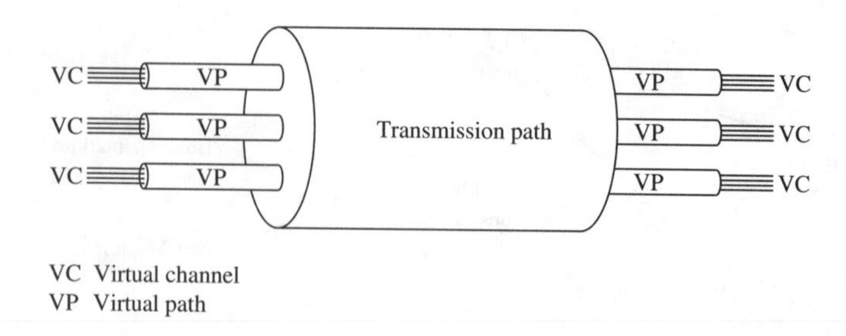

Figure 4.3 Relationship between virtual channel, virtual path and transmission path.

Virtual channel link: 'A means of unidirectional transport of ATM cells between a point where a VCI value is assigned and the point where that value is translated or removed.'

Similarly, a **virtual path link** is terminated by the points where a VPI value is assigned and translated or removed.

A concatenation of VC links is called a **virtual channel connection** (VCC), and likewise, a concatenation of VP links is called a **virtual path connection** (VPC).

The relationship between different levels of the ATM transport network is shown in Figure 4.4.

A VCC may consist of several concatenated VC links, each of which is embedded in a VPC. The VPCs usually consist of several concatenated VP links. Each VP link is implemented on a transmission path which hierarchically comprises digital sections and regenerator sections.

4.2.2 Switching of virtual channels and virtual paths

VCIs and VPIs in general only have significance for one link. In a VCC/VPC the VCI/VPI value will be translated at VC/VP switching entities. When explaining these basic VC/VP concepts it is not necessary to make a distinction between switch and cross-connect. Therefore, only the term 'switch' is used.

VP switches (see Figure 4.5) terminate VP links and therefore have to translate incoming VPIs to the corresponding outgoing VPIs according to the destination of the VP connection. VCI values remain unchanged.

VC switches (see Figure 4.6) terminate both VC links and, necessarily, VP links. VPI and VCI translation is performed. As VC switching implies VP switching, in principle a VC switch can also handle mere VP switching.

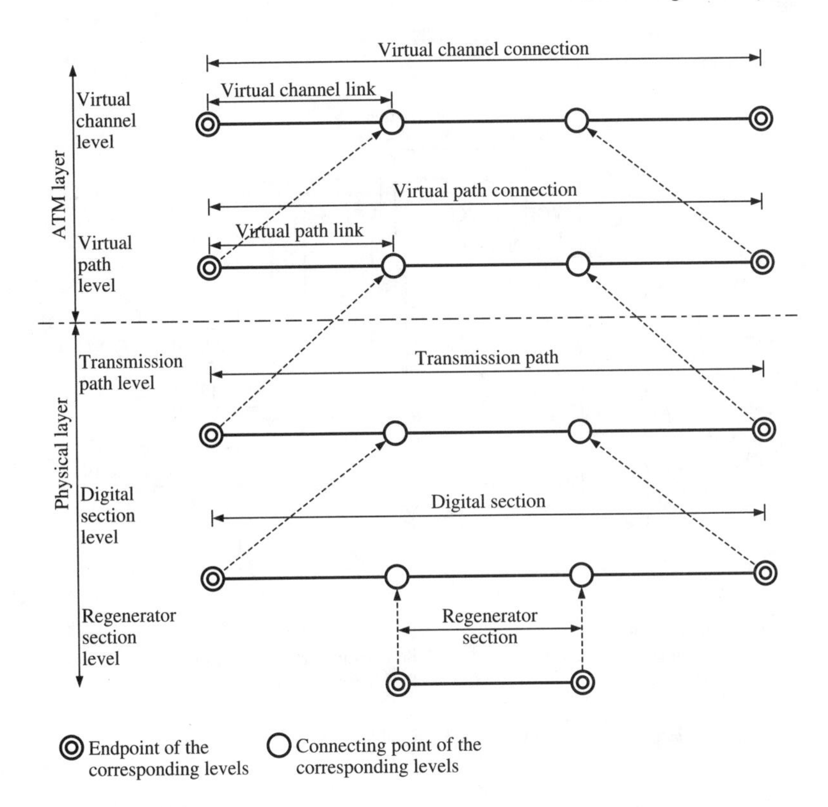

Figure 4.4 Hierarchical layer-to-layer relationship.

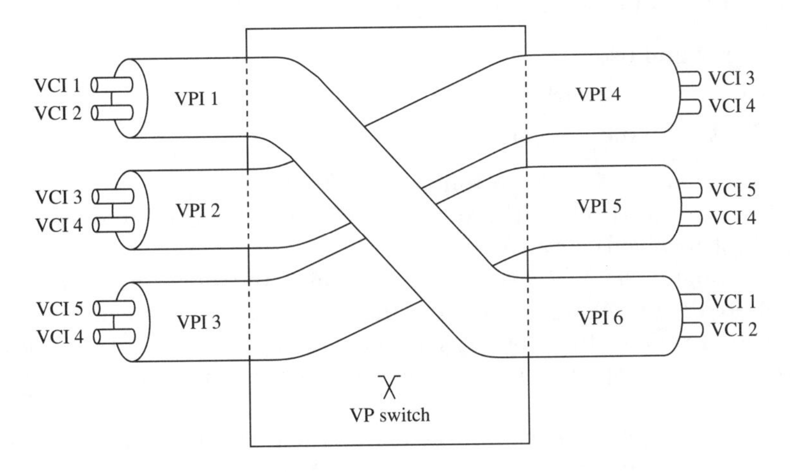

Figure 4.5 Virtual path switching.

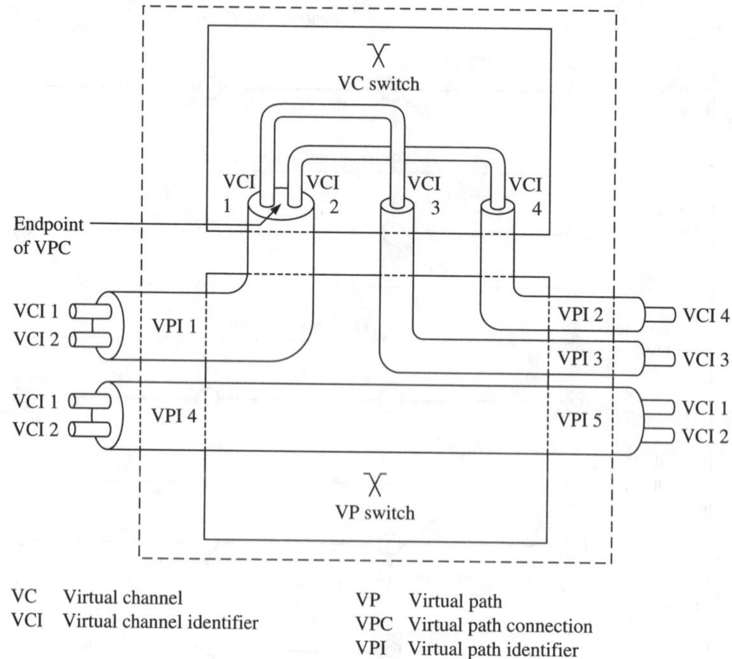

Figure 4.6 Virtual channel/virtual path switching.

4.2.3 Applications of virtual channel/path connections

VCCs/VPCs can be employed between:

- user and user
- user and network
- network and network.

All cells associated with an individual VCC/VPC are transported along the same route through the network. Cell sequence is preserved (first sent–first received) for all VCCs and VPCs.

User-to-user VCCs can carry user data and signaling information; user-to-network VCCs may, for instance, be used to access local connection-related functions (user–network signaling); and network-to-network VCC applications include network traffic management and routing.

A VPC between users provides them with a transmission 'pipe', the VC organization of which is up to them. This concept may, for example, be applied to LAN–LAN coupling.

A user-to-network VPC can be used to aggregate traffic from a customer to a network element such as a local exchange or a specific server.

Finally, network-to-network VPCs can be used to organize user traffic according to a predefined routing scheme or to define a common path for the exchange of routing or network management information. (More details on VCCs and VPCs can be found in Section 5.6.3.)

4.3 Signaling principles

4.3.1 General aspects

B-ISDN follows the principle of out-of-band signaling that has been established for the 64 kbit/s ISDN where a physical signaling D channel has been specified. In B-ISDN the VC concept provides the means logically to separate signaling channels from user channels.

A user may now have multiple signaling entities connected to the network call control management via separate ATM VCCs. The actual number of signaling connections and the bit rates allocated to them can be chosen in B-ISDN in a way that optimally satisfies a customer's need.

4.3.2 Capabilities required for B-ISDN signaling

B-ISDN signaling must be able to support:

- 64 kbit/s ISDN applications
- new broadband services.

This implies that existing ISDN signaling functions according to ITU-T Recommendation Q.931 (ITU-T, 1993v) must be included in B-ISDN signaling capabilities; on the other hand, the nature of B-ISDN – the ATM transport network – and the increasing desire for advanced communication forms, such as multimedia services, require specific new signaling elements. In the following, an overview of necessary B-ISDN signaling capabilities is given. (More details on such functions will be discussed in Chapter 8.)

ATM network-specific signaling capabilities have to be realized in order to:

- establish, maintain and release ATM VCCs and VPCs for information transfer;
- negotiate (and perhaps renegotiate) the traffic characteristics of a connection.

Other signaling requirements are basically not ATM related but reflect the fact that more powerful service concepts have become feasible. Examples are the support of multi-connection calls and multi-party calls.

For a multi-connection call, several connections have to be established to build up a 'composite' call comprising, for example, voice, image and data. It must also be possible to remove one or more connections from a call or to add new connections to the existing ones. Thus the network requires a means of correlating the connections of a call. In any case, it must be possible to release a call as a whole. These correlation functions should be performed in the origination and destination B-ISDN switches only, since transit nodes should not be burdened with such tasks.

A multi-party call consists of several connections between more than two endpoints (conferencing). Signaling to indicate establishment/release of a multi-party call and adding/removing one party is required. (A communication that is part of a multi-party call may also be of the multi-connection type.)

In a broadband environment, asymmetric connections (that is, low or zero bandwidth in one direction and high bandwidth in the other) will become important, so it is necessary to establish signaling elements to support such connections.

Another broadband issue affecting signaling is **interworking**, for example B-ISDN with non-B-ISDN services, or between video services with different coding schemes.

4.3.3 Signaling virtual channels

In B-ISDN, signaling messages will be conveyed out-of-band in dedicated **signaling virtual channels** (SVCs). Different types of SVC are provided (ITU-T, 1996f) as shown in Table 4.1.

There may be one **meta-signaling virtual channel** (MSVC) per interface. This channel is bidirectional and permanent. It is a sort of interface management channel used to establish, check and release point-to-point and selective broadcast SVCs.

Whereas the meta-signaling virtual channel is permanent, a point-to-point signaling channel is allocated to a signaling endpoint only while it is active. Point-to-point signaling channels are bidirectional. They are used to establish, control and release VCCs to carry user data (VPCs as well as VCCs may also be established without using signaling procedures, for example by subscription).

Table 4.1 Possible signaling virtual channels at B-ISDN user–network interface.

SVC type	*Directionality*	*Number of SVCs*
Meta-signaling channel	Bidirectional	1
General broadcast SVC	Unidirectional	1
Selective broadcast SVC	Unidirectional	Several possible
Point-to-point SVC	Bidirectional	One per signalling endpoint

Broadcast SVCs (BSVCs) are unidirectional (network-to-user direction only). They are used to send signaling messages either to all signaling endpoints in a customer's network or to a selected category of signaling endpoints. The general broadcast SVC reaches all signaling endpoints; it is always present. Selective broadcast SVCs may be provided as a network option so that all terminals belonging to the same service profile category can be addressed (a B-ISDN service profile contains information which is maintained by the network to characterize the services offered by the network to the user).

In **point-to-point** signaling access configurations, one pre-established signaling virtual channel can be used. In contrast, in a **point-to-multipoint** signaling access configuration, meta-signaling is required for managing the signaling virtual channels. More details on these configurations are described in Section 8.2.

Meta-signaling is not used for network-to-network signaling. In the case of a network–network VP containing signaling VCs, additional VCI values for signaling in this VP are pre-established. The method of pre-establishment is for further study.

To illustrate the SVC concept of B-ISDN, an example (based on ITU-T Recommendation I.311 (ITU-T, 1996f)) is given in Figure 4.7. This highlights different possibilities for carrying signaling information from the customer to the network and vice versa.

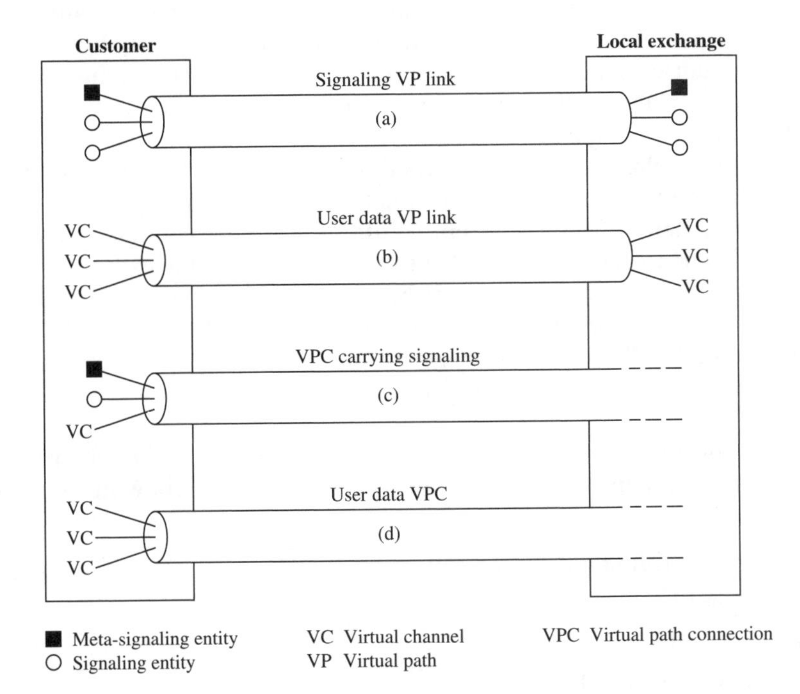

Figure 4.7 Allocation of signaling virtual channels to a customer.

Four different VP links/connections are depicted in the figure. The first (a) is a signaling VP link which transports all the signaling information to be exchanged between the customer and the local exchange, including meta-signaling. When a signaling capability to a point in the network other than the local exchange is required (for example, to communicate with a special service provider located elsewhere), such signaling can be done on an extra VPC (c) which may carry signaling and user data. This VPC goes through the local exchange and is terminated at the appropriate place. (The other two VPs (b) and (d) are shown for completeness; they only carry user data. In case (b), the corresponding VCs are switched in the local exchange while in (d) the VP as a whole passes transparently through the local exchange.)

4.4 Broadband network performance

Broadband networks based on ATM cell transfer must meet certain performance requirements in order to be accepted by both potential users and network providers. ATM-related performance parameters and measures need to be specified in addition to the performance parameters already introduced for existing networks. In this section, we deal only with ATM layer-specific network performance. The quality of service as perceived by the user may be influenced not only by the ATM transport network performance but also by higher layer mechanisms. In some cases, these will be able to compensate for shortcomings in the ATM transport network.

Cells belonging to a specified virtual connection are delivered from one point in the network to another, for example from A to B. A and B may denote the actual endpoints of a virtual connection, or may delimit a certain portion of the cell transport route (for example, A and B may indicate national network boundaries of an international ATM connection). For more details see ITU-T Recommendations I.353 (ITU-T, 1996q) and I.356 (ITU-T, 1996g). Because there is some transfer and switching delay, cells sent from A arrive at B within $\Delta t > 0$ (see Figure 4.8).

In order adequately to describe the quality of ATM cell transfer, ITU-T Recommendation I.356 first defines the following outcome categories:

- successfully transferred cell
- errored cell
- lost cell
- misinserted cell
- severely errored cell block.

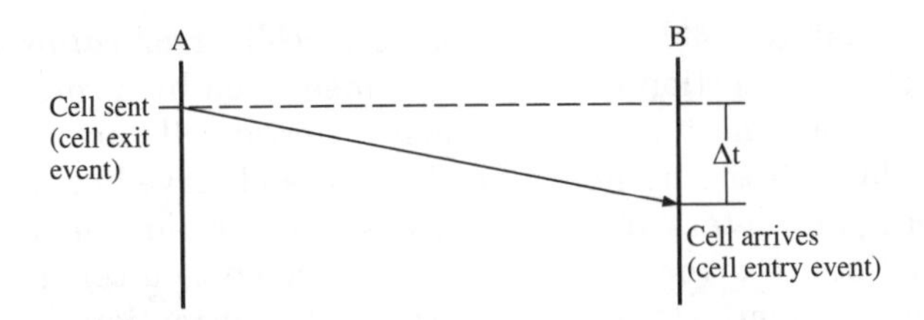

Figure 4.8 Cell transfer (schematic).

If Δt is less than a maximum allowed time T, the information field of the cell is not affected by bit errors, and the cell is received with a valid header (see Section 5.6.2), then the cell has been successfully transferred.

If the cell arrives in due time but there are one or more bit errors in the received cell information field or the cell has an invalid header, the cell is errored.

A lost cell outcome occurs if the cell arrives after time T (or never reaches B). Errors in the ATM cell header that cannot be corrected (see Section 5.3.1) or cell buffer overflows in the network (for example, in an ATM switch) lead to lost cells.

If a cell that has not been sent from A on this virtual connection arrives at B, then this misdelivered cell produces a misinserted cell outcome. Header errors that are not detected or are erroneously corrected may produce misinserted cells.

A severely errored cell block occurs if more than M errored cells, lost cells or misinserted cells are observed in a block of N cells transmitted consecutively on a given connection.

By making use of the above considerations, it is possible to define the performance parameters. The parameters and their definitions are listed in Table 4.2.

Table 4.2 ATM performance parameters.

Parameter	*Definition*
Cell loss ratio	Ratio of lost cells to transmitted cells
Cell misinsertion rate	Number of misinserted cells per connection second
Cell error ratio	Ratio of errored cells to the number of delivered cells
Severely errored cell block ratio	Ratio of the number of severely errored cell blocks to total number of cell blocks
Cell transfer delay	Δt (see Figure 4.8)
• Mean cell transfer delay	Arithmetic average of a specified number of cell transfer delays
• Cell delay variation (CDV)	Difference between a single observation of cell transfer delay and the mean cell transfer delay on the same connection

Note that in Table 4.2 cell loss and cell error *ratios* are given, whereas the cell insertion outcomes are measured by a *rate* (events per time unit). As the mechanism by which inserted cells are produced has nothing to do with the number of cells on the observed connection, this performance parameter cannot be expressed as a ratio, only as a rate.

Bit errors in errored cells can be corrected to a certain extent by error protection methods applied to the cell information field contents (see Section 5.7).

Lost and misinserted cells can cause severe problems if they are not detected. For example, when using a constant bit rate, the real-time services synchronism between sending and receiving terminals may be disturbed. Lost and misinserted cell events can be detected (in many cases) by monitoring a sequence number in the cell information field or an equivalent mechanism (see Section 5.7).

Both cell transfer delay and cell delay variation must be kept within a limited range in order to meet service requirements. Cell delay variation is introduced, for example, by ATM multiplexers or ATM switches and cross-connects (see Chapters 12 and 13) due to changing buffer fill levels.

ATM network performance requirements in terms of parameter values are not yet fully specified. This issue will be addressed again in the following chapters whenever appropriate (for example, in Chapters 12 and 13 on ATM switching and transmission). For cell transfer performance measurement methods see ITU-T Recommendation I.356, annex C (ITU-T, 1996g).

4.5 Traffic management aspects

There is only one common resource, the network, which has to be shared by several services and connections each with a different characterization. The required bit rate may vary from some kbit/s towards the Mbit/s range. Some services have stronger real-time constraints than others; some services might tolerate a few cell losses, others do not. However, ATM promises to support all these different requirements with a common network. Within such a network all connections may impact on each other. It is the task of traffic control to bound these effects and to achieve these two main goals:

- to provide specified and guaranteed levels of **quality of service** (QoS)
- to use available network resources efficiently.

In the early days of ATM, traffic control was mainly focused on dealing with the statistical cell flow fluctuations incurred because of the asynchronous nature of the traffic. These early considerations dealt with issues like buffer dimensioning at multiplexing and/or switching points. In order to fulfill stringent real-time requirements these buffers were kept as small as possible. However, soon it became evident that with these small buffers the second goal was almost unachievable. Furthermore, data traffic is considered to be 'self-similar'. This means that even if an aggregation of multiple connections is considered the same (unpredictable) traffic behavior still exists; that is, a statistical traffic smoothing through aggregation does not occur. Except in a very few cases (see Section 7.2.2), good network utilization is hard to achieve.

All these effects made it necessary to introduce large buffers in ATM switches, capable of storing large data bursts.

4.5.1 Overview of functions

To ensure the desired network performance outlined in the previous section, an ATM-based network will have to provide a set of traffic control and congestion control capabilities. ITU-T Recommendation I.371 (ITU-T, 1996i) identifies the following:

- network resource management
- connection admission control
- usage parameter control and network parameter control
- priority control and selective cell discard
- traffic shaping
- fast resource management
- congestion control.

Besides ITU-T, the ATM Forum is also working on traffic management issues. Its TM4.0 (ATM Forum, 1996c) laid the basis for many traffic control features specifications which were also adopted by ITU-T. Their definitions and descriptions are given in later sections. All these mechanisms considerably influence modern switch architectures.

4.5.2 ATM traffic parameters and transfer capabilities

To be able to optimize ATM network operation, ATM traffic is grouped into various categories of connections with different traffic characteristics. A source traffic descriptor – which is a set of traffic parameters such as peak cell rate, average cell rate, **cell delay variation** (CDV) tolerances, burstiness, peak duration – carries the information that is used by the network to match a source's connection request.

Traffic parameters must be understandable by the user so that a 'traffic contract' can be established between the user and the network.

Conformance to the negotiated traffic contract can then be enforced by the network (for example, through UPC and NPC, see Section 7.2.3). The complete traffic contract which is negotiated between the user and the network at connection establishment (by either signaling or subscription) comprises, besides the source traffic descriptor, an ATM transfer capability (see second subsection below), associated CDV tolerances (CDVT) and a quality-of-service class (see Section 4.5.3).

Peak cell rate and sustainable cell rate

The **peak cell rate** (PCR) of an ATM connection can be seen as:

> the inverse of the minimum inter-arrival time between two requests to send an ATM cell.

ITU-T Recommendation I.371 (ITU-T, 1996i) – which gives an exact definition of PCR, taking into account CDV too – allows 16 384 different peak cell rates in the range of 1 cell per second up to $4.290\,77 * 10^6$ cells per second.

The PCR is the maximum cell rate at which a source may send its cells during the connection's lifetime.

The **sustainable cell rate** (SCR), which is always smaller than the peak cell rate, is a parameter that is used for burst-type traffic. It is defined, together with the intrinsic **burst tolerance** (BT), by a **generic cell rate algorithm** (GCRA). The GCRA mechanism is described in ITU-T Recommendation I.371, where it is shown that two equivalent algorithms can be used to implement GCRA: the virtual scheduling algorithm and the continuous-state leaky bucket algorithm (they are equivalent in the sense that, for any sequence of cell arrivals, they determine the same cells to be conforming or non-conforming to the traffic contract). The BT is equivalent to the CDVT but with much higher values. The GCRA takes SCR, BT and CDVT into account.

An intuitive interpretation of the sustainable cell rate is the mean cell rate of a bursty source which will send its traffic at rates below and above this mean value. The duration of those periods when the source sends cells above the SCR is limited by the burst tolerance.

The **maximum burst size** (MBS), that is the amount of cells which may be sent at the PCR, can be derived from the SCR and BT values.

ATM transfer capabilities

ITU-T Recommendation I.371 (ITU-T, 1996i) defines ATM transfer capabilities for a number of services. An overview is given in Table 4.3 which also shows the corresponding definitions of the ATM Forum (1996c).

Table 4.3 Transfer capability classes according to ITU-T and the ATM Forum.

ITU-T	*ATM Forum*
Deterministic bit rate (DBR)	Constant bit rate (CBR)
Statistical bit rate (SBR)	Variable bit rate (VBR)
ATM block transfer (ABT)	*Not defined*
Available bit rate (ABR)	Available bit rate (ABR)
Not defined	Unspecified bit rate (UBR)

DBR/CBR is used by connections requesting a fixed amount of bandwidth to be continuously available during the connection's lifetime. A DBR/CBR service is specified by its peak cell rate and CDV value. One example of DBR/CBR traffic is 64 kbit/s speech connections.

SBR/VBR is suitable for applications with varying bit rate requirements where there exists prior knowledge of some traffic characteristics of the application. SBR/VBR uses peak cell rate, sustainable cell rate and intrinsic burst tolerance. (ATM Forum, 1996c) differentiates between **VBR-real time** (VBR-rt) for connections with real-time requirements and **VBR-non real time** (VBR-nrt) connections. One example of a VBR-rt connection is a video transmission with varying bandwidth.

ABT is employed in the case of traffic sources that send traffic sporadically and unpredictably but that require a guaranteed bit rate during each period of sending cells onto the ATM connection.

ABR supports LAN-like applications where users accept unreserved bandwidth but want to make optimum use of the bandwidth actually available. An ABR user will specify a maximum required bandwidth and a **minimum cell rate** (MCR) which may be zero. ABR keeps the end-terminal informed of the current load situation within the network. The end-terminal is expected to gear its cell emission rate accordingly. While there are no delay guarantees, limited cell loss is provided as long as the end-terminal reacts to the network flow control signals.

UBR resembles the best-effort service usually provided by the Internet. UBR does not guarantee any cell loss and cell delay performance; higher layer functions within end-terminals are expected to cope with these effects.

ABT, ABR and UBR are mainly designed for data applications.

In the following, only the ATM Forum's terms for the ATM transfer capabilities are used. More traffic management details are given in Chapter 7.

4.5.3 Service classes and quality of service

Since the ATM Forum defines only five traffic classes, namely CBR, VBR-rt, VBR-nrt, ABR and UBR, one can expect that these are the only

Table 4.4 ATM traffic classes according to ATM Forum.

	CBR	*VBR-rt*	*VBR-nrt*	*ABR*	*UBR*
CLR	Specified	Specified	Specified	Specified	Unspecified
CTD and CDV	Specified	Specified	Mean CTD only	Unspecified	Unspecified
PCR	Specified	Specified	Specified	Specified	Specified
SCR and BT	n/a	Specified	Specified	n/a	n/a
MCR	n/a	n/a	n/a	Specified	n/a
Reactive congestion control	No	No	No	Yes	No

CLR	Cell loss ratio	CBR	Constant bit rate
CTD	Cell transfer delay	VBR-rt	Variable bit rate – real time
CDV	Cell delay variation	VBR-nrt	Variable bit rate – non real time
PCR	Peak cell rate	ABR	Available bit rate
SCR	Sustainable cell rate	UBR	Unspecified bit rate
BT	Burst tolerance	n/a	Not applicable
MCR	Minimum cell rate		

classes being supported by end-terminals. Hence they will predominate over the other, ITU-T defined, transfer capabilities. Table 4.4 provides a classification of these traffic classes and shows their most relevant attributes.

4.6 Operation and maintenance aspects

4.6.1 General principles

ITU-T Recommendation M.60 (ITU-T, 1993q) defines maintenance as:

> 'The combination of all technical and corresponding administrative actions, including supervision actions, intended to retain an item in, or restore it to, a state in which it can perform a required function.'

The general principles for the maintenance of telecommunication networks which are also relevant to B-ISDN are contained in ITU-T Recommendation M.20 ('Maintenance Philosophy for Telecommunications Networks') (ITU-T, 1992a) and ITU-T Recommendation M.3600 ('Principles for the Management of ISDNs') (ITU-T, 1992b).

Table 4.5 gives a brief overview of **operation and maintenance** (OAM) actions (more details are given in ITU-T Recommendations M.20 and I.610 (ITU-T, 1995g).

Other recommendations relevant to the OAM of B-ISDN are ITU-T Recommendation M.3010 (ITU-T, 1996k) entitled 'Principles for a Telecommunications Management Network' (TMN), and M.3610 (ITU-T, 1996t) which applies TMN to B-ISDN. An example TMN architecture for B-ISDN customer access will be given in Section 6.1.

Table 4.5 Overview of OAM actions.

Action	*Description*
Performance monitoring	Normal function of the managed entity is monitored by the continuous or periodic checking of functions. As a result, maintenance event information will be produced.
Defect and failure detection	Malfunctions or predicted malfunctions are detected by continuous or periodic checking. As a result, maintenance event information or various alarms will be produced.
System protection	Effect of failure of managed entity is minimized by blocking or change-over to other entities. As a result, the failed entity is excluded from operation.
Failure or performance information	Failure information is given to other management entities. As a result, alarm indications are given to other management planes. Response to a status report request will also be given.
Fault localization	Determination by internal or external test systems of failed entity if failure information is insufficient.

4.6.2 OAM levels in B-ISDN

The OAM levels of the ATM transport network coincide with the levels introduced in Section 4.2.1 on network layering (see Figures 4.2 and 4.4). The ATM transport network comprises the physical layer (subdivided into regenerator section, digital section and transmission path level), and the ATM layer (subdivided into VP and VC levels) (see Figure 4.9).

The corresponding OAM information flows of each level – denominated F1 to F5 – are also shown in Figure 4.9. These OAM flows are bidirectional. As an example of an OAM flow, consider the monitoring of a VPC by supervising cells sent out at one endpoint of the VPC and mirrored at the other endpoint, to be evaluated at the sending side.

More information on these OAM information flows is presented in Section 6.2.

4.7 Customer network aspects

One essential block of the B-ISDN is the **customer network** (CN). Sometimes it is called the **customer premises network** (CPN) or **subscriber premises network** (SPN).

4.7.1 Reference configuration of the B-ISDN UNI

The reference configuration for the 64 kbit/s ISDN user–network interface, which is described in ITU-T Recommendation I.411 (ITU-T,

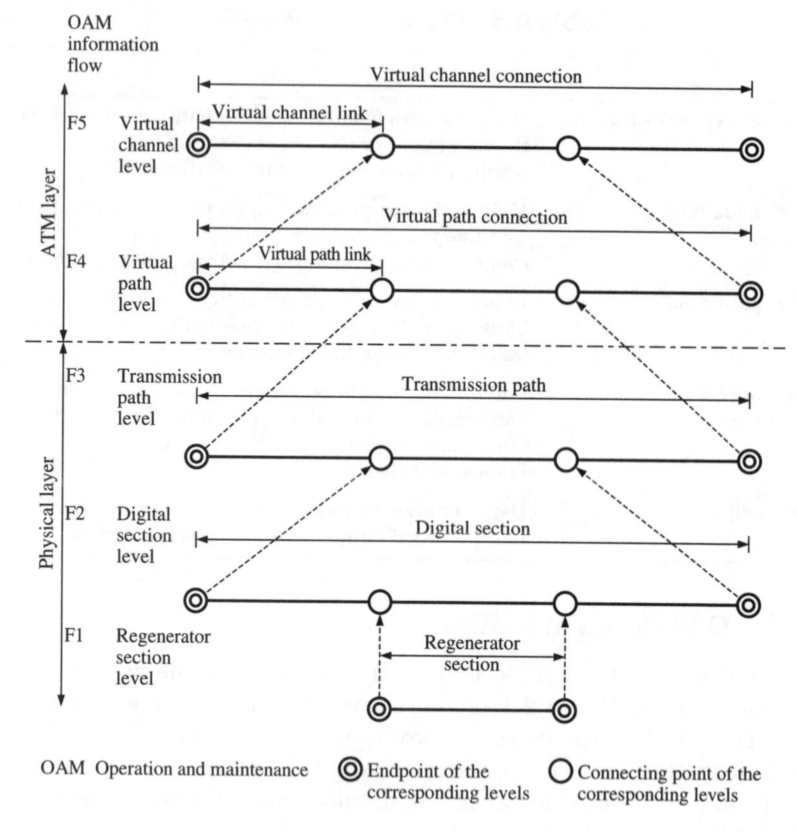

Figure 4.9 OAM hierarchical levels.

1993l), was accepted to be general enough for use in the B-ISDN environment (ITU-T, 1993m). Figure 4.10 shows the principles of the reference configuration for the B-ISDN UNI. It contains the following:

- Functional groups: Broadband network termination 1 (B-NT1), broadband network termination 2 (B-NT2) and broadband terminal equipment 1 (B-TE1).
- Reference points: T_B and S_B.

B-TE1 denotes a broadband terminal with standard interface. Physical interfaces may or may not occur at the reference points T_B and S_B. If they are realized, they must comply with the specified standard (see Section 5.2). While B-NT1 performs only line transmission termination and related OAM functions, B-NT2 may be, for example, a private branch exchange or LAN which performs multiplexing and switching of ATM cells.

The CN covers the area where users have access to the public network via their terminals. This is the part of the telecommunication network located at the user side of the B-NT1.

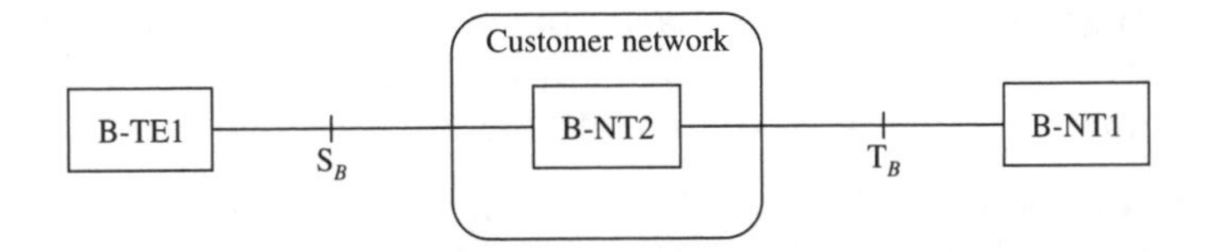

Figure 4.10 Reference configuration of the B-ISDN UNI.

The reference configuration shown in Figure 4.10 can be used for the functional description of the CN. The interface between the CN and the public network is usually at the reference point T_B. Thus the CN coincides with the functional group B-NT2.

4.7.2 Customer categories

Different aspects for the classification of CNs are possible (Baireuther, 1991; Besier, 1988; De Smedt et al., 1990), including the environment (residential, business), number of users or topology.

The **residential** environment is characterized by a small number of people (for example, a family) using broadband services mainly for entertainment. This category is considered to be more or less homogeneous and will be restricted to a single apartment or house. In many cases, no internal switching capabilities are necessary within a residential CN.

The counterpart of the residential CNs are the **business** CNs, which are subdivided according to their size:

Small business CNs: Small business CNs have a lot in common with residential CNs. Such a CN will be installed in a small area like an office or a shop with up to about 10 employees. Often the office or shop is combined with private housing and therefore the requirements of the residential CN (for example, entertainment services) must also be met. One major difference from the residential CN is that there may be a need for internal switching.

Medium business CNs: In medium and large business CNs, as well as in factory CNs, no distribution service need be supported for entertainment. Normally, only interactive services, such as telephony, videoconferencing or high-speed data transmission, will be used. One of the main characteristics is the internal switching capability. A medium business CN typically has between 10 and 100 users and spreads over several floors of a building.

Large business CNs: Large business CNs have more than 100 users. They spread over several buildings or floors of a building. Sometimes distances of up to 10 km must be spanned.

Factory CNs: Factory CNs can have the size of a medium or large business CN. They often have to satisfy exceptional physical requirements such as being robust against extreme heat, dust or electromagnetic interference.

The common characteristic of medium, large and factory CNs is the provision of internal communication. This facility is not often required in residential and small business environments.

4.7.3 General requirements

Two types of requirements are expected to be met by a CN (Vorstermans and De Vleeschouwer, 1988):

- service requirements
- structural requirements.

Service requirements deal with the service mix as well as the consequences of having to support these services. The service mix depends on the customer category. This type of requirement includes the bit rate to be supported as well as the information transfer characteristics, like mean and maximum delays, delay variation, error performance and throughput. It is possible to characterize realistic service mixes for each customer category, but it is difficult to estimate the evolution of coding techniques, such as variable bit rate video-coding, which will influence the bit rate required for a specific service.

The second type of requirements are the **structural requirements**, which include aspects of flexibility, modularity, reliability, physical performance and cost.

Flexibility is the ability to cope with system changes. This can be subdivided into four parts:

- Adaptability measures how the CN deals with changes that do not alter the global scale of the CN (for example, new wiring). This is very important in the terminal area for the residential as well as the business environment.
- Expandability indicates how the CN can grow (for example, by introducing new services, increasing the bit rate to be supported, installing new terminals, or expanding the scale of the CN).
- Mobility is the ability to interchange terminals. This requires a universal terminal interface.
- Interworking describes how the CN can interface to other networks. This is very important in areas which are already

covered by existing networks, such as **Local area networks** (LANs) and private **metropolitan area networks** (MANs).

Modularity is the provision of a flexible structure. The network should not be limited to a few applications. Therefore, a modular system must be used for the provision of CN capabilities.

Reliability deals with the sensitivity of the CN to errors (for example, bit errors, terminal failures and user-induced errors). This requirement is extremely important when a large number of people or terminals are affected by the error, or in all cases where highly reliable operation of the CN must be guaranteed, as is the case in hospitals and fire departments. Reliability requires redundancy and therefore even the terminal connection is duplicated within large CNs.

Physical performance concerns the optimum use of the physical medium. It includes aspects of coding efficiency and cable length and influences the hardware cost. Installation and maintenance are covered by the operating performance. This must be very simple in the terminal environment so that changes can be carried out rapidly and cheaply.

Cost is one of the most important requirements influencing the acceptance of the CN. In the residential area low costs are essential, whereas in the business area it is necessary to have reasonable costs during the introduction phase. Fast system growth can only be achieved if the incremental costs can be kept low.

4.7.4 Topologies

ITU-T Recommendation I.413 illustrates different physical configurations for the realization of a CN (ITU-T, 1993m). Figure 4.11 shows a few examples of CN configurations. (Other CN configurations are not precluded.)

The first configuration is the well-known star configuration in which each terminal is directly connected to the B-NT2 by a dedicated line.

A B-NT2 can be realized as a centralized system or as a distributed system. The latter can have LAN-like structures (for example, bus or ring) in which the terminals are all connected to a common medium via special medium adaptors in the general case.

However, ITU-T has also agreed to some new configurations in which terminals are directly connected to a common shared medium (for example, dual bus), as shown in Figures 4.11(b) and (c). The major motivations for these new configurations are simplicity of deployment, economy and evolutionary aspects of B-ISDN (Dobrowski et al., 1990). A shared medium configuration can be extended easily by adding a new terminal. When using a star configuration, extension may result in higher costs caused by the need for an additional or larger multiplexer.

Combinations of the star and shared medium configurations are

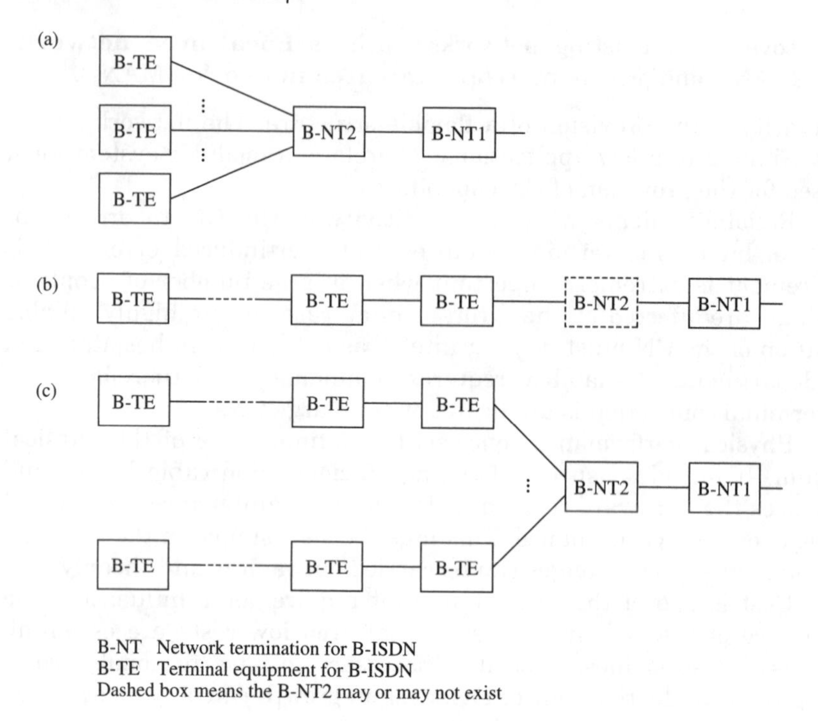

Figure 4.11 Physical configurations of the CN.

also possible (see Figure 4.11(c)). The configuration depicted in this figure is the starred bus system.

However, it is evident that terminals used in the shared medium configurations of Figure 4.11 have to include a medium access function. Since even with legacy LANs nowadays, shared medium configurations are being replaced by dedicated lines, that is, a star structure, one may expect that shared media will not play an important role.

The influence of these new configurations on the definition of the interface between a terminal and the B-NT2 will be discussed in Section 5.2.

4.7.5 Interworking with existing networks

In medium and large business environments, two different types of communication systems are used at present: voice and low-speed data communication like facsimile are handled by **private branch exchanges** (PBXs), while the need for pure data communication is satisfied by LANs.

LANs are primarily employed for in-house data communications. They are used for the interconnection of terminals, workstations, printers, databases, hosts and manufacturing systems. The traffic carried is characterized by short data bursts requiring high transmission speeds.

The number of LANs used within business premises has increased dramatically throughout the past decade. Various protocols are available which use different principles of operation and which have been standardized by a number of bodies, in particular the American National Standards Institute (ANSI), the Institute of Electrical and Electronic Engineers (IEEE), and the International Standards Organization (ISO).

Current LANs usually have an extent of up to 10 km and their transmission speed is in the range 1–16 Mbit/s. Typically, coaxial cable will be used for transmission. LANs can use a ring or bus topology, although star configurations dominate today and will continue to do so in the near future.

LANs can be classified by their topology (bus, ring, star), the transmission medium (twisted pair, coaxial cable, optical fiber) and the media access control (MAC) procedure. (In higher layers, identical protocols may be used.)

Today's most frequently used LANs are:

- carrier sense multiple access with collision detection (CSMA/CD) (ISO, 1990a)
- token bus (ISO, 1990b)
- token ring (ISO, 1991).

CSMA/CD was the first standardized MAC procedure. It is based on a development by Xerox called **Ethernet**. CSMA/CD uses a bus system with a specified transmission rate of 10 Mbit/s. A station with a packet to send listens to the carrier before sending. If the channel is idle the station begins sending, otherwise it waits till the channel becomes idle. Collision may occur when two or more stations begin sending at the same time. This will be detected by the sending stations and they will stop their packet transmission. The collision is resolved by the back-off algorithm (each station tries to send after a random time). In low loaded systems, this protocol works very well (low packet delay). However, the occurrence of collisions means that the performance deteriorates with increasing load.

The **token bus** system uses a coaxial cable with specified transmission rates of 1, 5 or 10 Mbit/s. All stations are passively coupled to the medium. Access to the channel is controlled by a token-passing protocol. The token is delivered from one station to its neighbor, where the 'neighbor' is defined by address rather than by physical location. The number of packets that can be sent during the interval when a station possesses the token is determined by the token-passing protocol. In order to fulfill different performance requirements, each station uses four internal priority classes.

The **token ring** network is made up of point-to-point unidirectional links interconnecting adjacent stations (active coupling) to form a closed loop. Transmission rates of 4 and 16 Mbit/s are specified for shielded twisted pair cables. Access to the ring is again controlled by a token-passing protocol. A station that is ready to send a packet has to wait for the token, which it receives from its physical neighbor. Different priorities can be attached to tokens but only one token can circulate at a time. During high load situations, token protocols prevent packet collisions but the delay times still increase with the network load.

Further developments in this area led to the so-called **high-speed local area networks** (HSLANs) which have a transmission rate of more than 100 Mbit/s. For example, Ethernet has been enhanced so that it can now offer transmission speeds of 100 Mbit/s. These networks are used to interconnect LANs as well as for the high-speed data communication needed by workstations and file-servers. HSLANs are not restricted to small areas. They may be installed in a metropolitan area, in which case they are called metropolitan area networks (MANs). They can cover a region of more than 100 km in diameter and up to 1000 stations can be attached to one network. The transmission medium is optical fiber. In principle LAN topologies can be used. Different types of HSLANs/MANs are discussed, for example:

- fiber-distributed data interface (FDDI) (ISO, 1989/90)
- DQDB (IEEE, 1991)
- high-performance parallel interface (HIPPI) (Tolmie and Renwick, 1993).

FDDI is available from different suppliers, and for DQDB commercially operated networks exist. HSLANs in the Gbit/s range are currently under study (see Section 4.7.6). HIPPI offers a transmission rate of 800 Mbit/s with IEEE-based framing (IEEE, 1989).

Increasing communication requirements demand the interconnection of LANs. This can be achieved by MANs (see Section 11.4) as well as ATM networks. An **interworking unit** (IWU) is required to perform protocol conversion and bit rate adaptation (necessary for the interconnection of different networks). The IWU can be attached directly to the B-NT1, as shown in Figure 4.12(a), or the LAN/MAN can be connected by means of the IWU via S_B to a B-NT2, if present, as shown in Figure 4.12(b).

The IWU may become a bottleneck in the case of heavy traffic load. To obtain a good performance (high throughput, small delay) it is necessary that the protocols used in the LAN and in the ATM network have a lot in common. Interconnection should be done on the lowest possible level. For further details see Chapters 10 and 11.

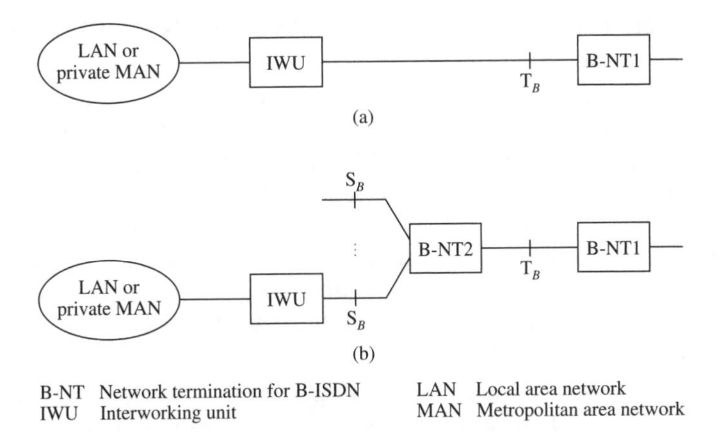

Figure 4.12 Interworking LAN – ATM network.

4.7.6 Gbit/s LANs

As mentioned above, LANs currently provide throughputs in the 100 Mbit/s range (for example FDDI (ISO, 1989/90). However, there are communication requirements, like the interconnection of super-computers with high-resolution graphic terminals and the need for a high-speed backbone network, which demand networks operating in the Gbit/s range.

At these high speeds, only optical fibers are suitable for the physical medium. Point-to-point systems can already operate at Gbit/s. However, shared medium LANs are much more complicated.

Research on LANs in the Gbit/s range has already started. Many problems exist which have to be resolved in the future. Some of the open questions are:

- optical power limitation (Henry, 1989)
- electronic bottleneck (Henry, 1989)
- network topologies (Acampora and Karol, 1989)
- medium access control (Müller et al., 1990)
- definition of protocols for high-speed applications
- compatibility with ATM.

Interconnection of such very high-speed LANs would obviously require high-speed links operating at 600 Mbit/s or higher.

Several approaches are currently being considered for these very high-speed LANs which are also applicable to the metropolitan area:

- ATM-LANs: This type of LAN offers a total throughput of several Gbit/s using a small central ATM switch to which the terminals are connected in a star fashion. More details on ATM-LANs can be found in Section 11.1.

- Shared medium network (ring or bus) with a suitable MAC mechanism. Possible MAC protocols are DQDB (IEEE, 1991), ATM ring (JTC, 1990), HANGMAN (Watson et al., 1992) or cyclic reservation multiple access II (CRMA-II) (van As et al., 1991). This list of MAC protocols is not complete. Good overviews on basic topologies and fundamental algorithms for the shared medium access protocols are given in Lemppenau et al. (1993) and Partridge (1994). Most of these LANs have MAC-PDUs the size of an ATM cell.
- Optical LANs using the WDM principle (see Section 13.6). The throughput of these LANs can be in the range of several tens of Gbit/s.

Review questions

1. What do the following terms specify:
   - digital section
   - transmission path
   - ATM virtual channel/virtual path
   - virtual channel/path link
   - virtual channel/path connection?
2. How can virtual channels/paths be identified?
3. What is the difference between VP switching and VC switching?
4. Which channels are used in ATM networks to convey signaling messages?
5. What are the main network performance parameters defined for ATM networks?
6. What traffic congestion control functions have been defined for ATM networks? What do they perform?
7. What is meant by QoS?
8. What are the five ATM transfer capabilities defined by the ATM Forum?
9. What are the service characteristics of the CBR traffic class?
10. What are F1, ..., F5 OAM information flows?
11. Which configurations for customer networks are possible?
12. How can the customer networks be classified?
13. What is the reference point between the customer network and the public network?

5

B-ISDN user–network interfaces and protocols

This chapter deals with the B-ISDN user–network interfaces and with ATM-based protocols. In Section 5.1 the protocol reference model developed by ITU-T for B-ISDN is presented. Sections 5.2 to 5.4 describe the user–network interfaces in general and their physical layer properties. Section 5.5 describes equipment-internal interfaces. Then in Sections 5.6 and 5.7, functions, codings and procedures for the adjacent ATM layer and ATM adaptation layer are described.

5.1 B-ISDN protocol reference model

5.1.1 General aspects

In modern communication systems, a layered approach is used for the organization of all communication functions. The functions of the layers and the relations of the layers with respect to each other are described in a **protocol reference model** (PRM).

A description of the PRM for the existing ISDN is given in ITU-T Recommendation I.320 (ITU-T, 1993h). In particular, it introduces the concept of separate planes for the segregation of the user, control and management functions. This PRM is the basis for the PRM of the broadband aspects of ISDN (B-ISDN PRM) which is described in ITU-T Recommendation I.321 (ITU-T, 1991e). The new recommendation takes into account the functionalities of B-ISDN, so expansion of the PRM contained in ITU-T Recommendation I.320 was necessary.

5.1.2 Layered architecture

According to the **open system interconnection** (OSI) reference model of the ISO (ISO, 1984), each open system can be described as a set of subsystems arranged in a vertical sequence (see Figure 5.1).

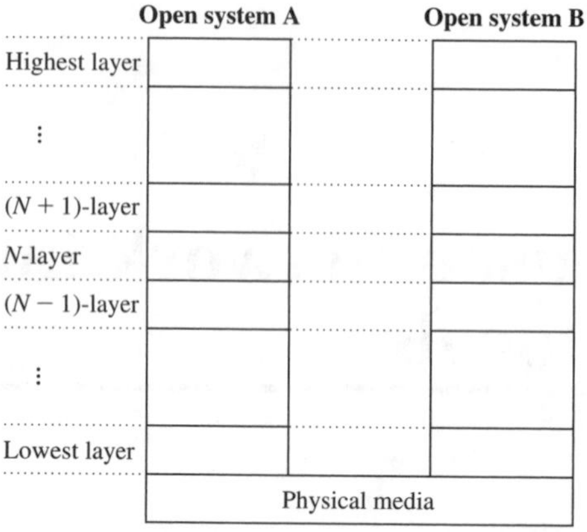

Figure 5.1 Layered structure of the OSI reference model.

An N-subsystem which consists of one or more N-entities only interacts with the subsystem above or below. The N-entity performs functions within layer N. Communication between peer N-entities (entities of layer N) uses an N-peer-to-peer protocol. The unit of data in an N-peer-to-peer protocol is called the N-protocol data unit (N-PDU). Peer N-entities communicate using the services provided by the layer below. The services of layer N are provided to layer $(N + 1)$. The point at which the N-services can be accessed by the layer above is called the N-service access point (N-SAP).

N-primitives are introduced to describe the interface between adjacent layers N and $(N + 1)$. Together with the N-primitive, the associated N-service data unit (N-SDU) is delivered from layer N to layer $(N + 1)$ and vice versa. For this purpose the N-service protocol (adjacent layer protocol) is used. Figure 5.2 illustrates this service concept.

Figure 5.3 shows the relationships among the various types of data unit. An N-PDU consists of N-protocol control information (N-PCI) and N-user data. The N-PCI is the information which is exchanged between N-entities.

5.1.3 Relationship between the B-ISDN PRM and the OSI reference model

The OSI reference model for ITU-T applications is defined in ITU-T Recommendation X.200 (ITU-T, 1994d). OSI is a logical architecture which defines a set of principles, including protocol layering, layer service definition, service primitives and independence. These principles are appropriate for the definition of the B-ISDN PRM. However, not all of these principles (for example, independence) have been fully applied in the B-ISDN PRM.

The OSI reference model uses seven layers (see Table 5.1). Each layer has its own specific functions and offers a defined service to the

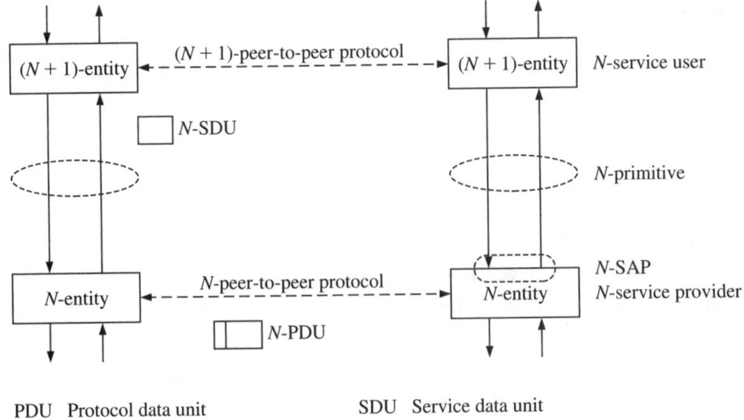

Figure 5.2 OSI service concept.

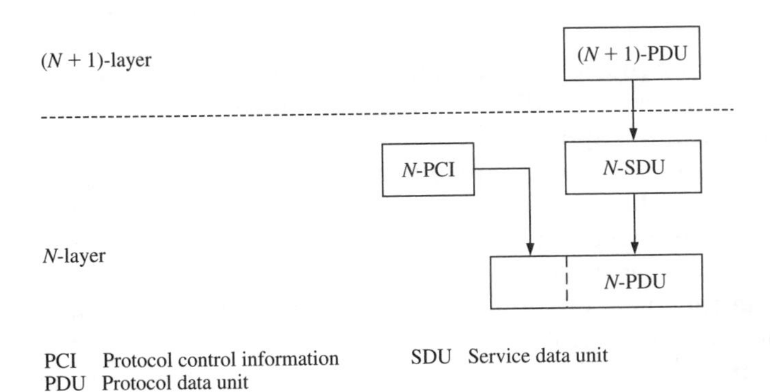

Figure 5.3 Relationships among the various types of data unit.

Table 5.1 OSI reference model.

Layer	*Name*
7	Application layer
6	Presentation layer
5	Session layer
4	Transport layer
3	Network layer
2	Data link layer
1	Physical layer

layer above, using the service provided by the layer below. This approach is also well suited to the B-ISDN PRM. Unfortunately, the exact relationship between the lower layers of the OSI reference model and those of the B-ISDN PRM is still not fully clarified.

5.1.4 B-ISDN PRM description

Figure 5.4 shows the B-ISDN PRM, which consists of three planes:

- user plane
- control plane
- management plane.

The **management plane** includes two types of function called **layer management** functions and **plane management** functions. All the management functions that relate to the whole system are located in the management plane which is responsible for providing coordination between all planes. No layered structure is used within this plane.

Layer management has a layered structure. It performs the management functions relating to resources and parameters residing in its protocol entities (for example, meta-signaling). Layer management handles the specific OAM information flows for each layer. More details about these management functions are presented in ITU-T Recommendation Q.940 (ITU-T, 1988f).

The **user plane** provides for the transfer of user information. All associated mechanisms, like flow control and recovery from errors, are included. A layered approach is used within the user plane.

A layered structure is also used within the **control plane**. This plane is responsible for the call control and connection control functions. These are all signaling functions which are necessary to set up, supervise and release a call or connection.

The functions of the physical layer (PL) and the ATM layer are the same for the control plane and the user plane. Different functions may occur in the ATM adaptation layer (AAL) as well as in higher layers.

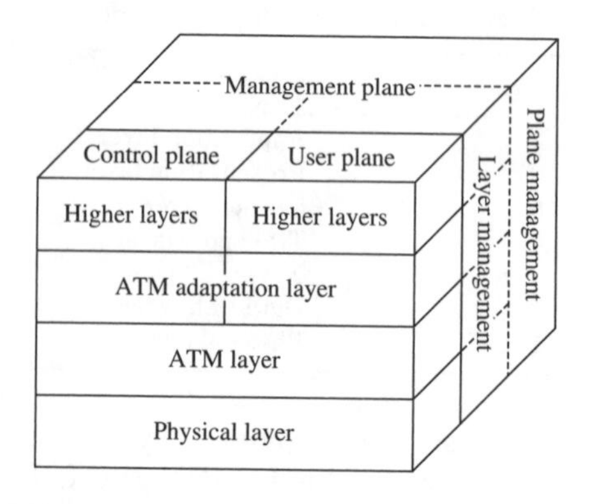

Figure 5.4 B-ISDN protocol reference model.

5.1.5 Layer functions

The functions of the physical layer, the ATM layer and the ATM adaptation layer are described in the following subsections. These descriptions are contained in ITU-T Recommendation I.321 (ITU-T, 1991e) and I.413 (ITU-T, 1993m). In Figure 5.5 the lower layers of the B-ISDN PRM and their functions are illustrated. Different fields within the cell header are used to implement some of these functions. These fields will be described in Section 5.6.2.

Cell terminology

Before the functions of the B-ISDN individual layers can be specified, it is necessary to clarify the notion **cell** because it is used for definitions in the ATM layer as well as in the physical layer.

The term cell is essential for B-ISDN, and therefore it is defined in ITU-T Recommendation I.113 (ITU-T, 1993e):

> 'A cell is a block of fixed length. It is identified by a label at the ATM layer of the B-ISDN PRM.'

More detailed definitions for the different kinds of cell are presented in ITU-T Recommendation I.321 (ITU-T, 1991e):

> **Idle cell** (physical layer): A cell that is inserted/extracted by the physical layer in order to adapt the cell flow rate at the boundary between the ATM layer and the physical layer to the available payload capacity of the transmission system used.
>
> **Valid cell** (physical layer): A cell whose header has no errors or has been modified by the cell header error control (HEC) verification process (for HEC mechanism see Section 5.3.1).
>
> **Invalid cell** (physical layer): A cell whose header has errors and has not been modified by the cell HEC verification process. This cell is discarded at the physical layer.
>
> **Assigned cell** (ATM layer): A cell that provides a service to an application using the ATM layer service.
>
> **Unassigned cell** (ATM layer): An ATM layer cell that is not an assigned cell.

Only assigned and unassigned cells are passed to the ATM layer from the physical layer. The other cells carry no information concerning the ATM and higher layers and therefore will be processed only by the physical layer.

Physical layer functions

The physical layer is subdivided into the **physical medium** (PM) sublayer and the **transmission convergence** (TC) sublayer.

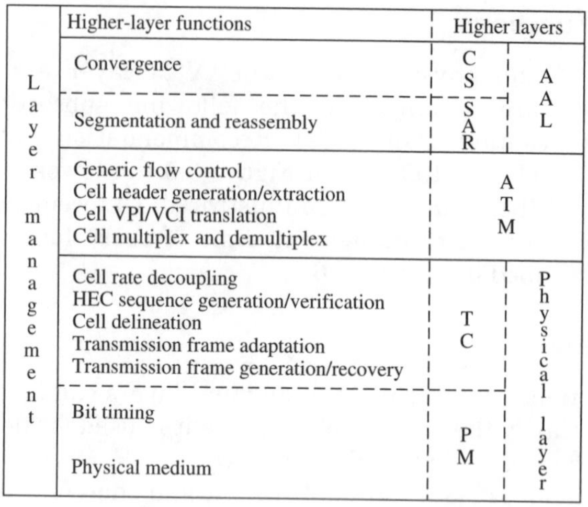

	Functions	Sublayer	Layer
Layer management	Higher-layer functions	Higher layers	
	Convergence	CS	AAL
	Segmentation and reassembly	SAR	AAL
	Generic flow control Cell header generation/extraction Cell VPI/VCI translation Cell multiplex and demultiplex	ATM	ATM
	Cell rate decoupling HEC sequence generation/verification Cell delineation Transmission frame adaptation Transmission frame generation/recovery	TC	Physical layer
	Bit timing Physical medium	PM	Physical layer

AAL ATM adaptation layer
CS Convergence sublayer
HEC Header error control
PM Physical medium
SAR Segmentation and reassembly
TC Transmission convergence
VCI Virtual channel identifier
VPI Virtual path identifier

Figure 5.5 The functions of B-ISDN in relation to the B-ISDN PRM.

The PM sublayer is the lowest sublayer and includes only the physical medium dependent functions. It provides the bit transmission capability, including bit alignment. Line coding and, if necessary, electrical/optical conversion is performed by this sublayer. In many cases, the physical medium will be an optical fiber. Other media, such as coaxial and twisted pair cables, are also possible. The transmission functions are medium specific.

Bit timing functions are the generation and reception of waveforms suitable for the medium, insertion and extraction of bit timing information, and line coding if required.

The TC sublayer performs five functions. The lowest function is the **generation and recovery of the transmission frame**.

Transmission frame adaptation is responsible for all actions necessary to adapt the cell flow according to the payload structure of the transmission system in the sending direction. In the opposite direction it extracts the cell flow from the transmission frame. This frame may be cell equivalent (no external envelope is used), a **synchronous digital hierarchy** (SDH) envelope or an envelope according to ITU-T Recommendation G.703 (ITU-T, 1991b). In the case of the B-ISDN UNI, ITU-T proposes an SDH envelope or the cell equivalent structure (ITU-T, 1996j). These alternatives are described in more detail in Section 5.3.

The functions mentioned so far are specific to the transmission frame. All other TC sublayer functions, which are presented in the following, can be common to all possible transmission frames.

Cell delineation is the mechanism that enables the receiver to recover the cell boundaries. This mechanism is described in ITU-T Recommendation I.432 (ITU-T, 1996j). The detailed procedure of the cell delineation mechanism is given in Section 5.3.1. To protect the cell delineation mechanism from malicious attack, the information field of a cell is scrambled before transmission. Descrambling is performed at the receiving side.

HEC sequence generation is done in the transmit direction. The HEC sequence is inserted in its appropriate field within the header. At the receiving side the HEC value is recalculated and compared with the received value. If possible, header errors are corrected, otherwise the cell is discarded. Details of the HEC mechanism are presented in Section 5.3.1.

In the sending direction, the **cell rate decoupling** mechanism inserts idle cells in order to adapt the rate of ATM cells to the payload capacity of the transmission system. In the receiving direction this mechanism suppresses all idle cells. Only assigned and unassigned cells are passed to the ATM layer.

ATM layer functions

The ATM layer is the layer above the physical layer. Its characteristic features are independent of the physical medium. Four functions of this layer have been identified.

In the transmit direction, cells from individual VPs and VCs are multiplexed into one resulting cell stream by the **cell multiplexing** function. The composite stream is normally a non-continuous cell flow. At the receiving side the **cell demultiplexing** function splits the arriving cell stream into individual cell flows appropriate to the VP or VC.

VPI and VCI translation is performed at ATM switching nodes and/or at cross-connect nodes. Within a VP node the value of the VPI field of each incoming cell is translated into a new VPI value for the outgoing cell. The values of the VPI and VCI are translated into new values at a VC switch.

The **cell header generation/extraction** functions are applied at the termination points of the ATM layer. In the transmit direction, after receiving the cell information field from the AAL, the cell header generation adds the appropriate ATM cell header except for the HEC value. VPI/VCI values could be obtained by translation from the SAP identifier. In the opposite direction, the cell header extraction function removes the cell header. Only the cell information field is passed to the AAL. This function could also translate a VPI/VCI into a SAP identifier.

The **GFC** function is only defined at the B-ISDN UNI. GFC supports control of the ATM traffic flow in a customer network. It can be used to alleviate short-term overload conditions (user-to-network direction) at the UNI (more details can be found in Section 5.6.2). GFC information is carried in assigned or unassigned cells.

ATM adaptation layer functions

The AAL is subdivided into the **segmentation and reassembly** (SAR) sublayer and the **convergence sublayer** (CS). The functions of the AAL are described in ITU-T Recommendation I.362 (ITU-T, 1993j).

The AAL is between the ATM layer and higher layers. Its basic function is the enhanced adaptation of services provided by the ATM layer to the requirements of the higher layer. Higher-layer PDUs are mapped into the information field of an ATM cell. AAL entities exchange information with their peer AAL entities to support AAL functions.

AAL functions are organized in two sublayers. The essential functions of the SAR sublayer are, at the transmitting side, segmentation of higher-layer PDUs into a suitable size for the information field of the ATM cell (48 octets) and, at the receiving side, reassembly of the particular information fields into higher-layer PDUs. The CS is service dependent and provides the AAL service at the AAL-SAP.

No SAP has yet been defined between these two sublayers. The need for such SAPs requires further study. Different SAPs for higher layers can be derived using different combinations of SAR and CS. For some applications, neither a CS nor an SAR is necessary, in which case they will be empty (see Section 5.7.1).

To minimize the number of AAL protocols, ITU-T proposed a service classification which is specific to the AAL. This classification was made with respect to the following parameters:

- timing relation
- bit rate
- connection mode.

Figure 5.6 depicts the AAL classes. Not all possible combinations make sense and therefore only four classes are distinguished.

Some examples for the different **service classes** are listed below:

Class A: Circuit emulation (for example, transport of a 2 Mbit/s or 45 Mbit/s signal), constant bit rate video

Class B: Variable bit rate video and audio

Class C: Connection-oriented data transfer

Class D: Connectionless data transfer.

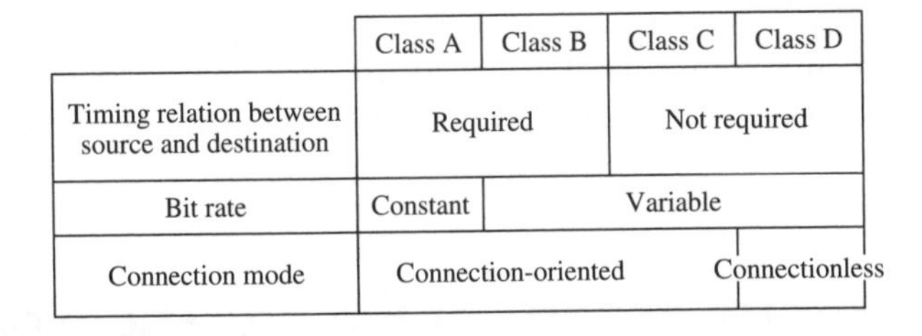

	Class A	Class B	Class C	Class D
Timing relation between source and destination	Required		Not required	
Bit rate	Constant	Variable		
Connection mode	Connection-oriented			Connectionless

Figure 5.6 Service classification for AAL.

Information flows of the physical layer

The following section looks at the information flows concerning the physical layer according to ITU-T Recommendation I.413 (ITU-T, 1993m). Different information flows exist:

- between the physical layer and the ATM layer
- between the physical layer and the management plane
- inside the physical layer between the sublayers.

From the PM sublayer to the TC sublayer, a flow of logical symbols (for example, bits) and associated timing information is exchanged. This information is also transferred in the opposite direction.

The physical layer provides the ATM layer with all cells belonging to this layer, with timing information and a clock derived from the line rate of the physical layer. In the opposite direction, assigned and unassigned cells, if any are available, and their associated timing are delivered. This layer informs the management plane about loss of the incoming signal and the indication of received errors or degraded performance. Bit errors may be detected by unexpected code violations or other mechanisms.

5.1.6 Relationship between OAM functions and the B-ISDN PRM

The relationship between OAM functions and the B-ISDN PRM is presented in ITU-T Recommendation I.610 (ITU-T, 1995g). A layered approach is used for the OAM functions which are allocated to the layer management. The different layer management functions are correlated with various layers. The independence requirement and the layered concept lead to the following principles:

(1) OAM functions related to OAM levels are independent of the OAM capabilities of other layers and have to be introduced at each layer.

(2) Each layer has its own OAM processing to obtain quality and status information. The results are delivered to the layer management or, if required, to the adjacent higher layer.

5.2 General aspects of the user–network interface

5.2.1 Transfer mode

As mentioned in Chapter 1, the B-ISDN UNI fully exploits the flexibility inherent in ATM. This means that the payload capacity of the interface (that is, the whole of the transmission capacity except for the small portion that is needed to operate the interface properly, see Section 5.3.1) is completely structured into ATM cells.

As at the UNI, there is no pre-assignment of cells to specific user applications; the actual use of cells for connections to be established across the interface can change dynamically. Different traffic mixes can easily be supported as long as the cell transfer capacity is not exceeded. (How this is related to the interface bit rate will be discussed later.)

5.2.2 Bit rates

As mentioned in Section 3.3, interface bit rates of about 150 Mbit/s and 600 Mbit/s at the physical layer have been specified. The exact figures are discussed below. It should be noted that the bit rate available for the end user's application is reduced at least by the overhead needed for operation of the physical layer, the ATM layer and the AAL.

The 150 Mbit/s interface is symmetric with respect to bit rate, offering 150 Mbit/s in both the network-to-user direction and the user-to-network direction. This sort of interface will predominantly be used for interactive services like telephony, videotelephony and data services. The extension of such an interface to a higher bit rate to meet the needs of users generating large traffic volumes seems quite natural. Thus, a bit rate-symmetric 600 Mbit/s interface was also conceived. (The reason for choosing 600 Mbit/s will soon become clearer.) Users who are expected to have a much higher traffic load from the network to themselves than in the other direction may get an asymmetric version of the 600 Mbit/s interface with a reduced upstream capacity (user-to-network direction) of only 150 Mbit/s. This would, for example, be suitable for the simultaneous transmission of several television programs to a residential customer who only needs the standard capacity for interactive services but a higher bit rate for distribution services like TV and sound programs.

The definition of two separate interface bit rates was a compromise between two diverging requirements, namely a very limited number of different interface types on the one hand and the cost-effectiveness of the interface on the other.

The exact interface bit rates at the physical layer are:

- 155.520 Mbit/s
- 622.080 Mbit/s.

These bit rates are identical to the two lowest bit rates of the SDH as defined in ITU-T Recommendation G.707 (ITU-T, 1996c). SDH is a transmission hierarchy which was adopted by ITU-T in 1988. It is based on the North American synchronous optical network (SONET) concept (ANSI, 1988) which was developed to:

- set a standard for optical transmission in order to react to the upcoming variety of manufacturer-specific implementations of optical transmission systems and interfaces;
- provide transmission facilities with flexible add/drop capabilities to allow for simpler multiplexing/demultiplexing of signals than in the existing plesiochronous digital hierarchy (PDH) (ITU-T, 1988a);
- grant generously dimensioned transmission overhead capacity to cater for various existing and assumed network operation and maintenance applications that were not, or at least only with difficulty, realizable in PDH.

More details on SDH can be found in Section 5.3.1 and Chapters 6 and 13 as well as in Breuer and Heelström (1990).

5.2.3 Interface structure

SDH has the inherent flexibility to transport quite different types of signals like ISDN channels, according to ITU-T Recommendation I.412 (ITU-T, 1988d), or ATM cells. Thus SDH – being a universal transmission concept – was proposed as the B-ISDN interface structure (ITU-T, 1987). Such a user–network interface implementation would have the advantage of full compatibility with the network–node interface. This avoids the otherwise necessary conversion of signals that are sent from the user through the network to other users. This property is extremely useful during the introductory phase of B-ISDN when a complete network infrastructure does not yet exist. Customers may easily be provided with access to a broadband network node (a cross-connect or a switch) which might be located in a different place from that of the 64 kbit/s ISDN local exchange that usually serves the customer, via SDH equipment. In some networks SDH will be implemented before B-ISDN.

The SDH-based B-ISDN UNI does have some drawbacks. A minor one is that the large overhead capacity provided by SDH is not actually needed at the user–network interface. However, this is not a strong argument against using SDH. Generation of the byte-structured SDH frame (see Figure 5.7) requires interface functions that would not be necessary if the interface was completely cell structured (this is a possible solution as no 'physical' channels are foreseen, as in the case of the 64 kbit/s ISDN – see ITU-T Recommendation I.412 (ITU-T, 1988d).

Insertion of ATM cells from several terminals into one SDH frame in a passive bus configuration (as standardized for the 64 kbit/s ISDN, see ITU-T Recommendation I.430 (ITU-T, 1995f) is almost impossible for realistic transmission lengths because of individually varying transmission delays and the high bit rates involved. (Though passive configurations are no longer foreseen in ITU-T Recommendation I.413 (ITU-T, 1993m), this was put forward as an argument in favor of the cell-based interfaces.)

For these reasons, two interface types were standardized (ITU-T, 1993m, 1996j): one based on SDH and the other based on mere cell multiplexing (see Figure 5.7).

Remarkably, although the interface bit rate of the cell-based interface is basically not determined by any frame structure but might be chosen freely, it was agreed within ITU-T to adopt the same bit rate (and payload capacity, see Section 5.3.1) for both interface types to facilitate interworking between the cell-based UNI and SDH in the network.

5.2.4 B-ISDN UNI reference configuration and physical realizations

As already mentioned in Section 4.7.1, the ISDN reference configuration for the basic access and primary rate access (ITU-T, 1993l) was applied

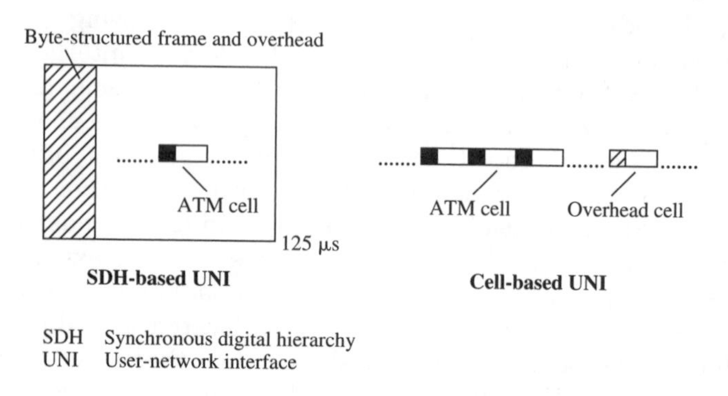

Figure 5.7 User–network interface options.

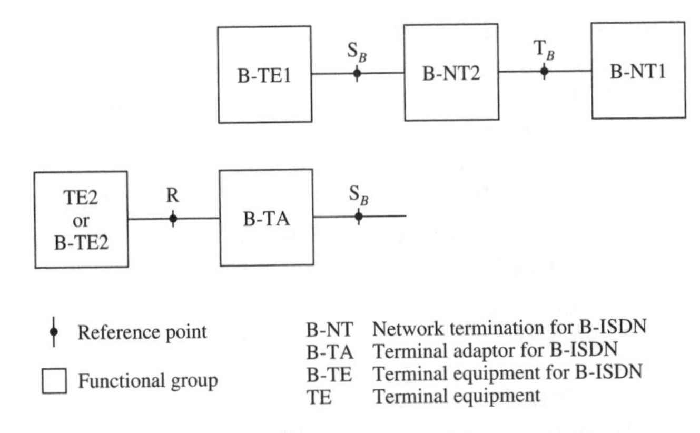

Figure 5.8 B-ISDN UNI reference configuration.

to B-ISDN with only minor modifications of notation. A reference configuration of the user–network access is a generic description based on two elements (see Figure 5.8):

- functional groups
- reference points.

Figure 5.8 mainly shows broadband functional groups and reference points. The corresponding entities for the 64 kbit/s ISDN are described in ITU-T Recommendation I.411 (ITU-T, 1993l).

The B-NT1 includes functions broadly equivalent to layer 1 of the OSI reference model (ITU-T, 1994d). Examples of B-NT1 functions are (according to ITU-T Recommendation I.413 (ITU-T, 1993m):

- line transmission termination
- transmission interface handling
- OAM functions.

The B-NT2 includes functions broadly equivalent to layer 1 and higher layers of the OSI reference model. Examples of B-NT2 functions are (ITU-T, 1993m):

- adaptation functions for different interface media and topologies
- multiplexing/demultiplexing/concentration of traffic
- buffering of ATM cells
- resource allocation; usage parameter control
- signaling protocol handling
- interface handling
- switching of internal connections.

The generic term 'B-NT2' covers a host of actual implementations: it may not actually exist ('null B-NT2' provided that the interface definitions allow direct connection of terminals to the B-NT1), may consist solely of layer 1 connections (wires), provide concentrating and/or multiplexing functions, or be a full-blown switch (private branch exchange). The B-NT2 functions may be concentrated or distributed, for example, in a bus or ring with its access nodes (see Section 4.7).

S_B and T_B denote reference points between the terminal and the B-NT2, and between B-NT1 and B-NT2, respectively. Physical interfaces need not occur at a reference point in any case. As an example, B-NT1 and B-NT2 functions might be combined as shown in Figure 5.9.

It is also possible for the terminal to include B-NT2 functionality (see Figure 5.10).

If the same interface standard applies to both S_B and T_B, these reference points may coincide (see Figure 5.11), thus permitting the direct connection of a terminal to the B-NT1.

B-ISDN will also offer 64 kbit/s ISDN services and interfaces. Terminal equipments 1 (TE1 complying with ITU-T Recommendation I.430 (ITU-T, 1995f), basic access) can be connected via such standard interfaces at the S reference point (as depicted in Figure 5.12).

Of course B-NT2 may provide multiple interfaces of each type at the S and S_B reference points.

Finally, to complete this brief description of the B-ISDN reference configurations, let us address the lower part of Figure 5.8 showing the functional groups broadband terminal adaptor (B-TA) and TE2/B-TE2, and the R reference point between these functional groups. Whereas there must be standardized broadband interfaces at S_B and T_B (according to ITU-T Recommendation I.432 (ITU-T, 1996j), at R any other (non-ISDN) interface can be used to connect a non-ISDN terminal

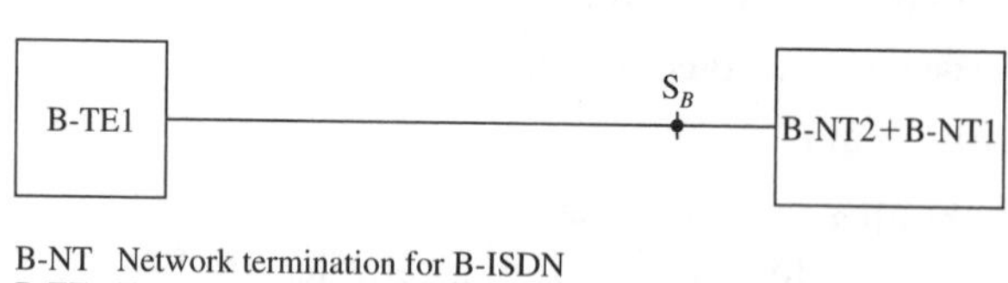

B-NT Network termination for B-ISDN
B-TE Terminal equipment for B-ISDN

Figure 5.9 Configuration with physical interface at S_B only.

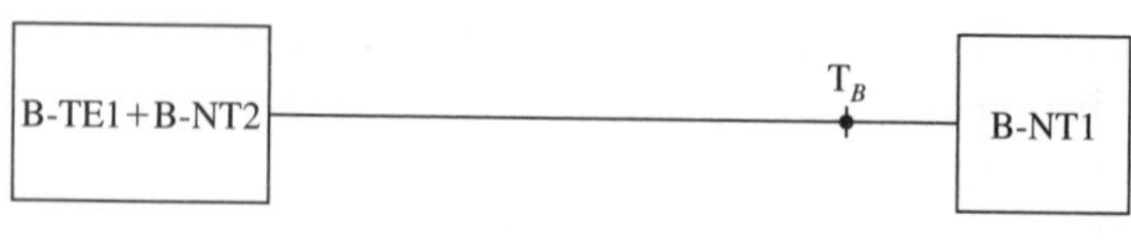

B-NT Network termination for B-ISDN
B-TE Terminal equipment for B-ISDN

Figure 5.10 Configuration with physical interface at T_B only.

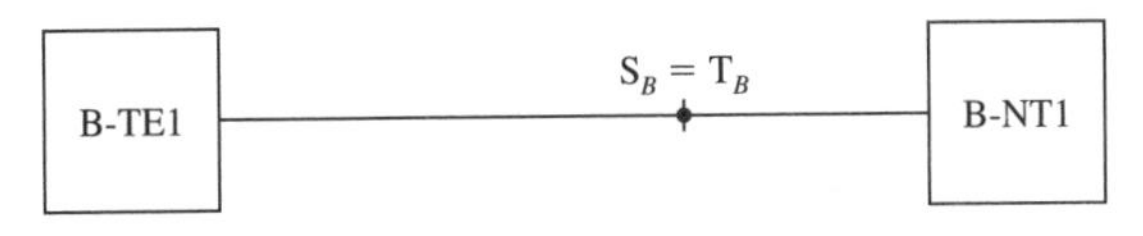

B-NT Network termination for B-ISDN
B-TE Terminal equipment for B-ISDN

Figure 5.11 Coinciding S_B and T_B.

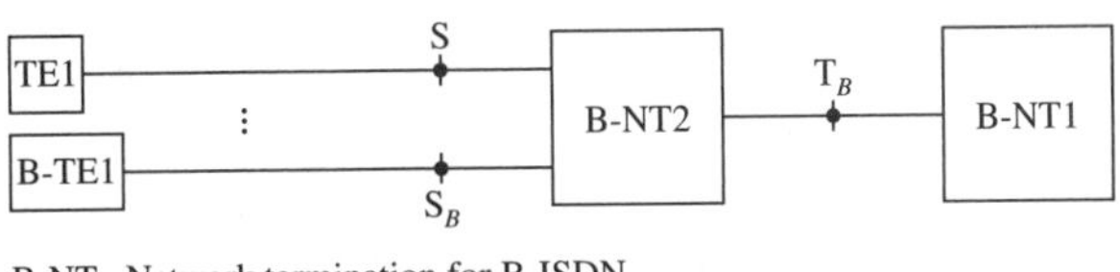

B-NT Network termination for B-ISDN
B-TE Terminal equipment for B-ISDN

Figure 5.12 B-NT2 offering 64 kbit/s interfaces and broadband interfaces.

(TE2 or B-TE2 in the figure, denoting a narrowband terminal or broadband terminal, respectively).

The provision of multiple terminal interfaces by the B-NT2, as indicated in Figure 5.12, is not restricted to a specific topology. Star, bus or ring configurations, or even mixtures of those topologies, such as starred bus, are possible (see Figure 5.13).

It may have become obvious from the previous figures that the B-ISDN standard allows many types of implementation according to the customers' needs. However, ITU-T Recommendation I.413 (ITU-T, 1993m) contains two restrictions:

(1) Only one interface per B-NT1 is allowed at the T_B reference point.

(2) The interfaces are point-to-point at the physical layer 'in the sense that there is only one sink (receiver) in front of one source (transmitter)' (ITU-T, 1993m). (This means that passive bus configurations are not supported. As such configurations are strongly restricted in terms of the number of connectable terminals and coverable transmission distance, they have been excluded as inappropriate for broadband.)

5.2.5 Special issues

Configurations like that in Figure 5.11 require a high degree of commonality between the interfaces at the S_B and T_B reference points. Of course, the terminal interface at S_B should be unique in order to support worldwide terminal interchangeability. However, whether the last requirement can be met is not certain as there are currently two

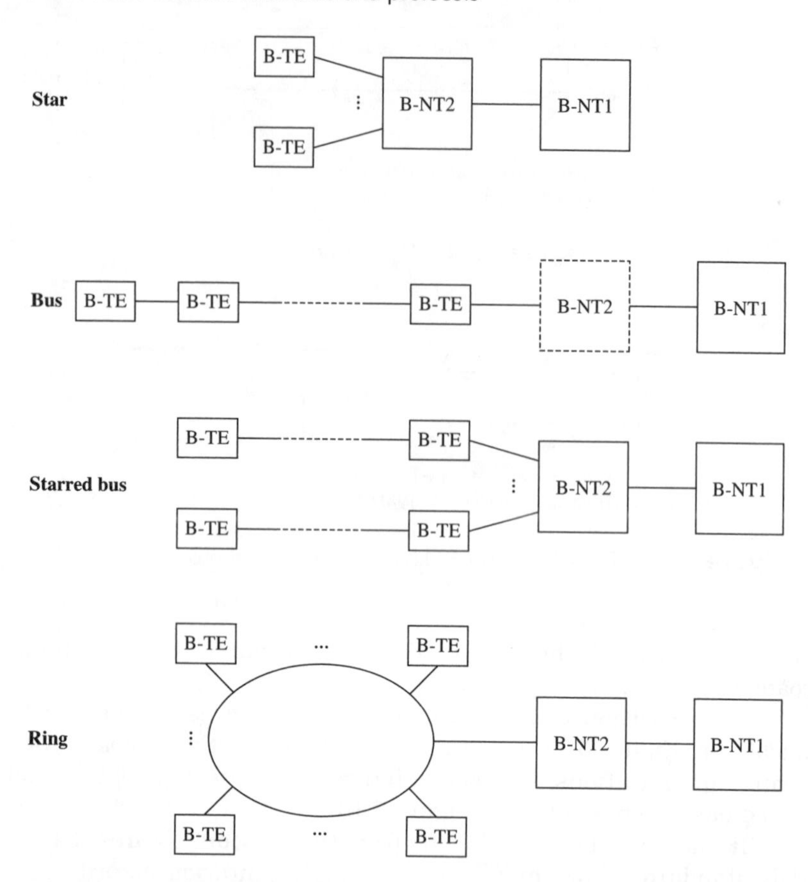

Figure 5.13 Multiple interface arrangements.

interface options at T_B (see Section 5.2.3) and there is strong support for the interfaces S_B and T_B being identical.

Another important issue is how, and to what extent, the standard interface should support shared medium configurations (bus and ring structures, as shown in Figure 5.13). Such configurations require medium access control functions which either can be part of the standard interface functionality, or otherwise have to be implemented whenever necessary in additional medium access control entities. (This problem, which affects the B-ISDN UNI definition, was introduced in Section 4.7.4.)

In a shared medium configuration, two different operation modes exist:

- **Distributed multiplexing distributed demultiplexing** (DMDD): This is a mechanism for collecting cells from multiple terminals for service at the B-NT2 or the local exchange. In the opposite direction it provides a mechanism for the distribution of cells to multiple terminals. In the downstream direction (network towards terminal) a terminal copies all the cells it has to handle. In the upstream direction (opposite direction) the terminal waits for an idle or unassigned cell which can be used to transport information towards the network. Such a terminal is called **unidirectional** because it can only send its own cells towards the network and receive cells pertaining to it from the network direction. All cells for internal communication are served by the B-NT2 or the local exchange.
- **LAN-like switching** (LLS): In this operation mode, **bidirectional** terminals are used. A terminal can receive and send cells in both directions. Therefore, cells used for internal communication can be served locally without involving the B-NT2 or the local exchange. This mode unburdens the B-NT2 or the local exchange from switching functions. However, LLS increases the complexity within the terminal.

In a star configuration as well as in a dual bus configuration using DMDD, only unidirectional terminals are necessary. The same type of terminal can be used as an **end terminal** in the dual bus system operating in the LLS mode. All other terminals in a dual bus with LLS are bidirectional. This does not contravene the requirement for terminal interchangeability since bidirectional terminals can also operate in the unidirectional mode. Thus the customer has a choice between the fully interchangeable bidirectional terminal and the simpler unidirectional one which has somewhat restricted functionality but may suffice in certain cases.

5.3 Physical layer of the user–network interface at 155/622 Mbit/s

According to the B-ISDN protocol reference model described in Section 5.1, the physical layer is split into two sublayers. These sublayers and their functions are shown in Figure 5.14.

The sublayer functions shown in Figure 5.14 were defined in Section 5.1. The following provides a detailed description of them based on ITU-T Recommendation I.432.1–5 (ITU-T, 1996j). OAM aspects are treated in Chapter 6.

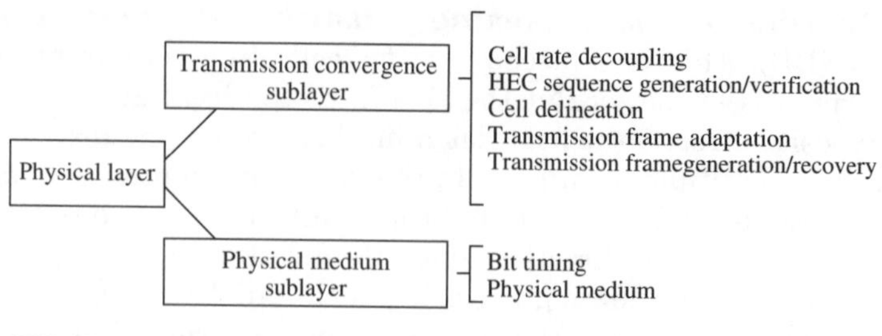

Figure 5.14 Physical layer structure.

5.3.1 Functions of the transmission convergence sublayer

At the physical bit level the SDH-based B-ISDN user–network interface has a gross bit rate of 155.520 Mbit/s or 622.080 Mbit/s. Its **interface transfer capability** is defined as (ITU-T, 1996j):

> 'the bit rate available for user information cells, signaling cells and ATM and higher layer OAM information cells, excluding physical layer frame structure bytes or physical layer cells.'

Its value of 149.760 Mbit/s for the 155.520 Mbit/s interface complies with SDH. The transfer capability of the 622.080 Mbit/s interface is 599.040 Mbit/s (four times 149.760 Mbit/s).

Additional user–network interfaces (which are especially important in the introduction phase of ATM) are explained in Section 5.4.

In the following, the transmission frame generation/adaptation aspects of the two interface options (SDH-based and cell-based) are described separately as they are rather different. The other transmission convergence sublayer functions can, in principle, be performed the same way in both options.

SDH-based interface at 155.520 Mbit/s

ITU-T Recommendation G.707 (ITU-T, 1996c) specifies the SDH. The transmission frame structure is shown in Figure 5.15. This frame is byte-structured and consists of nine rows and 270 columns. The frame repetition frequency is 8 kHz (9×270 byte $\times$ 8 kHz = 155.520 Mbit/s). The first nine columns comprise the section overhead (SOH) and administrative pointer 4 (AU-4). Another 9-byte column is dedicated to the path overhead (POH).

This structuring of transmission overheads complies with the OAM levels already introduced in Section 4.6.2. The use of these overhead bytes will be described in Chapter 6.

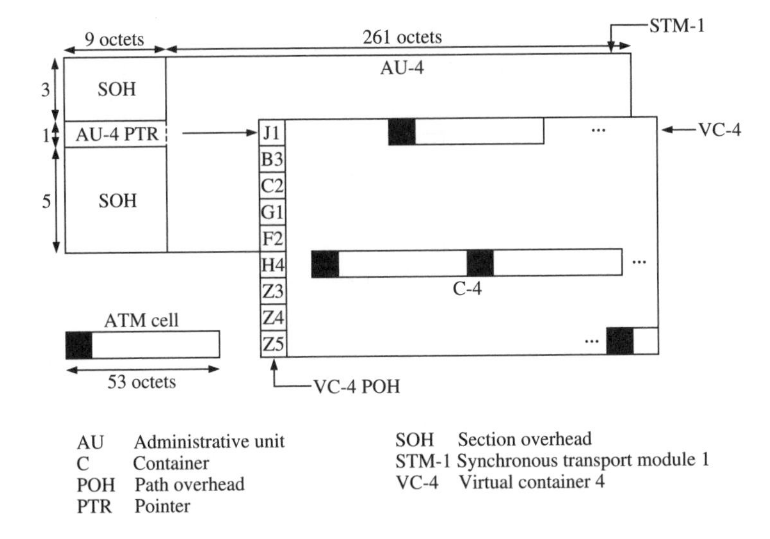

Figure 5.15 Frame structure of the 155.520 Mbit/s SDH-based interface.

Generation of the SDH-based user–network interface signal is as follows (ITU-T, 1996j). First the ATM cell stream is mapped into container 4 (C-4) (the SDH terminology is defined in ITU-T Recommendation G.707; here it is sufficient to say that C-4 is a 9 row × 260 column container corresponding to the transfer capability of 149.760 Mbit/s). Next C-4 is packed in virtual container 4 (VC-4) along with the VC-4 POH. The ATM cell boundaries are aligned with the byte boundaries of the frame. Note that an ATM cell may cross a C-4 boundary; as the C-4 capacity (2340 bytes) is not an integer multiple of the cell length (53 octets) and the C-4 capacity is entirely used for cell mapping, this will normally be the case.

Virtual container VC-4 is then mapped into the 9 × 270-byte frame (known as synchronous transport module 1, STM-1). The AU-4 pointer is used to find the first byte of VC-4. In principle, the first VC-4 byte can be located elsewhere in the STM-1 frame (excluding the first nine SOH columns).

POH bytes J1, B3, C2, G1 are activated (for the meaning of these bytes, see Chapter 6). Use of the remaining POH bytes is for further study.

SDH-based interface at 622.080 Mbit/s

There is a straightforward way of creating a 622.080 Mbit/s frame (STM-4) from four STM-1s according to ITU-T Recommendation G.707 (see Figure 5.16). However, the STM-4 payload can be structured in several ways, including simply as 4 × VC-4 or as one block. An

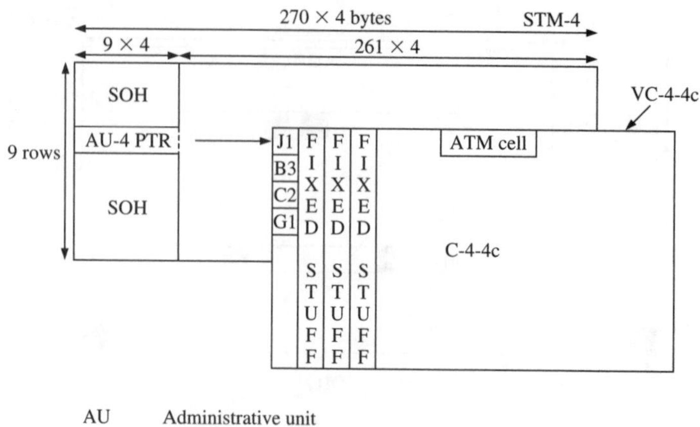

Figure 5.16 Frame structure of the 622.080 Mbit/s SDH-based interface.

advantage of the second option is the unrestricted use of about 600 Mbit/s for ATM cell mapping.

Possible future applications requiring more than the VC-4 capacity can easily be allocated bandwidth elsewhere in the 600 Mbit/s payload block. Therefore, ITU-T chose this option. The payload is exactly 4 × 149.760 Mbit/s = 599.040 Mbit/s. Although only one POH column is needed, the adjacent three columns in VC-4-4c are not usable as they are filled with stuffing octets ('fixed stuff'). This allows transmission of the STM-4 payload over four STM-1 links in cases where STM-4 transmission systems are not available in an ATM network.

Note that ATM cell mapping into STM-16 (2.5 Gbit/s) has been defined in a similar way. However, no UNI at a bit rate above STM-4 has been standardized as yet.

Cell-based interface

These interfaces (at both 155.520 Mbit/s and 622.080 Mbit/s) consist of a continuous stream of cells (see Figure 5.17) each containing 53 octets.

The maximum spacing between successive physical layer cells is 26 ATM layer cells. After 26 contiguous ATM layer cells the insertion of a physical layer cell is enforced in order to adapt the transfer capability to the interface rate. (The ratio 26:27 is the same as 149.760 Mbit/s to 155.520 Mbit/s or 599.040 Mbit/s to 622.080 Mbit/s.) When no ATM layer cells are available, physical layer cells are inserted.

The physical layer cells which are inserted at the transmitting side can be either idle cells (see next subsection and Table 5.3) or

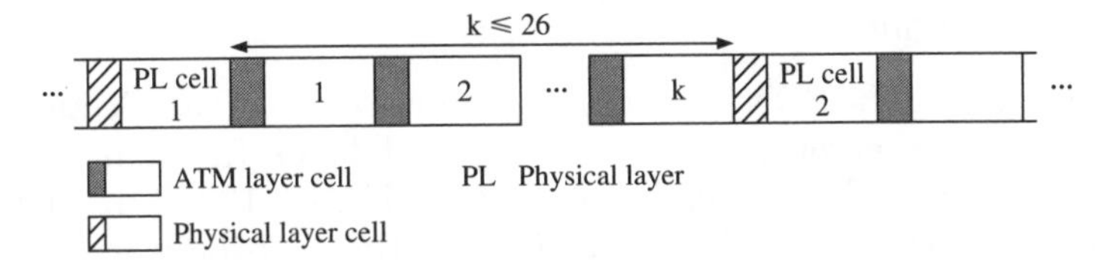

Figure 5.17 Structure of cell-based interfaces.

physical layer OAM cells, depending on operation and maintenance requirements.

Physical layer OAM information (which in the SDH case is allocated to SOH and POH) is here conveyed in specific physical layer OAM cells which are identified by unique cell header bit patterns reserved exclusively for these cell types. (This means that the ATM layer must not use these bit combinations as code points for the corresponding cell header fields, which are shown later on in Figure 5.31. The patterns are shown in Table 5.2.

Note that at the cell-based interfaces only the F1 and F3 OAM flows of Figure 4.9 occur. Levels F2 and F3 coincide for the cell-based interfaces, and the corresponding functions are supported by F3.

Cell rate decoupling

Whenever no assigned, unassigned or physical layer OAM cell is available for transmission, an **idle cell** will be inserted to adapt the cell stream to the transmission bit rate. Any idle cells will be discarded at the receiving side. The insertion and discarding of idle cells is called **cell rate decoupling**.

Idle cells are identified by a standardized pattern for the cell header, which is shown in Table 5.3. This is used throughout the ATM network to identify idle cells. Each octet of the information field of an idle cell is filled with 01101010.

Table 5.2 Header pattern for physical layer OAM cells.

Cell type	*Octet 1*	*Octet 2*	*Octet 3*	*Octet 4*	*Octet 5*
F1	00000000	00000000	00000000	00000011	HEC valid code
F3	00000000	00000000	00000000	00001001	HEC valid code

Table 5.3 Header pattern for idle cell identification.

Octet 1	*Octet 2*	*Octet 3*	*Octet 4*	*Octet 5*
00000000	00000000	00000000	00000001	HEC valid code

Header error control

According to the B-ISDN protocol reference model (see Section 5.1), ATM cell header error control is a physical layer function and it is described in ITU-T Recommendation I.432.1 (ITU-T, 1996j). It should be noted that the HEC method standardized for the user–network interface can be employed universally in the ATM network.

First the HEC generation algorithm is described and then its capabilities.

Every ATM cell transmitter calculates the HEC value across the first four octets of the cell header and inserts the result in the fifth (HEC field). The HEC value is defined as 'the remainder of the division (modulo 2) by the generator polynomial $x^8 + x^2 + x + 1$ of the product x^8 multiplied by the content of the header excluding the HEC field' (the transmitter device computing this remainder presets its register to all 0s before performing the division), to which the fixed pattern '01010101' will be added modulo 2.

This HEC code is capable of:

- correcting single-bit errors
- detecting certain multiple-bit errors

in the ATM cell header. Both error processing capabilities will be used by the equipment receiving ATM cells according to the state–event diagram depicted in Figure 5.18.

After initialization the receiver is in 'correction mode'. When a single-bit error is detected, it is corrected; when a multi-bit error is detected, the cell is discarded. In both cases, the receiver switches into 'detection mode'. In this receiver state, each cell with a detected single-bit or multi-bit error is discarded. If no further errored cell is detected the receiver switches back to 'correction mode'.

The above receiver operation has been chosen to take into account the error characteristics of fiber-based transmission systems. These exhibit a mix of single-bit errors and relatively large **error bursts** (on a low error level). The specified HEC method ensures recovery from single-bit errors, and a low probability of delivering cells with errored headers under bursty error conditions.

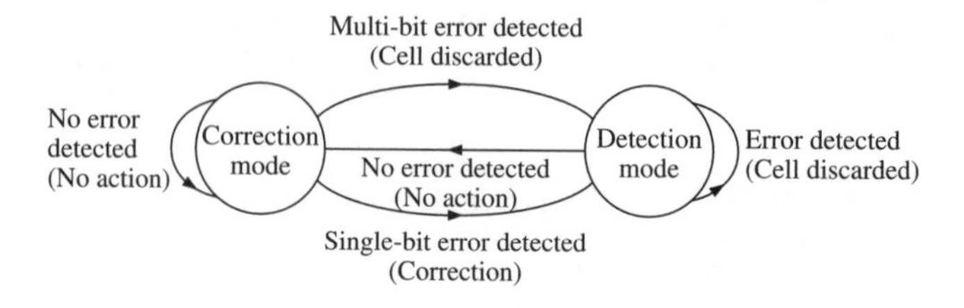

Figure 5.18 HEC receiver actions.

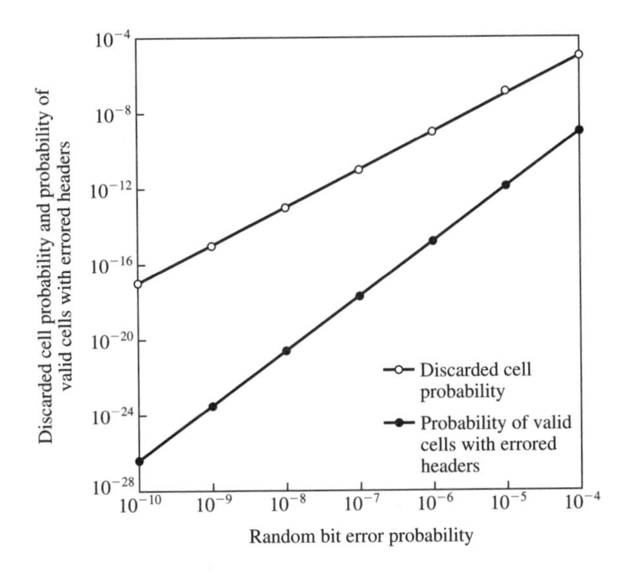

Figure 5.19 HEC performance.

Figure 5.19 shows the performance of the HEC mechanism as a function of bit error probability.

For a bit error probability of, say, 10^{-8} (a realistic value for fiber-based transmission systems) the probability that cells will be discarded (as a result of header errors which have been recognized but could not be corrected) is about 10^{-13}. The probability of valid cells with errored headers (cells with unrecognizable header errors) is about 10^{-20}.

Cell delineation

According to ITU-T Recommendation I.432.1 (ITU-T, 1996j) 'cell delineation is the process which allows identification of the cell boundaries'. The method recommended by ITU-T for cell delineation is based on the correlation between the header bits to be protected (the first four octets of the cell header) and the relevant control bits (1-octet HEC field, see the previous subsection). Figure 5.20 shows the state diagram for HEC-based cell delineation.

In the HUNT state, a bit-by-bit check of the assumed header field is performed. When more information is available (for example, octet boundaries), this may optionally be used. When the HEC coding law (see previous subsection) is seen to be respected, that is, syndrome equals zero, one assumes that a header has been found and goes to the PRESYNCH state. Now the HEC correlation check is performed cell by cell. If δ consecutive correct HECs are found, the SYNCH state is entered; if not, the system goes back to the HUNT state. The system

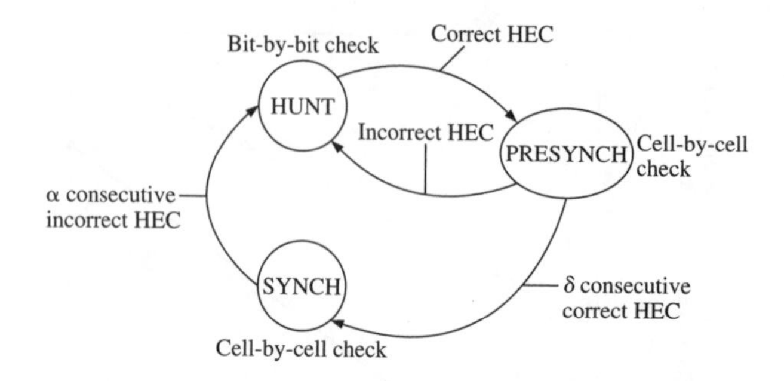

Figure 5.20 Cell delineation state diagram.

leaves the SYNCH state for the HUNT state if α consecutive incorrect HECs are identified.

The values for α and δ obviously influence the performance of the cell delineation process. Robustness against false misalignment caused by bit errors depends on α, while robustness against false delineation in the resynchronization phase depends on δ. The following values of α and δ have been fixed:

$\alpha = 7$, $\delta = 6$ for SDH-based interfaces
$\alpha = 7$, $\delta = 8$ for cell-based interfaces.

With $\alpha = 7$, a 155.520 Mbit/s ATM system will be in SYNCH for more than one year even when the bit error probability is about 10^{-4}. With $\delta = 6$, the same system with the same bit error probability will need about 10 cells or 28 μs to re-enter SYNCH after loss of cell synchronization (ITU-T, 1996j).

This cell delineation method could fail if the header HEC correlation were imitated in the information field of ATM cells. This might be effected by a malicious user or might be caused inadvertently by an application that accidentally uses the same generator polynomial. To overcome such difficulties, the information field contents will be scrambled using a self-synchronizing scrambler with the polynomial $x^{43} + 1$ in the case of SDH-based interfaces. The scrambler is effective only in PRESYNCH and SYNCH and is disabled in HUNT.

For cell-based interfaces, a distributed sample scrambler of 31st order has been specified in Recommendation I.432.1.

5.3.2 Physical medium characteristics

This section addresses topics such as:

- physical medium to be deployed at the user–network interface at 155/622 Mbit/s;

- bit timing and interface code;
- power feeding;
- modes of operation of the interface.

Physical medium

The broadband integrated services digital network will largely be based on the deployment of optical fiber transmission (see Section 3.3) in the trunk network and the user access network. However, it is not so evident whether optical transmission should also be used at the interfaces at the S_B and T_B reference points.

As the interface range at S_B and T_B is usually much shorter than the distances that have to be bridged in the access network, electrical media can be employed at S_B and T_B (at least for the 155.520 Mbit/s interface). Such an electrical interface was assessed to be cheaper (at least as a short- and medium-term solution) and easier to handle in terms of installation and maintenance. For the 155.520 Mbit/s interface, a range of up to about 200 m can be covered with an electrical interface; for the 622.080 Mbit/s interface the range would decrease considerably to about 100 m (RACE, 1989).

Use of the same medium for 155.520 Mbit/s and 622.080 Mbit/s would support upgradability of the 155.520 Mbit/s interface to the higher bit rate. The same medium at S_B and T_B would allow for terminal portability from S_B to T_B. The optional use of both media at the interface at either reference point would require medium adaptors in all cases where equipment is to be used that does not comply with the deployed medium.

It should be noted that in the case of electrical interfaces, longer interface ranges than 100 m to 200 m can be achieved by, for example, inserting an optical transmission system. This solution, however, requires electrical/optical conversion twice.

ITU-T has specified both an electrical and an optical 155.520 Mbit/s interface to be applied at T_B and S_B. The maximum range of the electrical interface depends on the attenuation of the transmission medium. A maximum range of about 100 m can be achieved for microcoax (4 mm diameter) and 200 m for cable TV (CATV) type (7 mm diameter). For the optical interface, the attenuation of the optical path has been specified to be in the range of 0 to 7 dB.

The feasibility and range of application of an electrical 622.080 Mbit/s interface needs further study. For the optical 622.080 Mbit/s interface the same attenuation range of 0 to 7 dB as for the 155.520 Mbit/s interface has been specified.

Detailed information on the physical medium characteristics of the electrical 155.520 Mbit/s interface and optical interfaces (at both 155.520 Mbit/s and 622.080 Mbit/s) is presented in Tables 5.4 and 5.5.

Table 5.4 155.520 Mbit/s electrical interface characteristics.

Item	*Specified solution*
Transmission medium	Two coaxial cables, one for each direction
Wiring configuration	Point-to-point
Impedance	75 Ω with a tolerance of ±5% in the frequency range 50–200 MHz
Attenuation of the electrical path	Approximate $\sqrt{f}$ law with a maximum insertion loss of 20 dB at a frequency of 155.520 MHz
Electrical parameters	According to ITU-T Recommendation G.703 (ITU-T, 1991b)

Table 5.5 Optical interface characteristics for 155.520 Mbit/s and 622.080 Mbit/s.

Item	*Specified solution*
Attenuation range	0–7 dB
Transmission medium	Two single mode fibers according to ITU-T Recommendation G.652 (ITU-T, 1993a), one for each direction
Operating wavelength	1310 nm (second window)
Optical parameters	According to ITU-T Recommendation G.957 (ITU-T, 1995c)
Safety requirements	Parameters for IEC 825 (IEC, 1993) Class 1 devices should not be exceeded even under failure conditions

Bit timing and interface code

In normal operation, *timing* for the transmitter is locked to the timing received across the interface. Timing may alternatively be provided locally by the clock of the customer equipment in the case of the cell-based interface option. Locally provided timing will be used under fault conditions; then the interface works in free-running clock mode. For this mode, a tolerance ± 20 ppm has been defined.

Coded mark inversion (CMI) as described in ITU-T Recommendation G.703 (ITU-T, 1991b) has been chosen as the interface code for the electrical 155.520 Mbit/s interface. CMI has several advantages (see RACE, 1989):

- Simple implementation (for example, easy clock extraction).
- Zero direct current (DC) and low frequency content.
- Guaranteed signal transitions: the number of transitions in the encoded data stream is independent of the applied data stream. Bit sequence independence required for the interface (ITU-T, 1993m) is easily achieved (and malicious attack to prevent timing extraction is made impossible).
- No bit error multiplication.
- Ability to trace discrete bit errors through code violations.

However, CMI has the drawback of doubling the transmission rate of the CMI-coded signal (baud rate = 2 × bit rate). This is not critical in the case of the 155.520 Mbit/s interface but is clearly disadvantageous for the 622.080 Mbit/s interface.

Therefore the commonly used line coding will be non-return to zero (NRZ) for the optical interfaces at 155.520 Mbit/s and 622.080 Mbit/s. The convention used for the optical logic levels is:

- emission of light for a binary ONE
- no emission of light for a binary ZERO.

The extinction ratio must be in accordance to the definitions given in ITU-T (1995c).

Power feeding

An optional power feeding mechanism similar to that used for the primary rate access (ITU-T, 1993o) has been specified in I.432.2. This method is briefly described.

A separate pair of wires at the interface at the T_B reference point is to be used to power the B-NT1 via this interface. The power sink is fed:

- either by a source under the responsibility of the user when requested by the network provider;
- or by a power supply unit, under the responsibility of the network provider, connected to the mains electric supply on the customer's premises.

When power is provided by the user, the source may be an integral part of the B-NT2 or B-TE, or it may be physically separated from B-NT2 or B-TE as an individual power supply unit.

The power available at the B-NT1 via the user–network interface (at T_B) should be at least 15 W, with the feeding voltage in the range of −20 V to −57 V relative to ground.

The power source must be protected against short circuits and overload, and the power sink of B-NT1 should not be damaged by interchanging the wires.

Modes of operation

The B-ISDN user–network interface is normally in the 'fully active' state (ITU-T, 1993m). Other modes of operation have been under discussion, including an emergency mode in the case of power failure and a deactivated mode to save power. If activation/deactivation is to be implemented it will become necessary to define:

- activation/deactivation signals
- activation/deactivation procedures.

Deactivation of the interface would be used to minimize power consumption during idle periods when no connections are established. The emergency mode is necessary to guarantee minimum communication facilities in the case of power failure. For example, at least one telephone set should work. (This mode requires battery backup in B-NT1 and/or B-NT2.)

Some considerations regarding economizing on power consumption by establishing a deactivated interface state can be found in ETSI (1990). Deactivation allows for:

- extension of the time of the battery-powered operation under power failure conditions (deactivation is especially important in the emergency state);
- reduction of the heat management problem.

These benefits could be increased if activation/deactivation of the interface were accompanied by activation/deactivation of the line. In this case the network operator would also get relief in terms of power consumption and heat management at the local exchange or remote switching/multiplexing/concentrating unit.

In ETSI (1990) it is reported that in the deactivated state the power consumption of the network termination and line termination is about one-half of the corresponding value in normal (fully active) mode. The time taken to switch from the deactivated state back to the active state is estimated to be about 10 ms to 50 ms. The same handshake principle as described in ITU-T Recommendation I.430 (ITU-T, 1995f) for the basic access could be used as a basis for the B-ISDN activation/deactivation procedure.

It should be noted that the transition of the interface from deactivated state to active state is not only a physical layer task. It also affects higher layers, such as the ATM layer where recovery procedures may be necessary.

The present interface specifications by ITU-T and the ATM Forum no longer consider such a deactivated state.

5.4 Additional user–network interfaces

5.4.1 Objectives and overview

In many ATM-based networks, one of the main applications will be data communications between business users. Often the full range of possible broadband services is not required, leading to reduced bit rate requirements at the user–network interface. ATM interfaces at the following bit rates below 155 Mbit/s have been introduced to cater for such applications:

- 100 Mbit/s (at S_B)
- 51.84 Mbit/s (at S_B)
- 34/45 Mbit/s based on PDH bit rates ("E3"/"T3") (ITU-T, 1988a)
- 25.6 Mbit/s (at S_B)
- 1.5/2/6 Mbit/s based on PDH bit rates ("T1"/"E1"/"T2") (ITU-T, 1988a) as 'poor man's ATM interfaces'.

The rationale for these additional interface types is simply cost-effectiveness, and sometimes the non-availability of SDH-based transmission systems in the access network. The reason for choosing terminal interfaces at 52 Mbit/s and 25.6 Mbit/s is the desire to utilize low-cost (preferably existing) in-house cabling; 1.5/2 Mbit/s ATM interfaces can be provided via ordinary copper wiring for telephony using the high-speed digital subscriber line (HDSL) technique (Koljonen, 1992). In the following some more technical details about these interfaces are given.

5.4.2 ATM interfaces based on PDH bit rates

ITU-T Recommendation G.804 (ITU-T, 1993b) defines ATM cell mapping into PDH signals. In fact only the PDH bit rates of ITU-T Recommendation G.702 (ITU-T, 1988a) have been retained completely; the old frame structures of the PDH signals have been partly replaced with new ones (described in ITU-T Recommendation G.832 (ITU-T, 1995b)) which are capable of supporting ATM cell transport and SDH element transport.

The 1.5/2 Mbit/s signal frame of ITU-T Recommendation G.704 (ITU-T, 1995a) is used for ATM cell transport by mapping ATM cells into time slots 1 to 24, or 1 to 15 and 17 to 31, respectively. Cell delineation is performed via the HEC mechanism as for the standard 155 Mbit/s interface (see Section 5.3.1), in the case of the 1.5 Mbit/s interface; however, no scrambling is used.

The frame structure for the 34 Mbit/s signal is shown in Figure 5.21.

The functions supported by the above overhead bytes are similar to the SDH functions (see Section 6.3.1); for a detailed description see ITU-T (1995b). The ATM cells are mapped into the 530 payload bytes.

In contrast to the 34 Mbit/s cell mapping, at 45 Mbit/s (using the multiframe format of ITU-T Recommentation G.704) the physical layer convergence protocol (PLCP) mapping is employed. PLCP has been adopted from the DQDB-slot mapping.

Whereas the ATM Forum has already specified interfaces at all of the above bit rates as UNIs (see UNI 3.0 (ATM Forum, 1993c) and UNI 3.1 (ATM Forum, 1994c)), ITU has only included the 1.5 Mbit/s and 2 Mbit/s UNIs in its specification set (see ITU Recommendation I.432.3 (ITU, 1996j)) so far.

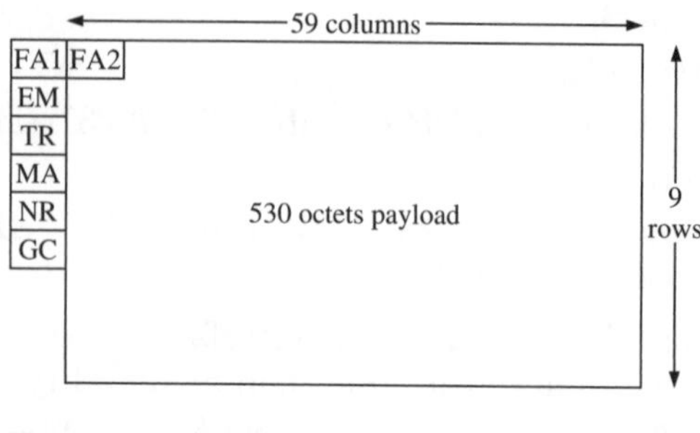

EM Error monitoring
FA Frame alignment signal
GC General purpose communications channel
MA Maintenance and adaptation byte
NR Network operator byte
TR Trail trace

Figure 5.21 Frame structure at 34.368 Mbit/s.

Both the 1.5 Mbit/s and the 2 Mbit/s UNI apply to the T_B as well as the S_B reference point. The exact interface bit rates are 1.544 Mbit/s and 2.048 Mbit/s respectively, and their transfer capabilities usable for ATM cell transport are 1.536 Mbit/s and 1.920 Mbit/s. ATM cells are directly mapped into the frame structures and octet-aligned. OAM functions are similar to those defined for the SDH-based interfaces (see Section 6.2); for further details see ITU Recommendation I.423.3 (ITU-T, 1996j).

The 6 Mbit/s UNI (ATM Forum, 1995c) has an exact interface bit rate of 6.312 Mbit/s and a cell transfer capability of 6.144 Mbit/s. Mapping of ATM cells into the frame is accomplished in the same way as for the 1.5/2 Mbit/s interfaces.

5.4.3 'Private' ATM interfaces

The ATM Forum has specified a **100 Mbit/s multimode fiber ATM interface** based on the FDDI physical layer. This interface is meant to be used as a private UNI which connects customer premises equipment, such as computers, bridges, routers and workstations, to a port on an ATM switch. According to the ATM Forum (1993c), 'the private UNI does not need the OAM complexity or link distance provided by telecom lines'. Details concerning the physical medium, bit timing and line coding are described in ATM Forum (1993c).

There is no frame with this interface, and not even a continuous stream of cells. Unless cells are actively being sent, the line contains idle codes. Cells can occur any time the line is idle. Each cell is preceded with the start-of-cell code (1 byte). (This byte, together with the 53 octets of a cell, must be contiguous on the line.)

HEC-generation/verification complies with ITU-T Recommendation I.432 (ITU-T, 1996j).

Two other, special, low-cost UNIs applying at the S_B reference point only have been defined by the ATM Forum and by ITU; their interface bit rates are 51.840 Mbit/s and 25.600 Mbit/s.

51.840 Mbit/s is a subrate of SDH; this interface (see ITU Recommendation I.432.4 (ITU-T, 1996j) and ATM Forum (1994e) specification either employs the SDH frame format and a subset of the SDH OAM functions and fields, or is cell-based. The cell transfer capability is 48.384 Mbit/s. Category 3 UTP *(unshielded twisted pair)* cabling is used enabling a maximum distance to be covered of about 100 m. As to the PMD layer characteristics of this interface, the following items are specified:

- bit error ratio should not exceed 10^{-10};
- timing is SDH-based (under normal conditions derived from the network clock; under fault conditions a tolerance of $\pm$ 100 ppm is permitted);
- symbol encoding with 16-QAM code (symbol rate of 12.96 Mbaud);
- two different scrambler polynomials are used to ensure that the signal in one direction is uncorrelated to the signal in the other direction.

The 25.6 Mbit/s UNI is defined in ITU Recommendation I.432.5 (ITU-T, 1996j) and ATM Forum (1995d). It deploys either 100 Ω UTP (*unshielded twisted pair*) or 120 Ω/150 Ω STP (*shielded twisted pair*) cables covering a distance of about 100 m. The signals are 4B5B encoded yielding a line symbol rate of 32 Mbaud. Bit error and timing criteria are the same as for the 52 Mbit/s interface.

As the 4B5B block code causes multiple bit errors for each corrupted bit, only the error detection mode is used for ATM cell header error control (see Section 5.3.1). Cell delineation is accomplished by prefixing a specific *start-of-cell* command to each ATM cell before transmission.

5.4.4 Inverse multiplexing ATM (IMA)

As outlined in Section 5.1, the underlying transport mechanism for ATM cells is provided by the physical layer. A key role is played by the existing and widely deployed interfaces for which an entire transport infrastructure exists, for example T1/E1 and T3/E3. Network operators still charge a considerable amount of money for the physical bandwidth offered. One can expect that only a very few users can or will afford an STM-1 ATM interface. Since T1/E1 interfaces are much cheaper and widely installed, many subscribers will presumably use these interface types. However, for some applications the bandwidth of a single T1/E1 line will not be sufficient and the next higher interface (T3/E3) will be too expensive. This 'bandwidth-price' gap can be filled by a function called **inverse multiplexing for ATM** (IMA) as defined by the ATM Forum (ATM Forum, 1997e). IMA allows several physical T1/E1 lines to be combined in such a way that they constitute a single transmission pipe which can be used for ATM cells. The combination of n lines builds an interface with a virtual bandwidth of n times T1/E1, called a *link group*. Thus, for example, a 4.5 Mbit/s ATM connection can be transported over three combined T1 lines.

As illustrated in Figure 5.22, the IMA function at the sending side distributes the cells equally over the available lines and the receiver combines the cells back into a single cell stream. The cells are distributed in a round robin manner on a cell-by-cell basis across the links of a link group. It is possible to combine up to 32 T1/E1 lines to form a single transmission pipe. However, economic reasons will reduce this number for actual networks. If, for example, the price for a T3/E3 interface is about six to eight times higher than for a single T1/E1 line, only link groups with a similar maximum number of links can be economically used.

It should be noted that it is absolutely vital to maintain cell sequence integrity. Therefore, the IMA function makes use of specific OAM cells, called **IMA control protocol** (ICP) cells, which are periodically inserted into the cell flow on each link. These ICP cells form a so-called *IMA frame*.

In general, each T1/E1 line will have a different length and thus different transmission delays. The IMA function is capable of equalizing a maximum differential delay of 25 ms. Additional IMA functions are defined for maintenance purposes. Link failures and link recoveries are handled automatically. This allows some level of redundancy to be added by assigning more physical link capacity than is actually used by the ATM layer. In case of a single (or a few) link failure(s), the remaining capacity might be sufficient to carry all ATM connections without adverse effects. ICP cells also are used to support these maintenance functions, to maintain synchronization and to determine the differential delay between IMA links.

A second new cell type was introduced for cell rate decoupling purposes; instead of idle cells, specific *filler cells* are used which are inserted on each individual link if no ATM cell is available.

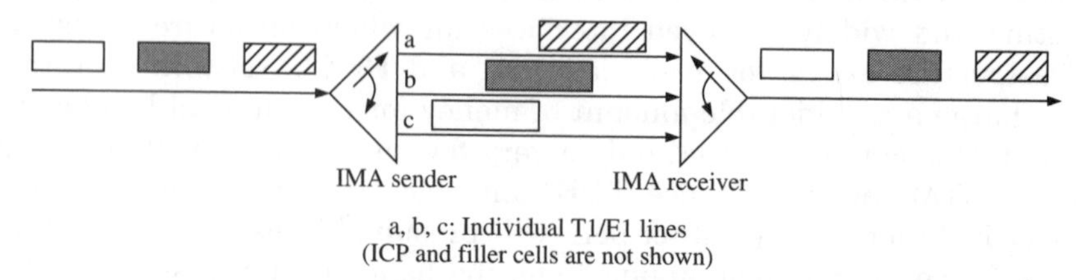

Figure 5.22 The IMA principle (unidirectional).

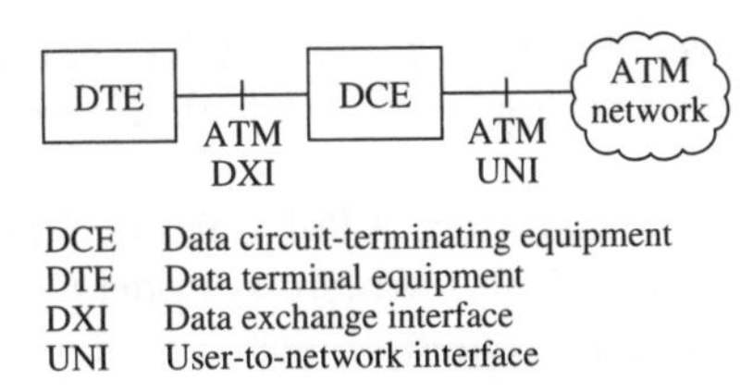

Figure 5.23 ATM data exchange interface (DXI).

The same IMA principles can be applied to other interfaces, for example T3/E3, as well. However, all links within a group must be of the same type.

5.4.5 Frame-based interfaces

Frame-based UNIs provide a means of offering ATM services in cases where the deployment of a native ATM UNI is not possible or not desirable. The **data exchange interface** (DXI) was defined in 1993 by the ATM Forum (ATM Forum, 1993b) mainly for connecting non-ATM-capable devices to an ATM network.

The **frame-based user-to-network interface** (FUNI) was defined for ATM for access rates of $n \times 64$ kbit/s. At these low speeds the ATM cell overhead is perceived as being an unnecessary burden compared with frame-based transport mechanisms with considerably less overhead. The DXI is the older frame interface which presumably will be replaced by the FUNI.

ATM data exchange interface (DXI)

The DXI allows a **data terminal equipment** (DTE) such as a router to be connected via **data circuit terminating equipment** (DCE) to an ATM switch. The DCE provides the conversion of the DXI format to an ATM UNI as shown in Figure 5.23. This allows a non-ATM-capable DTE to communicate with a peer station via an ATM network. The DXI can be run over physical layers that comply with V.35 (ITU-T, 1988i) or a **high-speed serial interface** (HSSI).

The DXI operates via AAL 5 and AAL 3/4 (see Section 5.7) only; there is no support for CBR services. Three modes of operation, as outlined in Table 5.6, are supported. The maximum possible number of virtual connections ranges from 1023 to $2^{24} - 1$ and the maximum **service data unit** (SDU) size varies from 9224 octets to 65 535 octets. Figure 5.24 shows the DXI frame format for mode 1; mode 2 uses a slightly different frame format. The DXI header contains fields to carry **cell loss priority** (CLP) and **congestion notification** (CN) infomation.

For addressing purposes a 10- or 24-octet **DXI frame address** (DFA) is used which is converted to (or from) a VPI/VCI combination within the DCE.

For management purposes (ATM Forum, 1993b) defines a **local management interface** (LMI) which supports SNMP and/or ILMI-based management. The managed objects are defined in the DXI LMI MIB.

Frame-based user-to-network interface (FUNI)

Whereas the main purpose of the DXI is to connect non-ATM-capable devices to an ATM network, the FUNI focuses on offering an efficient transport mechanism for data traffic. Its main purpose is to reduce the relatively large ATM overhead incurred from the ATM cell header as compared to a frame-based transport. A segmentation and reassembly of user frames takes place on the network side of the FUNI in an **interworking function** (IWF) which provides the frame-to-cell conversion (see the reference configuration in Figure 5.25). Thus the transport overhead via the access link between user equipment and IWF can be reduced.

Table 5.6 DXI operation modes.

Mode	*AAL*	*Maximum SDU size*	*Number of VCs*	*FCS*
1a	AAL 5 only	9232 octets	1023	16 bit
1b	AAL 3/4 for at least one VC	9224 octets	1023	16 bit
	AAL 5	9232 octets		
2	AAL 5 and AAL 3/4	65 535 octets	16 777 215 ($2^{24}-1$)	32 bit

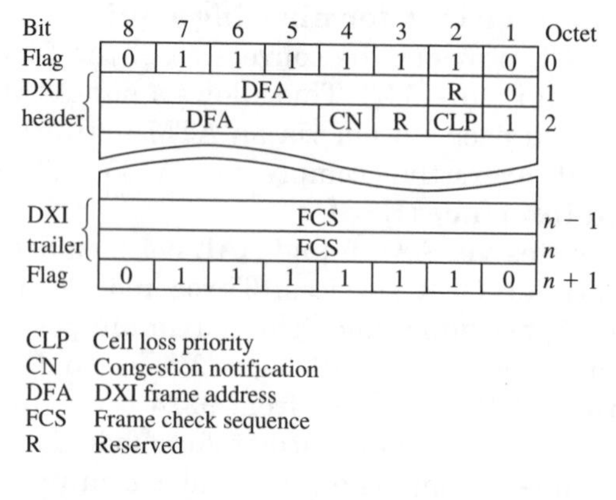

Figure 5.24 DXI frame format for mode 1a/b.

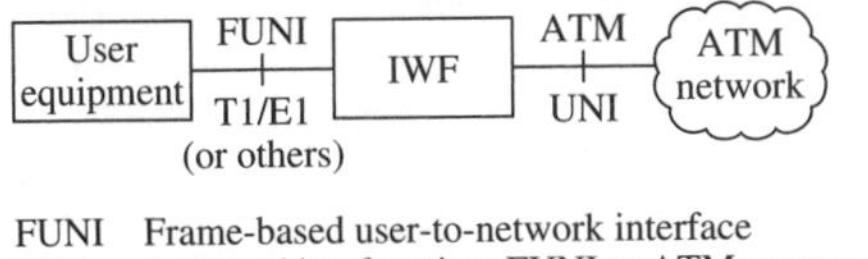

Figure 5.25 FUNI reference configuration.

FUNI is mainly designed to operate via $n \times 64$ kbit/s over E1/T1 links which constitute a concatenated transport container rather than individual time slots. Interconnection with a peer FUNI or with any other ATM UNI is possible. The FUNI is based on the DXI and compatible with it. A FUNI version 2 was defined by the ATM Forum (1996a).

Figure 5.26 shows the FUNI frame format for mode 1. The frame identifier is now known as the **frame address** (FA) instead of the DFA and an additional identification of OAM frames is possible. Compliant equipment must support DXI mode 1a, it optionally may support mode 1b and it shall not support mode 2 (see previous subsection). Table 5.7 shows the possible FUNI modes of operation.

Compared to an ATM UNI the FUNI provides the same features but with some restrictions. The only AAL types supported are 3/4 and 5, and only the VBR-nrt and UBR traffic classes are allowed. However, VPI/VCI multiplexing, UNI signaling, network management (OAM, ILMI) and traffic shaping are defined as in the ATM UNI case.

Signaling over the FUNI is based on UNI 4.0 (ATM Forum, 1996d) without the need to support all UNI 4.0 capabilities; for example, AAL 1 connection set-up requests are not possible. On the FUNI user side the protocol stack contains a slightly modified Q.2119 'B-ISDN

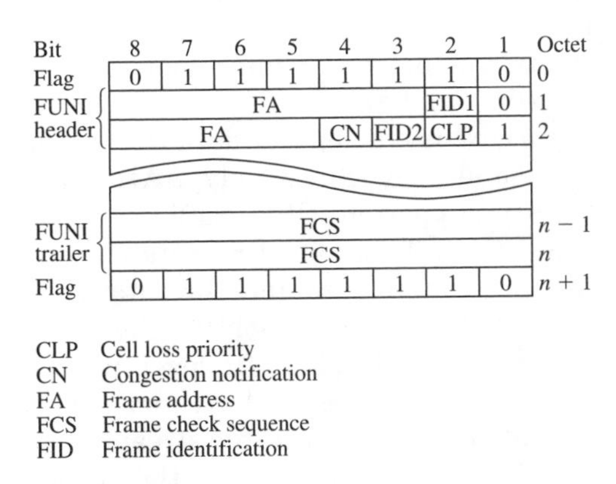

Figure 5.26 FUNI frame format for mode 1.

Table 5.7 FUNI operation modes.

Mode	*AAL*	*Access link speed*	*Maximum SDU*	*Number of VCs*	*FCS*
1a	AAL 5 only	E1/T1 and above below E1/T1	9232 octets 4096 octets (65 535 octets*)	512	32 bit 16 bit (32 bit*)
(1b*)	AAL 3/4 for at least one VC	E1/T1 and above below E1/T1	9224 octets 4088 octets	512	32 bit 16 bit (32 bit*)
	AAL 5	E1/T1 and above below E1/T1	9232 octets 4096 octets (65 535 octets*)		32 bit 16 bit (32 bit*)
2 (n.s.)	n.s.	n.s.	n.s.	n.s.	n.s.

n.s. Not supported
* Optional

ATM adaptation layer convergence function for SSCOP above the frame relay core service' (ITU-T, 1996r). The same protocol stack is used for SNMP/ILMI as well.

Normally, the first ATM network element controls the user cell stream by means of a UPC function. This requires the IWF to perform traffic shaping in order to meet PCR, SCR, MBS and CDVT parameters (as explained in Section 4.5.2). Policing in the ATM-to-frame (from network to user) direction does not occur, so shaping in that direction is not necessary.

The use of OAM cells (see Section 6.2.2) via the FUNI is optional and accomplished by carrying a full 53-octet OAM cell within the FUNI data link frame. An OAM frame can be differentiated from a user frame by setting the FID bits accordingly. Carrying the entire OAM cell allows differentiation among end-to-end and segment cells since the PTI field can be evaluated.

More information on network management and the FUNI MIB can be found in ATM Forum (1996a).

Although the FUNI is efficient for data traffic it does not allow for complete access consolidation. Since only AAL 3/4 and AAL 5 and VBR-nrt and UBR are supported, CBR traffic such as voice cannot be transported via the FUNI. Thus, for corporate customers with a local network comprising, for example, routers for data traffic and a PBX for voice traffic, network access consolidation via the FUNI is not possible. Only by using an additional TDM multiplexer it becomes possible to share the T1/E1 access link between FUNI and, for example, N-ISDN traffic. Within the central office both traffic streams are separated again (demultiplexing function) and connected to the ATM network via the IWF and to the N-ISDN network, respectively (see Figure 5.27).

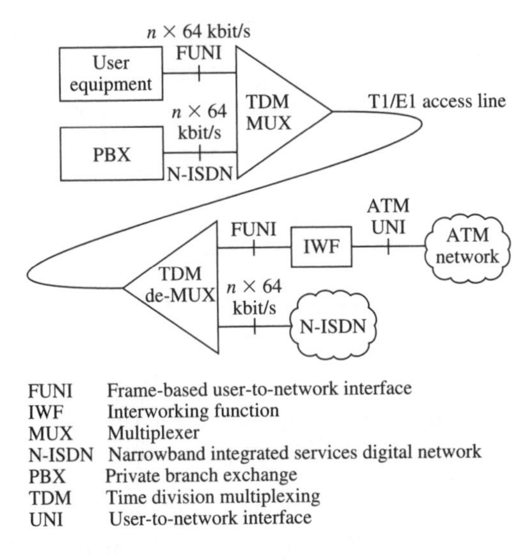

Figure 5.27 Using a T1/E1 access line for FUNI and other traffic.

5.5 Equipment-internal interfaces

All the interfaces so far described have been defined for interworking and interoperability reasons. This allows network operators to connect networking equipment from different vendors. In addition to these external interfaces the ATM Forum has specified two equipment-internal interfaces. The main purpose of this 'standardization' effort is to enable switch vendors to obtain integrated circuits from different manufacturers for building ATM switches.

5.5.1 UTOPIA interface

Initially, the **universal test and operations physical interface for ATM** (UTOPIA) was defined as a standardized interface between the physical layer and the ATM layer. However, nowadays this interface is also used between other parts of the protocol stack, for example between the ATM layer and the AAL. It is the physical realization of the logical boundaries between layers of the ATM protocol reference model as illustrated in Figure 5.28. Since normally the different functions are realized within the same networking device, for example an ATM switch, this interface will in almost all cases constitute an equipment-internal interface.

UTOPIA allows the building of ATM products based on commercially available components from different IC vendors. Today

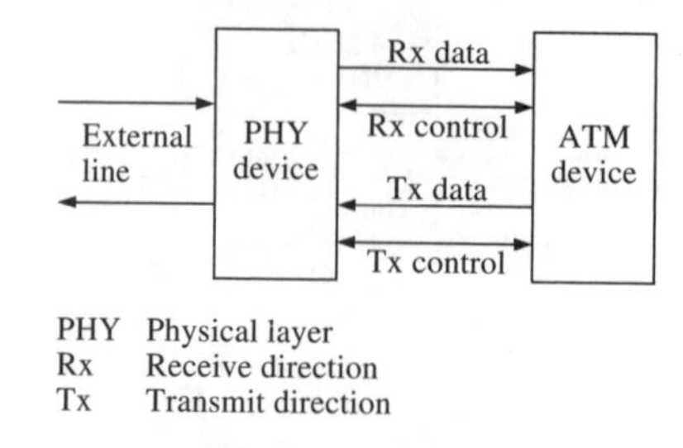

Figure 5.28 The UTOPIA interface.

there is a wide variety of components for the various functions such as physical layer handling, ATM cell processing or support of the different AAL types.

Two variants of the UTOPIA interface need to be differentiated: UTOPIA level 1 as specified by the ATM Forum in 1994 (ATM Forum, 1994b) and its successor UTOPIA level 2 (ATM Forum, 1995a).

For the sake of simplicity only the interface between physical layer devices and ATM layer devices is considered in the following.

UTOPIA level 1

UTOPIA level 1 is an 8-bit-wide interface operated at a maximum of 25 MHz allowing for a maximum throughput of 155 Mbit/s in each direction. It allows exactly one physical layer device to be connected with one ATM layer device. Via this interface 53-octet UTOPIA ATM cells are transmitted. These are normal ATM cells but instead of the HEC field, the fifth header octet contains a **user-defined field** (UDF) which can be used for non-standardized functions between devices. However, although all HEC operations are normally executed by the physical layer device, the HEC can be transferred to the ATM layer by placing it in the UDF.

The cell transfer itself is controlled via hardware handshaking signals (Rx/Tx control in Figure 5.28).

The entire transfer is based on a master–slave relationship between the two devices. The master regularly polls the slave and if the slave has a cell available for transmission the actual transfer starts. This requires a FIFO cell buffer in slave devices. Normally, the physical layer device constitutes the UTOPIA slave but some devices allow the UTOPIA interface to be configured as either master or slave.

UTOPIA level 2

UTOPIA level 2 is an 8-bit- or 16-bit-wide interface operated at 25 MHz, 33 MHz or 50 MHz allowing for a maximum throughput of 622 Mbit/s.

The 33 MHz frequency was chosen to facilitate deployment in PCI-bus based environments like PCs. The UTOPIA level 2 cell format for the 16-bit-wide option is shown in Figure 5.29. Since a second UDF field was introduced the cell length is now 54 octets. The most notable enhancement, besides the higher throughput which allows OC-3/STM-4 ATM interfaces to be built, is the capability to interconnect more than one physical layer device with a single (or multiple) ATM layer IC(s). Hence, at UTOPIA level 2 multiplexing can occur. This allows the building of more cost-effective low bit rate interfaces which share a common ATM layer device. Up to eight physical layer devices can be controlled by one ATM IC. Each of the physical layer devices may have several line interfaces. A maximum of 31 line interfaces is supported by UTOPIA level 2. The UTOPIA master controls the cell transfers. It sequentially polls all connected devices/ports within a single cell cycle and selects the next device/port that has a cell ready for transmission.

5.5.2 WIRE

Workable interface requirements example, better known as WIRE, is an ATM Forum specification for another equipment-internal interface. It specifies the interface beteen the PMD part and the TC part of the physical layer (see Section 5.3). Again this specification allows ICs for TC or PMD operations, respectively, to be obtained from different vendors. Even before WIRE was created, the logical split within the physical layer was implemented in two different devices since the PMD requires a media-dependent, mostly high-speed circuitry, whereas the TC could be built of digital technology with relatively low clock rates. WIRE is an attempt to 'standardize' this split. Serial, 8/16/32-bit-wide parallel and quad 8-bit parallel variants are specified. The throughput can be as high as 2.4 Gbit/s. More details on this interface can be found in ATM Forum (1996b).

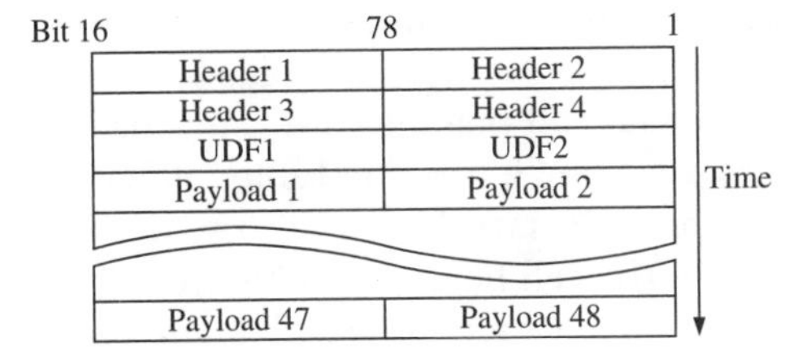

Figure 5.29 UTOPIA level 2 format in 16-bit mode.

5.6 ATM layer

ITU-T Recommendation I.150 (ITU-T, 1995d) includes the functional characteristics of the ATM layer; its specification is given in ITU-T Recommendation I.361 (ITU-T, 1995e).

5.6.1 Cell structure

The cell is the basic element of the ATM layer. As mentioned in Section 5.1.5, the term 'cell' is also used at the physical layer. A cell consists of a 5-octet header and a 48-octet information field. Its structure is shown in Figure 5.30.

The following numbering conventions are defined in ITU-T Recommendation I.361:

- Octets are sent in increasing order starting with octet 1. Therefore, the cell header will be sent first, followed by the information field.
- Bits within an octet are sent in decreasing order starting with bit eight.
- For all fields, the first bit sent is the most significant bit (MSB).

5.6.2 Cell header

The cell header at the UNI differs from that at the network–node interface (NNI) in the use of bits 5–8 of octet 1. The NNI is the interface between network nodes. At the NNI these bits are part of the VPI, whereas at the UNI they constitute an independent unit, the **generic flow control** (GFC) field. Figure 5.31 depicts the cell header used at the UNI and the NNI.

The different fields defined within the cell header have no meaning for physical layer cells. Their significance is restricted to ATM cells.

Pre-assigned cell header values

In order to differentiate cells for the use of the ATM layer from those cells only used at the physical layer as well as to identify unassigned cells (see Section 5.1.5), pre-assigned cell header values are used. These values are shown in Table 5.8. As shown in the table, physical layer cells and unassigned cells are characterized by an all-0 pattern in the header bits 5–28. All other values of the cell header can be used by assigned cells.

The differentiation between physical layer cells and unassigned cells is based on the use of the least significant bit (LSB) of octet 4 of the cell header. This bit is thus not used for cell loss priority indication as in the case of assigned cells.

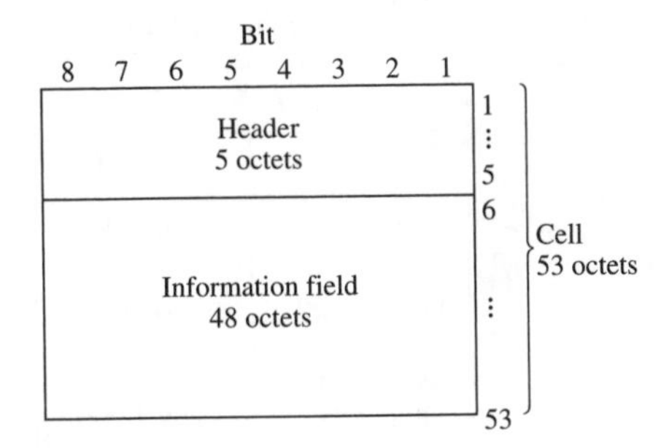

Figure 5.30 Cell structure.

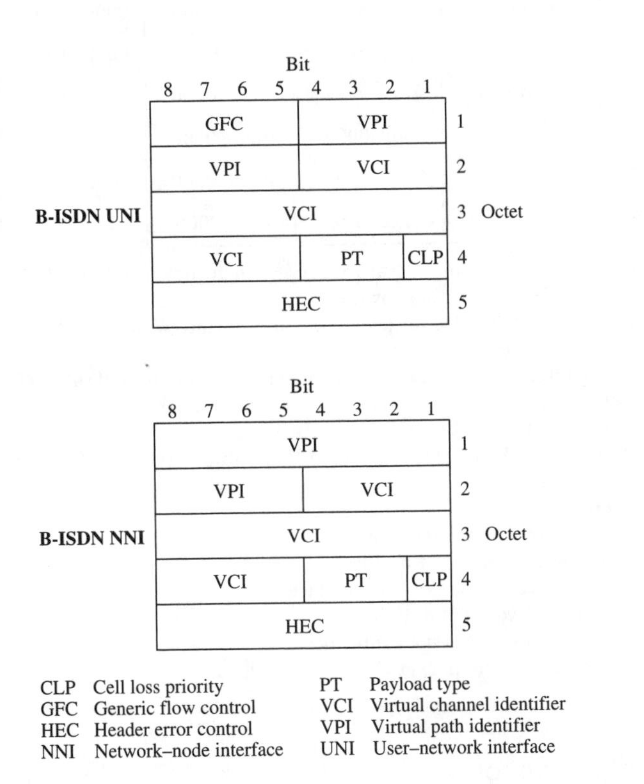

Figure 5.31 Cell header at the B-ISDN UNI and the B-ISDN NNI.

Several types of physical layer cells have already been defined by pre-assigned header values, as shown in Tables 5.2 and 5.3. The use of these cells has been demonstrated in Section 5.3.1. Other cell header values are pre-assigned for signaling, OAM, and other functions (see Table 5.8).

Table 5.8 Pre-assigned cell header values at the UNI/NNI.

	VPI		*VCI*	*PT C*
	Octet 1	*Octet 2*	*Octet 3*	*Octet 4*
Reserved for use of the physical layer	pppp 0000	0000 0000	0000 0000	0000 ppp 1
Idle cell	0000 0000	0000 0000	0000 0000	0000 000 1
Unassigned cell	gggg 0000	0000 0000	0000 0000	0000 xxx 0
Meta-signaling	gggg yyyy	yyyy 0000	0000 0000	0001 0a0 c
General-broadcast signaling	gggg yyyy	yyyy 0000	0000 0000	0010 0aa c
Point-to-point signaling	gggg yyyy	yyyy 0000	0000 0000	0101 0aa c
F4 segment OAM flow	gggg zzzz	zzzz 0000	0000 0000	0011 0a0 a
F4 end-to-end OAM flow	gggg zzzz	zzzz 0000	0000 0000	0100 0a0 a
VP Resource Management	gggg zzzz	zzzz 0000	0000 0000	0110 110 a
LANE configuration direct	yyyy yyyy	yyyy 0000	0000 0001	0001 aaa c
ILMI	0000 0000	0000 0000	0000 0001	0000 aaa 0
IFMP/GSMP	0000 0000	0000 0000	0000 0000	1111 aaa 0
PNNI PTSP	zzzz zzzz	zzzz 0000	0000 0011	0000 aaa 0

All other VCI values to decimal 31 are reserved for future use. Octet 5 of the cell header contains the valid HEC code (not shown here)

a Indicates the bit is available for use by the appropriate ATM layer function
c shall be set to 0 by sender
g Indicates the bit is available for use by the GFC protocol or VPI (at NNIs)
p Indicates the bit is available for use by the physical layer
x Indicates the bit is a 'don't care' bit
y Any VPI value. For VPI = 0 the specified VCI value is reserved for user signaling with the local exchange
z Any VPI value
GSMP General switch management protocol
IFMP Ipsilon flow management protocol
ILMI Integrated local management interface
LANE Local area network emulation
PNNI Private network-to-network interface
PTSP PNNI topology state packet

Generic flow control

The **generic flow control** (GFC) field consists of four bits. Its default value is 0000 as long as the GFC function is not used. GFC information is carried in either **assigned** or **unassigned** cells (ITU-T, 1991e).

For a very long time ITU-T was unable to decide on a concrete protocol for GFC. Although it has now been defined, it is rarely used.

The functional description of the GFC is included in ITU-T Recommendation I.150 (ITU-T, 1995d). The GFC mechanism helps to control the traffic flow from ATM connections at the UNI.

The GFC mechanism can be used in point-to-point configurations of the customer network. Support of shared medium topologies is for further study.

Two modes of operation are defined:

- the controlled access
- the uncontrolled access.

To differentiate between **uncontrolled** and **controlled** operation the following applies: any equipment which receives 10 or more non-zero GFC fields within 30 000 cell times should consider the other ATM entity to be executing the 'controlled transmission' set of procedures (ITU-T, 1995e).

Furthermore, controlled and controlling equipment is differentiated. As an example, the B-NT2 in Figure 4.11(a) may control the attached B-TEs.

Using the controlled access mode allows:

(a) the bandwidth of an end-station to be limited to a fixed fraction of its interface rate, or

(b) the network access for 'controlled ATM connections' to be controlled and a cell to be explicitly marked as being on a controlled connection.

In case (a) the B-NT2 would issue a periodic HALT signal to the B-TEs. Whenever such a signal is received cell transmission must be stopped. This applies to all active virtual connections. In case (b) only the rate of particular VCCs or VPCs is controlled. The detailed GFC procedure and the appropriate setting of GFC bits is described in ITU-T (1995e).

Virtual path identifier

The **virtual path identifier** (VPI) field at the UNI consists of eight bits and is used for routing. The VPI at the NNI comprises the first 12 bits of the cell header, thus providing enhanced routing capabilities. Pre-assigned VPI values are used for some special purposes. As mentioned above, all bits of the VPI field are set to zero in an unassigned cell. Table 5.8 shows the currently predefined VPI/VCI values. Other uses of pre-assigned VPIs/VCIs are not excluded.

Virtual channel identifier

Together with the VPI field, the **virtual channel identifier** (VCI) field constitutes the routing field of a cell. A field of 16 bits is used for the VCI at the UNI as well as the NNI. It also has some pre-assigned values. The VCI value for unassigned cells is shown above. With this assignment in mind it is evident that the VCI value of zero is not available for user VC identification. Other pre-assigned VPI/VCI values can be found in Table 5.8, which is still open for further amendments.

Payload type

Three header bits are used for the **payload type** (PT) identification. Table 5.9 describes the **payload type identifier** (PTI) coding.

The payload of user information cells contains user information as well as service adaptation functions. In network information cells, the payload is used to carry information the network needs for its operation and maintenance. One example is the F5 information flow (see Section 4.6.2) which supports the OAM of VCCs. Here the PTI is used to distinguish between user cells and F5 cells pertaining to the same VCC. The **ATM-layer-user-to-ATM-layer-user** (AUU) indication will be used by AAL type 5 (see Section 5.7.5).

The **congestion indication** (CI) bit within the PTI of user cells may be modified by any network element that is congested to inform the end user about its state.

The use of **resource management** cells is described in Chapter 7.

Cell loss priority

The **cell loss priority** (CLP) field consists of one bit which is used explicitly to indicate the cell loss priority. If the value of the CLP bit is '1' the cell is subject to discard, depending on the network conditions. However, the agreed quality of service (QOS) parameters will not be violated. In the other case (CLP = '0'), the cell has high priority and therefore sufficient network resources have to be allocated to it. The CLP bit may be set by the user or the service provider. Cells belonging to a CBR connection always have high priorities. Many VBR services

Table 5.9 PTI values.

PTI	*Interpretation*
000	User data cell, congestion not experienced, ATM-layer-user-to-ATM-layer-user indication = 0
001	User data cell, congestion not experienced, ATM-layer-user-to-ATM-layer-user indication = 1
010	User data cell, congestion experienced, ATM-layer-user-to-ATM-layer-user indication = 0
011	User data cell, congestion experienced, ATM-layer-user-to-ATM-layer-user indication = 1
100	OAM F5 segment associated cell
101	OAM F5 end-to-end associated cell
110	Resource management cell
111	Reserved for future functions

require a guaranteed minimum capacity as well as a peak capacity. Some of these services may take advantage of the CLP bit to distinguish between cells of high and low loss sensitivity.

When a VBR connection is established, the rate of higher priority cells is determined. However, it can be renegotiated during the connection phase. Cells of higher priority which exceed the agreed parameters are subject to the normal UPC/NPC (see Section 7.2.3).

Header error control

This field is part of the cell header but it is not used by the ATM layer. It contains the **header error control** (HEC) sequence which is processed by the physical layer. The HEC mechanism is specified in ITU-T Recommendation I.432 (ITU-T, 1996j). A description of the mechanism is given in Section 5.3.1.

5.6.3 ATM layer connections

An ATM layer connection is the concatenation of ATM layer links in order to provide an end-to-end transfer capability to access points.

The VPI is used to distinguish different VP links which are multiplexed at the ATM layer into the same physical layer connection at an interface in one direction. Different VC links within a VPC are identified by their individual VCIs.

Two VCs belonging to different VPs at the same interface may have identical VCI values. Consequently both the VCI and VPI are necessary to correctly identify a VC.

A VPI is changed at points where a VP link is terminated (for example, cross-connect, concentrator and switch), while the VCI is changed at points where VC links are terminated. In this way, VCI values are preserved within a VPC.

Active connections at the UNI

At the UNI, 24 bits are available for routing. However, the actual number of bits used is negotiated between the user and the network (for example, on a subscription basis). The lower requirements of the user or the network determine the number of active routing bits. The following rules have been agreed by ITU-T (1995e) for the determination of the position of active routing bits within the VPI/VCI field:

- used bits of the VPI field and the VCI field will be contiguous;
- bit allocation will always begin with the LSB of its appropriate field;
- unallocated bits of the routing field, that is, bits not used by either the user or the network, will be set to zero.

Virtual channel connections

The virtual channel connection (VCC) is defined in ITU-T Recommendation I.113 (ITU-T, 1993e) as follows (see Section 4.2.1):

> 'A concatenation of virtual channel links that extends between two points where the adaptation layer is accessed.'

One of four methods can be used to establish a VCC at the B-ISDN UNI:

- **Semi-permanent** or **permanent** VCCs (PVC) are established at subscription time. No signaling procedure is necessary.
- A VCC is established/released by using a **meta-signaling** procedure. This approach is applied for establishing a signaling VC only (see Section 4.3.3).
- Establishment/release of a **switched end-to-end VCC** (SVC) can be done by a user-to-network signaling procedure.
- If a VPC already exists between two UNIs, a VCC within this VPC can be established/released by employing a user-to-user signaling procedure.

Cell sequence integrity is preserved within a VCC. Traffic parameters are individually negotiated between the user and the network at VCC establishment. These parameters can be renegotiated during the connection phase. All cells originating from the user are monitored by the network to ensure that the agreed parameters are not violated. This mechanism is called usage parameter control (details can be found in Section 7.2.3).

The QoS for the connection is negotiated between the user and the network when the VCC is established.

The VCI can be determined by the network or the user, or by negotiation between network and user, otherwise a standardized value will be used. In general the value of the VCI field is independent of the service provided over that VC. However, to simplify terminal interchangeability and initialization, the same VCI value will be used at all UNIs for some fundamental functions, such as meta-signaling.

Virtual path connections

The definition of the **virtual path connection** (VPC) is given in ITU-T Recommendation I.113 (ITU-T, 1993e) (see Section 4.2.1):

> 'A concatenation of virtual path links that extends between the point where the virtual channel identifier values are assigned and the point where those values are translated or removed.'

One of the following methods can be used to establish/release a VPC between VPC endpoints:

- **Semi-permanent** or **permanent** VPCs are established/released on a subscription basis and therefore no signaling procedure is necessary.
- VPC establishment/release may be done by a user-to-network or network-node-to-network-node signaling procedure.
- A VPC can also be established/released by the network using **network management** procedures.

Cell sequence integrity within a VPC is preserved for each VCC carried and for the VPC as a whole. During VPC establishment, the traffic parameters for the VPC are negotiated between the user and the network. If necessary, these parameters can subsequently be renegotiated. Changing these traffic parameters can improve traffic performance (Huber and Tegtmeyer, 1993). All input cells from the user to the network are monitored to supervise the agreed traffic parameters.

At VPC establishment, the QoS is selected from the range of QoS classes supported by the network. A VPC can carry VCCs requiring different QoS classes. Therefore, the QoS of the VPC must meet the most demanding QoS of the VCCs carried.

Some VCIs within a VPC may be reserved for, say, network OAM functions (for example, to implement the F4 flow, see Section 4.6.2) and will not be available to the user.

5.7 ATM adaptation layer

The AAL functions have already been described in Section 5.1.5. The AAL provides for the mapping of higher-layer PDUs into the information field of cells and the reassembly of these PDUs. Other AAL functions support different applications. The detailed specification of the AAL (see also ITU-T Recommendation I.363 (ITU-T, 1996h)) is presented in the following.

Several AAL protocol types are defined. Each type consists of a specific SAR sublayer and CS. This classification fits the AAL service classes described in Section 5.1.5 (for example, CBR services of class A will use AAL type 1). However, no strict relationship between the AAL service classes and the AAL protocol types is requested. Other combinations of the described SAR and CS protocols or other SAR and/or CS protocols may also be used to support specific services. The definitions and relationships between PDU, SDU and PCI are described in Section 5.1.2.

5.7.1 AAL type 0

Although 'AAL type 0' is not an officially used term it can be considered important, being an AAL with empty SAR and CS. This means that no AAL functionality is required and that the content of the cell information field is directly and transparently transferred to the higher layer. However, a detailed description of such a service is not yet available in ITU-T standards.

5.7.2 AAL type 1

Normally, **CBR services** (class A) use AAL type 1 because it receives/delivers SDUs with a constant bit rate from/to the layer above. **Timing information** is also transferred between source and destination. If necessary, information about the **data structure** can be conveyed too. Indication of lost or errored information is sent to the higher layer if these failures cannot be recovered within the AAL.

The functions performed by the AAL are as follows:

- segmentation and reassembly of user information
- handling of cell delay variation
- handling of cell payload assembly delay
- handling of lost and misinserted cells
- source clock frequency recovery at the receiver
- recovery of the source data structure at the receiver
- monitoring of AAL-PCI for bit errors as well as handling those errors
- monitoring of the user information field for bit errors and possible corrective actions.

In the case of **circuit emulation**, monitoring of the end-to-end QoS is necessary. This will be located at the CS. For this purpose, a cyclic redundancy check (CRC) may be calculated for the information carried in one or more cells. The result is transferred to the receiver within the information field of a cell or in a special OAM cell.

Circuit emulation is believed to be an important feature of B-ISDN as it allows existing circuit-based signals (for example, 1.5 Mbit/s or 2 Mbit/s) to be transported, meeting the requirements on delay, jitter, bit error rate and so on for such signals. The user will not even be aware of the transfer mechanism involved.

Segmentation and reassembly sublayer

The SAR-PDU consists of 48 octets. The first octet includes the PCI; all other octets are available for the SAR-PDU payload. The PCI is subdivided into a 4-bit **sequence number** (SN) and a 4-bit **sequence number protection** (SNP) field. The SN consists of a **convergence sublayer indication** (CSI) bit and a 3-bit **sequence count** field. The SNP field contains a 3-bit CRC which protects the SN field and an even parity bit which has to be calculated over the resulting 7-bit code word. Figure 5.32 shows the SAR-PDU format of AAL type 1.

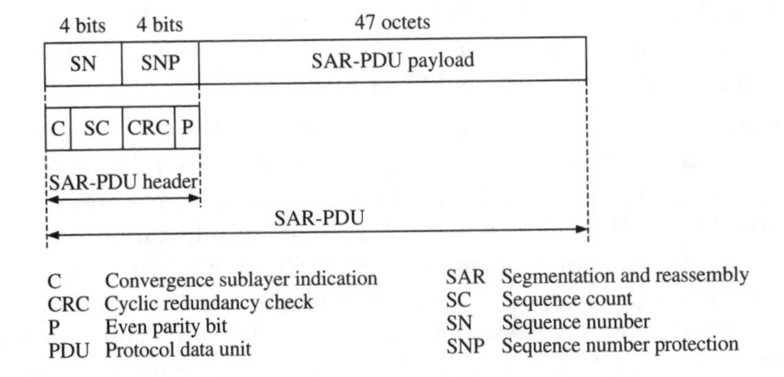

Figure 5.32 SAR-PDU format for AAL type 1.

The sequence count value of the SN makes it possible to detect the loss or misinsertion of cells. However, for systems with high cell loss ratios this method is not very robust since the 3-bit sequence count field is relatively short. The CSI bit can be used to transfer timing information and/or information about the data structure.

The SNP provides error detection and correction capabilities. The following two-step approach will be used which allows the correction of all single-bit errors and the detection of multiple-bit errors:

(1) The SN is protected by the polynomial $G(x) = x^3 + x + 1$.

(2) The resulting 7-bit code word is protected by an even parity check.

Convergence sublayer

The functions of the CS depend strongly on the service to be supported. Some of these functions are listed in the following. If required a clock may be derived from the terminal interface. Examples of how these functions can be performed are also given:

- Handling of **cell delay variation**. A buffer is used to support this function. Buffer underflow or overflow may lead to the insertion of dummy bits or the dropping of excessive bits, respectively.
- Further handling of lost and misinserted cells, such as the insertion of dummy information or discarding of gained cells.
- **Source clock frequency recovery** using the **synchronous residual time stamp method** (SRTS). A **residual time stamp** (RTS) is used to measure and convey information about the frequency difference between a common reference clock derived from the network clock and a service clock. The 4-bit RTS is

transferred by the CSI bit in successive SAR-PDU headers with an odd number in the sequence count field (SN = 1, 3, 5, 7). For details of this method see ITU-T (1996h).

- Transfer of **structure information** between source and destination, for example to support 8 kHz-based frame formats in circuit-mode services. Two modes of operation are possible, called **non-P** and **P** format. For the P format an additional pointer field (1 octet) is provided in the SAR-PDU payload. This pointer indicates the start of a structured block. For P format operations the CSI value in SAR-PDU headers with an even SN (that is, SN = 0, 2, 4, 6) is set to 1. For details of this method see ITU-T (1996h).
- **Forward error correction** (FEC) may be used to ensure high quality for some video and audio applications. This may be combined with bit interleaving to give more secure protection against errors. One example of an FEC for unidirectional video services uses a Reed–Solomon code; it is described in ITU-T (1996h). This method increases the overhead by 3.1% and the introduced delay is 128 cell cycles.

5.7.3 AAL type 2

AAL type 2 has been discussed from the very beginning of the ATM/AAL standardization efforts carried out within ITU-T. For a long period of time it was assumed to be an AAL for a class B type of service, that is, a VBR service with a timing relation between source and destination. It was focused on applications like VBR audio or VBR video. However, a detailed specification was never completed.

In the meantime an **AAL composite user** (AAL-CU) was proposed. This AAL-CU provides a method of multiplexing several sources which are carried over a common VCC. A specification of this AAL type was finalized and the AAL-CU was renamed AAL type 2 (I.363.2, ITU-T (1996h)). Hence the scope of the original AAL type 2 was completely changed.

This 'new' AAL type 2 can be used for a 'bandwidth efficient transmission for low-rate, short length packets in delay sensitive applications' (ITU-T, 1996h). An important example of such traffic is compressed voice as generated by mobile networks. Since even the packetization delay for a normal 64 kbit/s voice channel may lead to echo and delay problems (see also Section 14.1), this situation gets worse with, for example, 8 kbit/s compressed voice samples. Filling an ATM cell with 8 kbit per second requires about 48 ms. This value was considered to be much too high. One alternative to overcome this problem is to fill an ATM cell only partially. This, however, leads to a very inefficient use of network resources. A new mechanism was needed. With AAL type 2 it is now possible to multiplex several low-bandwidth sources into a single ATM cell. Thus the packetization delay can be reduced considerably (filling an ATM cell with two 8 kbit/s streams requires only 24 ms, and so on).

As shown in Figure 5.33 the AAL type 2 is subdivided into a **common part sublayer** (CPS) and a **service-specific convergence sublayer** (SSCS). The multiplexing of several AAL type 2 channels (or AAL-SDUs) occurs in the CPS. The multiplexing itself is performed on a per packet basis, called CPS-packets. At the AAL type 2 service access point (AAL2-SAP) the existence of an entire AAL-SDU is assumed.

The SSCS can support different higher-layer services or, for example, provide an assured data transfer. Different SSCS functions are currently being studied by ITU-T; however, standardization is still pending.

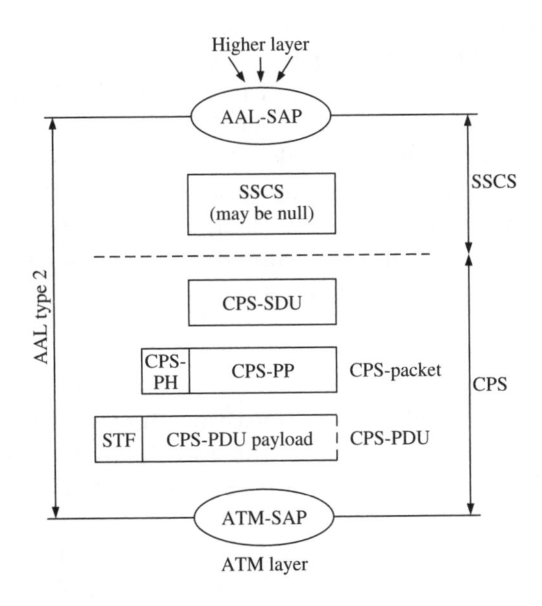

AAL ATM adaptation layer
CPS Common part sublayer
PDU Protocol data unit
PH Packet header
PP Packet payload
SAP Service access point
SDU Service data unit
SSCS Service-specific convergence sublayer
STF Start field

Figure 5.33 Structure of AAL type 2.

Common part sublayer

The AAL type 2 CPS offers the following service to the SSCS:

- non-assured data transfer of 45- or 64-octet SDUs (no bit error corrections, no retransmission in case of loss);
- multiplexing of several SDUs onto a single ATM VCC;
- maintaning the CPS-SDU sequence integrity.

As mentioned above, the CPS provides for the multiplexing of several information streams into a single ATM VCC. Figure 5.34 shows the CPS-packet format. It consists of a 3-octet **CPS-packet header** (CPS-PH) and a **CPS-packet payload** (CPS-PP) field which can have a length either of up to 45 octets or up to 64 octets. The maximum default length of the CPS-PP is 45 octets.

The **channel identifier** (CID) is used for identifying the different packets. The same value is used for both directions of a bidirectional connection. Some CID values are reserved for specific functions, hence the maximum possible number of user connections on a single ATM connection is 248. The **length indicator** (LI) specifies the length of the CPS-PP. The **user-to-user indication** (UUI) field is either transparently transmitted between peer SSCS entities or it can be used for layer management purposes. Like the ATM cell header, a **header error control** (HEC) field is used for protecting the CPS-PH. However, the CPS-PH HEC is only five bits long and is only used for error detection; that is, correction of corrupted headers does not take place. The generator polynomial is as follows:

$$G(x) = x^5 + x^2 + x$$

One or more of these CPS-packets constitute the CPS-PDU payload. In order to form an ATM-SDU, that is, the 48 octets which constitute the ATM cell's payload field, the entire CPS-PDU consists of a prepended 1-octet header and a padding field at the end if necessary. The PDU header is also called the **start field** (STF). The entire CPS-PDU is shown in Figure 5.35. The PDU header is made up of an **offset field** (OSF), a **sequence number** (SN) and a **parity** (P) field. The OSF measures the number of octets between the end of the PDU-header (or STF) and the start of the first CPS-packet. The SN is a modulo 2 counter of CPS-PDUs. An odd parity check, encoded in the P field, protects the STF.

The CPS-PDU payload may carry one or more CPS-packets. A CPS-packet may cross the ATM cell boundary and hence be transported in subsequent ATM cells (see Figure 5.36).

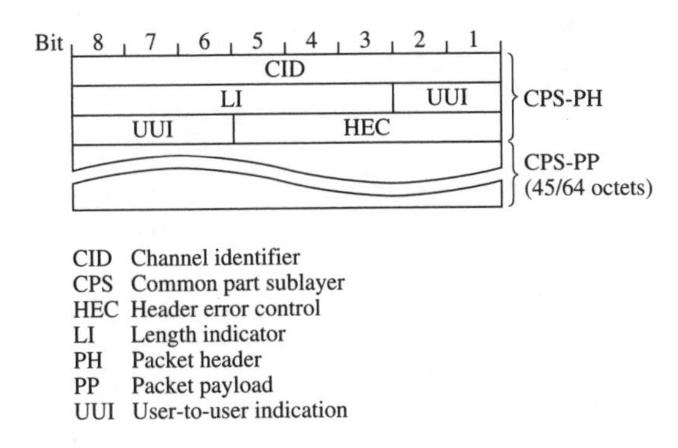

Figure 5.34 CPS-packet format for AAL type 2.

AAL type 2 negotiation procedure

The **AAL type 2 negotiation procedure** (ANP) has been defined to provide a mechanism which allows AAL type 2 channels between two endpoints to be dynamically allocated and de-allocated. The entire ANP procedure is carried out by AAL type 2 management entities. The ANP protocol information is carried via CID = 1 and the main ANP functions are:

- allocation of AAL type 2 channels over an existing VCC;
- removal of AAL type 2 channels from an existing VCC;
- exchange of status information for AAL type 2 channels.

More detailed information about the ANP can be found in ITU-T Recommendation I.363.2 (ITU-T, 1996h). An appendix to this recommendation summarizes some initial discussions on a specific OAM flow for AAL type 2 connections. However, the need for such a function has to be carefully evaluated against the additional complexity and transmission overhead. Furthermore, AAL 2 specific traffic management issues require further investigation.

5.7.4 AAL type 3/4

The name of this AAL reflects its development: as the service classes were specified, separate AALs were allocated for class C and class D services, namely AAL type 3 and AAL type 4. In the meantime both types have merged, thereby supporting both service classes.

Figure 5.37 shows the general structure of this AAL; the CS is split into a **common part** (CPCS) and a **service-specific part**. The **service-specific convergence sublayer** (SSCS) is application dependent and may be null; it is not discussed in this chapter.

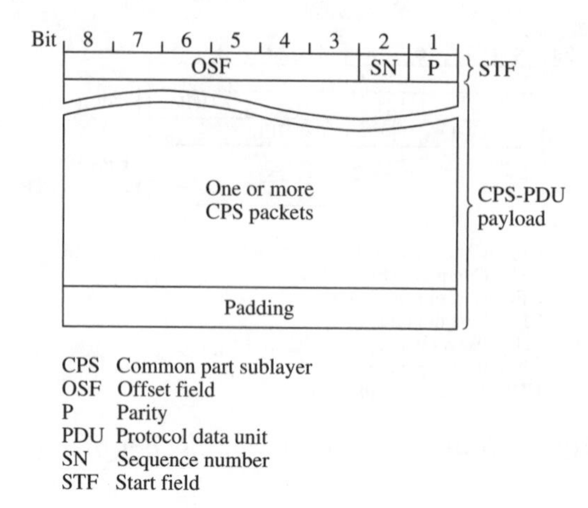

Figure 5.35 CPS-PDU of AAL type 2.

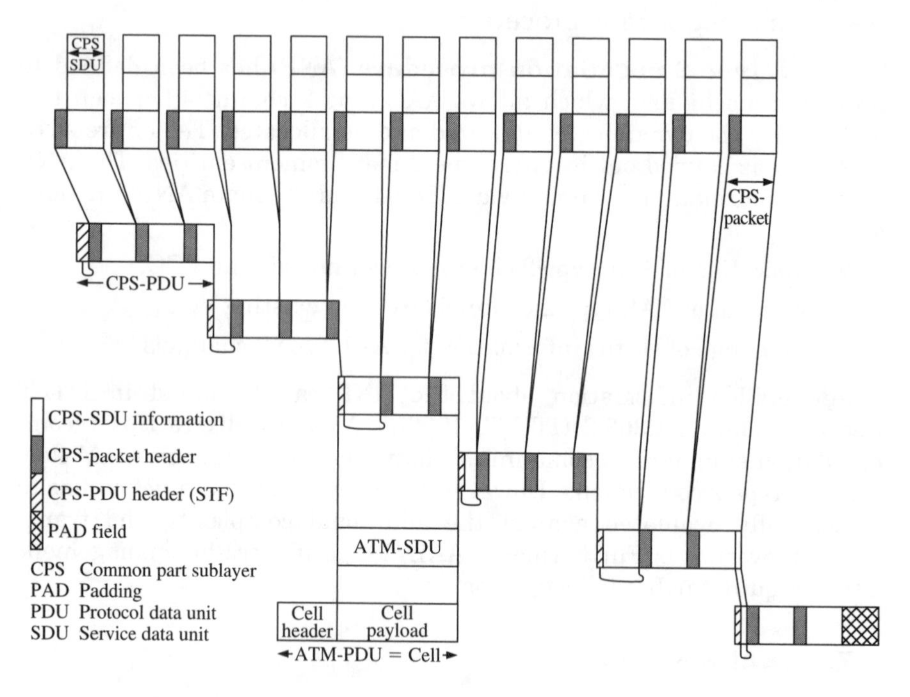

Figure 5.36 Filling ATM cells with CPS packets.

Data applications like CBDS and SMDS (see Section 10.3.4) make use of AAL type 3/4.

Two modes of service are defined for AAL type 3/4. The **message mode** service can be used for framed data transfer (for example, high-level data link control frame), and the **streaming mode** service is

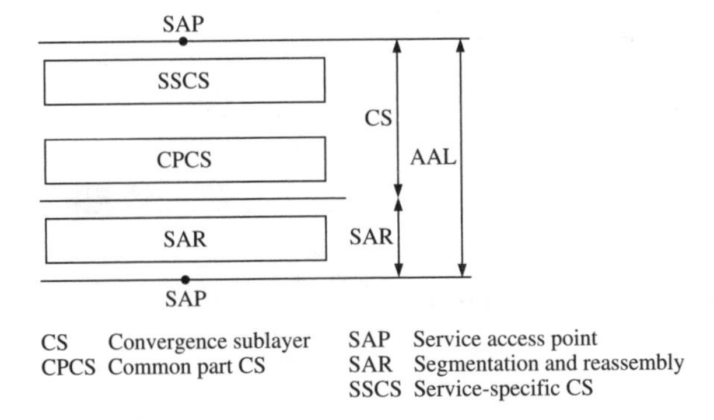

Figure 5.37 AAL type 3/4.

suitable for the transfer of low-speed data with low delay requirements. The application of these modes to a particular service depends on the service requirements.

The message mode service transports a single AAL-SDU in one, or (optionally) more than one, CS-PDU, which may build one or more SAR-PDUs. Figure 5.38 shows the operation of this mode.

In the streaming mode, one or more fixed-size AAL-SDUs are transported in one CS-PDU. The AAL-SDU may be as small as one octet and is always delivered as one unit, because only this unit will be recognized by the application (one SAR-SDU contains at most one AAL-SDU). Figure 5.39 illustrates the operation of the streaming mode service.

Two peer-to-peer operation procedures are offered by both service modes:

- **Assured operation:** The assured operation retransmits missing or errored AAL-SDUs, therefore flow control is provided as a mandatory feature. This operation mode may be restricted to point-to-point connections at the ATM layer.
- **Non-assured operation:** In this mode, lost or errored AAL-SDUs are not corrected by retransmission. The delivery of corrupted AAL-SDUs to the user may be provided as an optional feature. In principle, flow control can be applied to point-to-point ATM layer connections. No flow control will be provided for point-to-multipoint ATM layer connections.

Segmentation and reassembly sublayer

In general CS-PDUs are of variable length. When accepting such a PDU, the SAR sublayer generates SAR-PDUs containing up to 44 octets of CS-PDU data. The CS-PDU is preserved by the SAR sublayer. This requires a **segment type** (ST) indication and a SAR payload fill indication. The

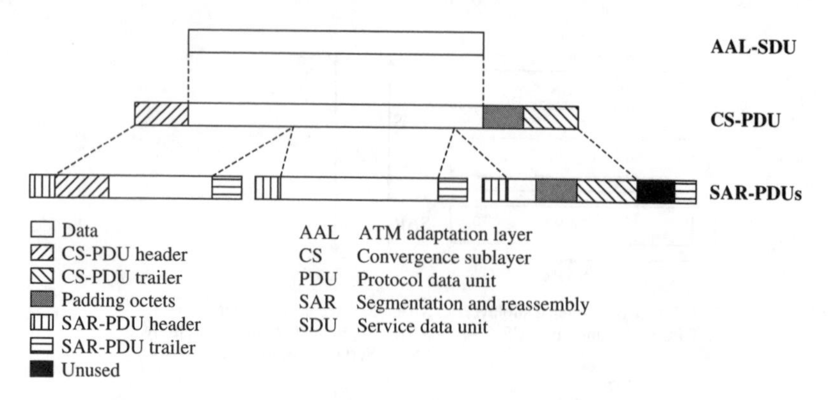

Figure 5.38 Message mode service.

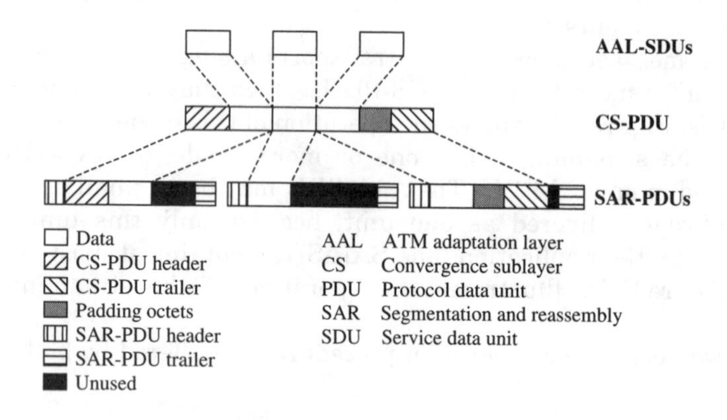

Figure 5.39 Streaming mode service.

ST indication identifies a SAR-PDU as being **beginning of message** (BOM), **continuation of message** (COM), **end of message** (EOM) or **single-segment message** (SSM). The payload fill indication represents the number of octets of a CS-PDU contained in the SAR-PDU payload. In the case of message mode service, the SAR-PDU payload of all BOMs and COMs contains exactly 44 octets, whereas the payload of EOMs and SSMs is of variable length. In streaming mode, the SAR-PDU payload of all segments depends on the AAL-SDUs.

Error detection is the second function of the SAR sublayer. This function includes detecting bit errors in the SAR-PDU as well as detecting lost or misinserted SAR-DPUs. An indication is sent to the CS if one of these errors occurs.

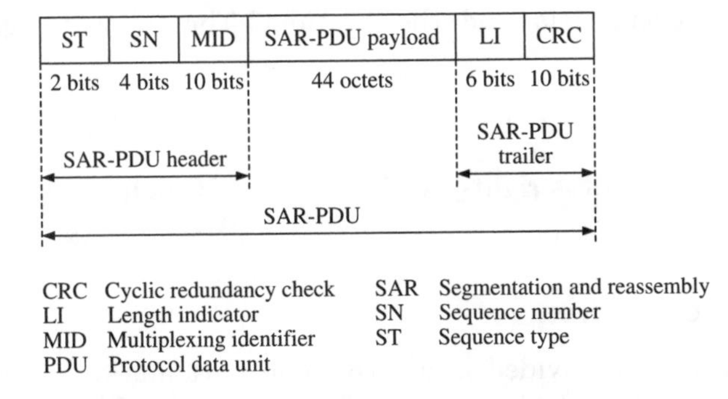

CRC Cyclic redundancy check
LI Length indicator
MID Multiplexing identifier
PDU Protocol data unit
SAR Segmentation and reassembly
SN Sequence number
ST Sequence type

Figure 5.40 SAR-PDU format for AAL type 3/4.

The third function of the SAR sublayer is the concurrent multiplexing/demultiplexing of CS-PDUs from multiple AAL connections over a single ATM layer connection.

To support all these functions, four octets are used (two for the SAR-PDU header and two for the SAR-PDU trailer). Therefore, of the 48 octets of the SAR-PDU only 44 remain for the payload. Figure 5.40 illustrates the SAR-PDU format. The coding of this PDU conforms to the conventions and rules described in Section 5.6.1.

The segment type consists of two bits which are used to identify a BOM, COM, EOM or SSM.

Four bits are available for the sequence number field. The SN or a SAR-PDU is incremented by '1' relative to the SN of the previous SAR-PDU belonging to the same AAL connection (numbering modulo 16).

The remaining 10 bits of the SAR-PDU header form the **multiplexing identifier** (MID) field. SAR-PDUs with an identical MID value belong to a particular CS-PDU. The MID field assists in the interleaving of ATM-SDUs from different CS-PDUs and reassembly of these CS-PDUs. If multiple AAL connections use the same ATM layer connection, these AAL connections must have identical QoS characteristics. Multiplexing/demultiplexing is done on an end-to-end basis. The ATM layer connection which is used by different AAL connections is administered as a single entity.

This is comparable to the multiplexing function of AAL type 2 where the CID corresponds to the MID. The SAR-PDU field (44 octets) is filled with CS-PDU data (left justified). If this field is not fully filled the remaining unused bits are coded as zero.

The **length indicator** (LI) field consists of six bits and contains the number of octets, binary coded, from the CS-PDU which are included in the SAR-PDU payload field (with a maximum of 44 octets). An LI value of 63 associated with an ST indicating an EOM leads to an abortion of partially transmitted CS-SDUs at the receiver.

The CRC field (10 bits) is filled with the result obtained from a CRC calculation which is performed across the SAR-PDU header, the

SAR-PDU payload field and the LI field. The following generating polynomial is used:

$$G(x) = x^{10} + x^9 + x^5 + x^4 + x + 1$$

The LSB of the result is right-justified in the CRC field.

Convergence sublayer

This sublayer is subdivided into a common part and a service-specific part. The CPCS transfers user data frames with any length from 1 to 65 535 octets. The CPCS functions require a 4-octet CPCS-PDU header and a 4-octet CPCS-PDU trailer. Additionally, a **padding** (PAD) field is provided for 32-bit alignment. Figure 5.41 shows the CPCS-PDU format.

The **common part indicator** (CPI) field is used to interpret the remaining fields in the CPCS-PDU header and trailer. Currently it is set to all zeros and the resulting interpretation of the other fields is described in the following (other settings might be used to identify AAL layer management messages and are for further study).

The **beginning tag** (Btag) and **ending tag** (Etag) fields allow the proper association of CPCS-PDU header and trailer. The same numerical value is set into both fields. The **BASize** field indicates to the receiving peer entity the maximum buffering requirements to receive the CPCS-SDU. The PAD field ensures that the CPCS-PDU payload is an integer multiple of four octets. It may be 0–3 octets long and does not convey any information. Similarly, the **alignment** (AL) field is used for 32-bit alignment of the CPCS-PDU trailer. The **length** field is used to encode the length of the CPCS-PDU payload field. It is also used by the receiver to detect the loss or gain of information.

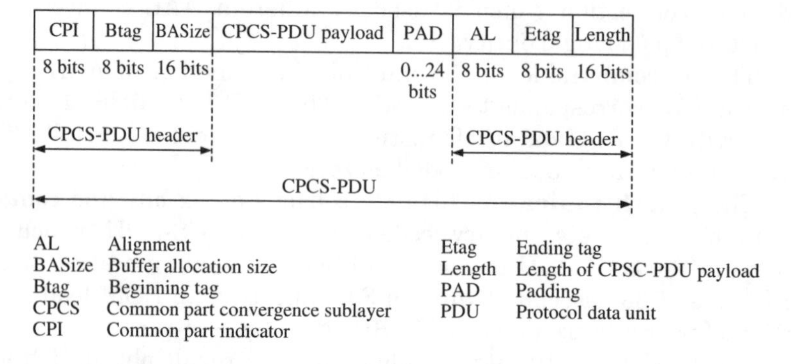

Figure 5.41 CPCS-PDU format for AAL type 3/4.

5.7.5 AAL type 5

AAL type 5 will be applied to variable bit rate sources without a timing relation between source and destination. It provides services similar to AAL type 3/4 and will mainly be used for data applications. The reason for defining this additional AAL type was its reduced overhead ('simple and efficient AAL'). Its message mode service, streaming mode service and assured/non-assured operation are identical to those defined in Section 5.7.4 for AAL type 3/4. However, one essential difference is that AAL type 5 does not support a multiplexing function, so there is no MID field. AAL type 5 usage examples are signaling frame relay and IP over ATM. It is anticipated that it will be used for other future applications as well.

AAL type 5 is subdivided into a SAR and CS, so the structure is the same as in Figure 5.37. The CS is further subdivided into a CPCS and an SSCS. This SSCS is application dependent and may be null. If needed, one possible function of the SSCS might be the multiplexing of different AAL connections (as known from the AAL 3/4 using the MID field). Another example of an SSCS can be found in Section 10.4.

Segmentation and reassembly sublayer

The SAR sublayer accepts SDUs which are an integer multiple of 48 octets from the CPCS. No additional overhead, that is, no more fields, are added to the received SDUs at the SAR sublayer. Only segmentation and, in the reverse direction, reassembling functions are performed. For the recognition of the beginning and end of a SAR-PDU, AAL type 5 makes use of the AUU parameter. This parameter is part of the PT field in the ATM header (see Section 5.6.2). An AUU parameter value of '1' indicates the end of a SAR-SDU, while a value of '0' indicates the beginning or continuation of a SAR-SDU. Thus an ST field (as provided in AAL type 3/4) is not used. The fact that this AAL type makes use of information conveyed in the ATM cell header can be considered as 'level mixing'. This means that the operations of AAL type 5 are no longer completely independent of the underlying ATM layer, which is an obvious infringement of the PRM specified for ATM (see Section 5.1). Nevertheless, it was adopted because of its simplicity and efficiency.

Convergence sublayer

The CPCS provides for the transfer of user data frames with any length from 1 to 65 535 octets. Additionally, one octet of user-to-user information is transparently transferred with each CPCS-PDU. A CRC-32 is used to detect bit errors. The CPCS-PDU format of AAL type 5 is shown in Figure 5.42.

The functions of the CPCS require an 8-octet CPCS-PDU trailer. In addition, a PAD field provides for a 48-octet alignment of the CPCS-PDU.

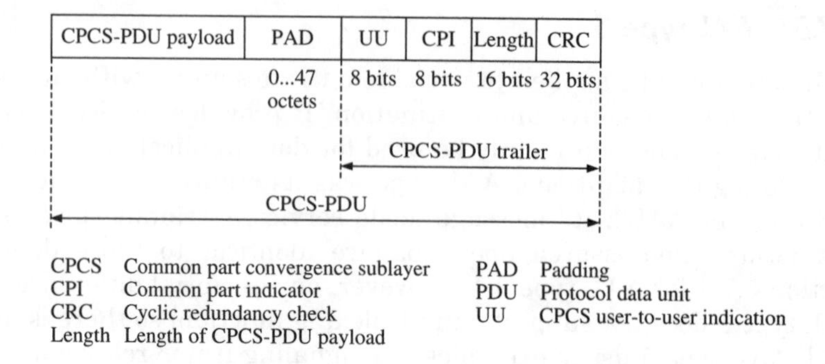

Figure 5.42 CPCS-PDU format for AAL type 5.

The CPCS user-to-user indication field is used for the transparent transfer of CPCS user-to-user information. Functions of the CPI fields may be similar to those described for AAL type 3/4. For the time being it is merely used to align the CPCS-PDU trailer to 64 bits. The length field is used to encode the length of the CPCS-PDU payload field. It is also used by the receiver to detect the loss or gain of information. Setting the length field to zero leads to partially transmitted CPCS-SDUs being aborted. The CRC-32 is used to detect bit errors in the CPCS-PDU.

Review questions

1. What is a protocol reference model (PRM)?
2. What do the terms N-SAP, N-PDU and N-SDU mean?
3. What are the names of the OSI layers and what are their tasks?
4. What planes and layers build up the B-ISDN/ATM PRM?
5. Which cell-related functions are performed at the *physical* layer of the ATM PRM?
6. What functions does the ATM layer provide? Explain what these functions do.
7. What is an idle cell?
8. At what bit rates are SDH-based ATM UNIs provided?
9. How are ATM cells mapped into SDH frames?
10. Which other UNIs have been defined by ITU-T and/or the ATM Forum?
11. What is meant by inverse multiplexing ATM?
12. Which frame-based interfaces are defined?
13. Between which layers of the B-ISDN reference model is the UTOPIA interface defined?

14. For which bit rates can the UTOPIA interface be used?
15. What does the ATM cell header structure look like?
16. What is the HEC field used for?
17. What is the PTI field used for?
18. What is the CLP bit used for?
19. What are the first four cell header bits at the UNI reserved for?
20. What is meant by AAL 0?
21. What is a typical example of using AAL type 1?
22. Why was there a need for AAL type 2?
23. Which AAL type is most frequently used for data applications?

6

Operation and maintenance of the B-ISDN UNI

This section covers **operation and maintenance** (OAM) aspects of the user–network interface and the customer access controlled by the network. The minimum functions required to maintain the physical layer and the ATM layer of the customer access, including the user–network interface, are described according to ITU-T Recommendation I.610 (ITU-T, 1995g). However, most of the mechanisms defined for the UNI will be used at NNIs as well. OAM of the layers above the ATM layer is not considered here.

First, the network configuration for OAM of the customer access is presented (Figure 6.1). Then OAM functions of the physical layer and ATM layer and applications of the hierarchically structured OAM information flows – see Section 4.6.2 – are described (Section 6.2). In Section 6.3, implementation of these flows in the physical layer – flows F1, F2, F3 – and in the ATM layer – flows F4 and F5 – is discussed. The last section (Section 6.4) presents the integrated local management interface (ILMI) as specified by the ATM Forum (1996h).

6.1 Network configuration for OAM of the customer access

The general arrangement for maintenance of the customer access – based on the principles set out in ITU-T Recommendation I.601 (ITU-T, 1988e) – is shown in Figure 6.1.

The customer installation in the figure comprises a maintenance entity that communicates with the customer access maintenance center located in the B-ISDN.

Comprehensive general maintenance services are also available in the network to support maintenance of the customer access.

B-ISDN will make use of functions provided by the TMN, the principles of which are described in ITU-T Recommendation M.3010 (ITU-T, 1996k).

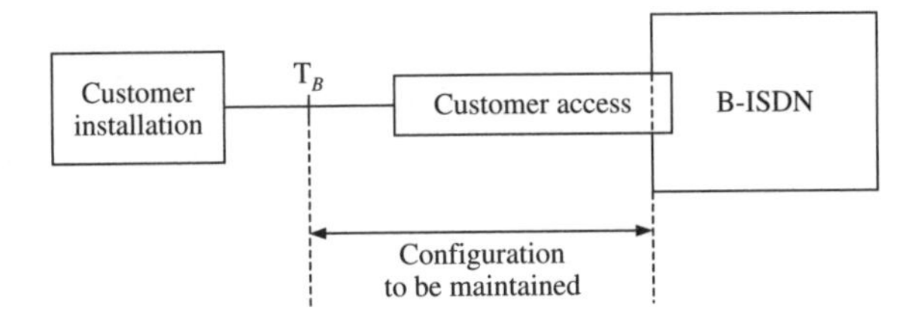

Figure 6.1 Maintenance configuration of the customer access.

The TMN is an independent information processing system. It enables network operators (of public as well as virtual private networks) to supervise components of the telecommunication network via defined interfaces and protocols.

Network management application functions supported by TMN include:

- performance management
- fault management
- configuration management
- accounting management
- security management.

The architectural relation of the B-ISDN customer access with the TMN is shown in Figure 6.2.

The protocols used for maintenance are specified through Q interfaces. Results of network-element internal monitoring will be passed to the TMN via these Q interfaces. The ATM Forum defines management interfaces at similar locations but calls them M interfaces.

6.2 OAM functions and information flows

Figure 4.9 shows the hierarchical level structure of the physical layer and ATM layer and the corresponding information flows. These levels and flows are summarized in Table 6.1.

Note that flows F1, F2 and F3 belong to the physical layer, and flows F4 and F5 to the ATM layer.

Two kinds of F4 and F5 flows can exist simultaneously on a VPC or VCC. One is an end-to-end flow used for end-to-end VPC/VCC operations communications. The other is a segment flow. A segment is a portion of a VPC or VCC which is typically under the control of a single network operator. Using these segment flows the network operator can monitor and operate the portions of VPCs/VCCs which are in its responsibility.

F4 and F5 flows, which are usually generated/terminated either at the endpoints of a VPC/VCC or at the connecting points terminating a VPC/VCC segment, may, optionally, also be monitored at intermediate points ('in flight').

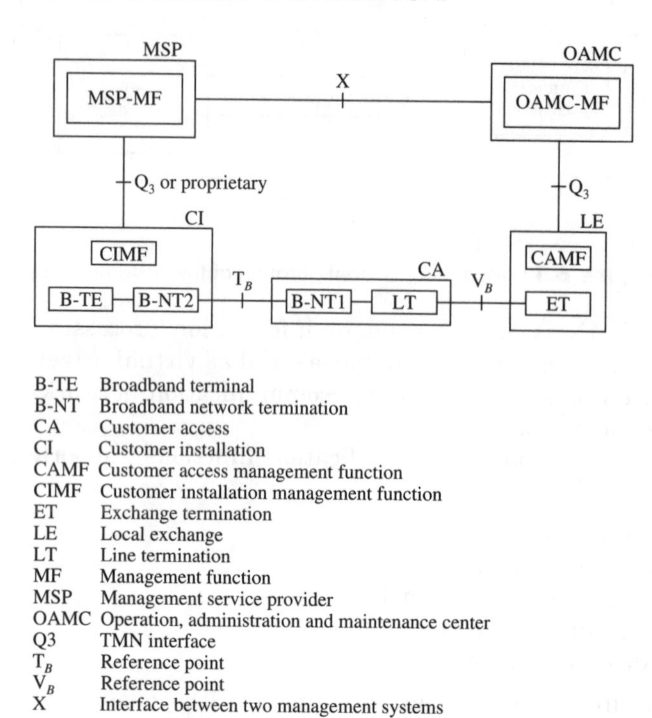

Figure 6.2 Example of TMN architecture for the customer access.

Table 6.1 OAM hierarchical levels and flows.

Level	*Flow*
Regenerator section	F1
Digital section	F2
Transmission path	F3
Virtual path	F4
Virtual channel	F5

6.2.1 Physical layer

Physical layer OAM functions are split into two categories:

- OAM functions which are dedicated to the detection and indication of unavailability states and which require real-time failure information transfer towards the affected endpoints for system protection. These functions are supported solely by the corresponding flows F1, F2 and F3.
- OAM system management functions dedicated to performance monitoring and reporting, or for localization of failed equipment.

These functions may be supported by the flows F1, F2 and F3 or by other means (for example, TMN via the Q interfaces, see Figure 6.2).

The functions described above are shown in Table 6.2 for the SDH-based interface and in Table 6.3 for the cell-based interface. The tables cover failures occurring on the B-NT2 to B-NT1 section (at reference point T_B) and on the transmission path beginning/terminating at B-NT2 (as later illustrated in Figure 6.4).

An **alarm indication signal** (AIS) is a maintenance signal of the physical layer which indicates the detection and location of a transmission failure. AIS is applicable at both the section and path levels. The same holds for the **remote defect indication** (RDI) signal (see Table 6.5). The two signals serve the following purposes:

'AIS is used to alert the associated downstream termination point and connection point that an upstream failure has been detected and alarmed.'

'RDI is used to alert the associated upstream termination point that a failure has been detected downstream.'

Table 6.2 Physical layer operation and maintenance functions of the SDH-based interface.

Level	*Function (category)*	*Failure detection.*
Regenerator section	Signal detection, frame alignment	Loss of signal, loss of frame
Multiplex section	Section error monitoring	Unacceptable error performance
Multiplex section adaptation	AU-4 pointer operation	Loss of AU-4 pointer or path-AIS
Transmission path	Cell rate decoupling	Defect in insertion/ suppression of idle cells
	Cell delineation	Loss of cell synchronization

Table 6.3 Physical layer operation and maintenance functions of the cell-based interface.

Level	*Function (category)*	*Failure detection*
Regenerator section	Signal detection, physical layer OAM cell recognition	Loss of signal, loss of F1 cell recognition
	Section error monitoring	Unacceptable error performance
Transmission path	Cell rate decoupling	Defect in insertion/ suppression of idle cells
	Physical layer OAM cell recognition	Loss of F3 cell recognition
	Cell delineation	Loss of cell synchronization

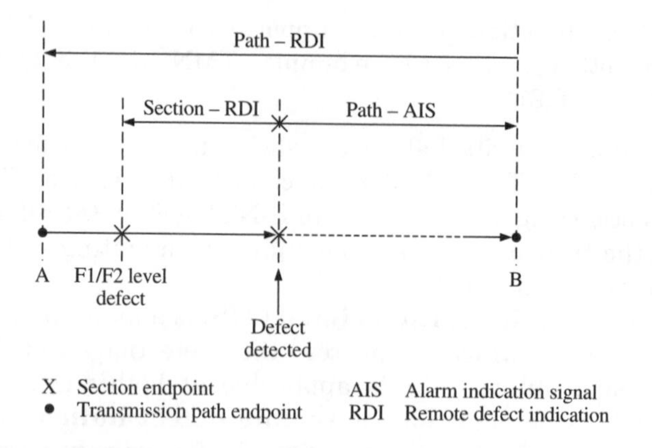

Figure 6.3 Illustration of the AIS/RDI concept.

If a fault is detected at a section endpoint, the section RDI is sent in the backward direction. A path AIS is sent in the forward direction to inform the path endpoint B (see Figure 6.3) about the defect affecting the transmission path. This path endpoint will then notify the path endpoint at A using a (backward) path RDI. This concept is illustrated in Figure 6.3.

Figure 6.4 shows the span of the flows F1 to F3 relating to physical layer OAM for different access configurations.

The upper part of the figure shows a customer directly connected to the local exchange. The transmission path (and the related F3 flow) extends between B-NT2 and the exchange termination (ET). This path comprises the two sections between B-NT2 and B-NT1, and between B-NT1 and the line termination (LT). Each section is assigned an F2 flow. In the example given in the figure, the second section comprises two regenerator sections with corresponding F1 flows. In the configuration shown in the middle of the figure, the section starting at B-NT1 is terminated by an LT in front of an STM multiplexer. Another section begins after this STM multiplexer. (Possible division of these sections into regenerator sections is not shown.) In the last configuration (lower part of the figure) a customer is connected with a VP cross-connect which terminates the transmission paths.

6.2.2 ATM layer

ATM layer OAM functions are monitoring of VP/VC availability and performance monitoring at the VP and VC levels (see Table 6.4).

The span of F4 and F5 flows is illustrated by the example in Figure 6.5. The VPC maintained by means of F4 here extends between B-NT2 and the ET, while the VCC extends between B-NT2 or a terminal and a termination point, which may be, for example, located behind the B-NT2 of another customer.

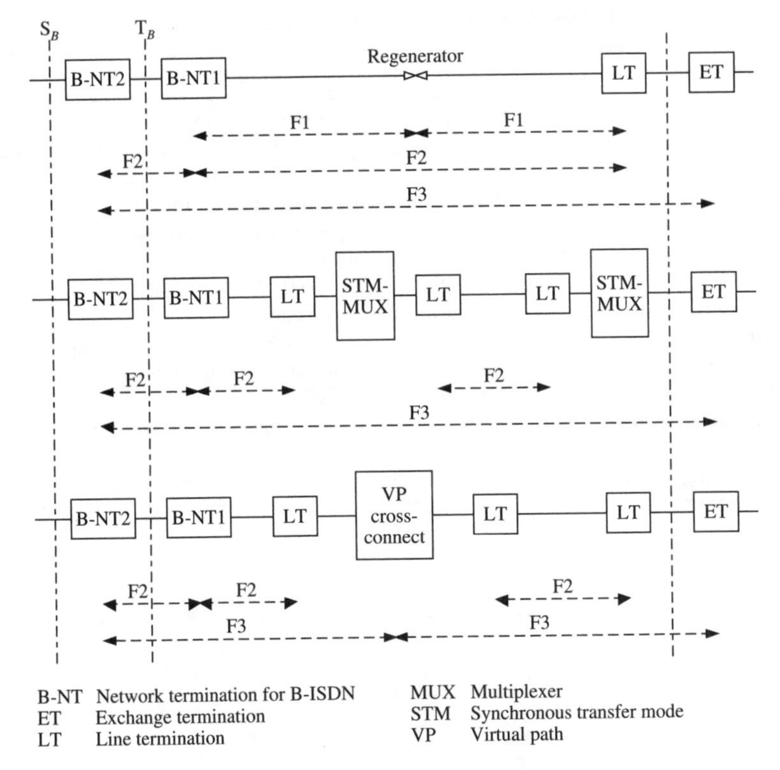

Figure 6.4 Physical layer OAM flows.

In the event of a defect the actions are similar to those at the physical layer. VP/VC-level alarms are sent to all affected active VPCs/VCCs from the connecting point of the VPC/VCC (for example, an ATM switch) which detects the defect. VP-AIS/VC-AIS and VP-RDI/VC-RDI are used at the ATM layer in an analogous way to the physical layer alarms.

VP/VC alarms are generated following the detection of defects pertaining to the physical layer or the ATM layer. This is shown in Figure 6.6.

VP/VC-AIS and VP/VC-RDI OAM cells are sent periodically during the failure condition with a frequency of one cell per second. The AIS/RDI status is declared after one AIS/RDI cell has been received. It is removed if no further AIS/RDI cell has been received for 2.5 ± 0.5 seconds or on receipt of one valid cell, that is, a user cell or continuity check cell.

To monitor VP/VC availability, continuity check cells may be sent downstream by a VPC/VCC endpoint when no user cell has been sent for a period of 1 second. If the endpoint of a VPC/VCC where continuity

Table 6.4 OAM functions of the ATM layer.

OAM function	*Main application*
AIS	For reporting defect indications in the forward direction
RDI	For reporting remote defect indications in the backward direction
Continuity check	For continuously monitoring continuity
Loopback	For on-demand connectivity monitoring For fault localization For pre-service connectivity verification
Forward performance monitoring	For estimating performance
Backward performance monitoring	For reporting performance estimations in the backward direction
Activation/deactivation	For activating/deactivating performance monitoring and continuity check
System management	For use by end-systems only (not further standardized)

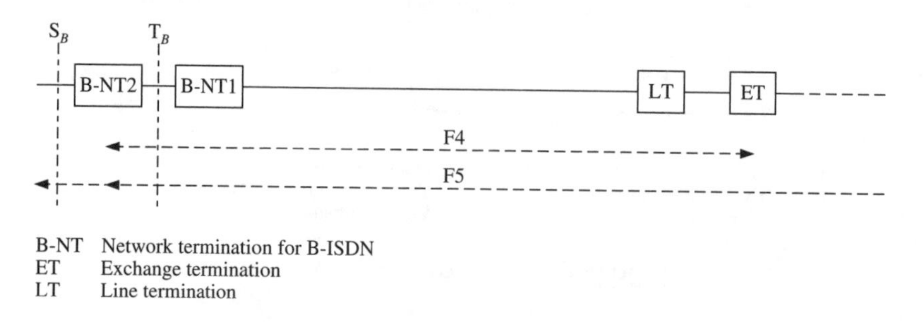

Figure 6.5 ATM layer OAM flows.

checking applies does not receive a cell within a time interval of 3.5 ± 0.5 seconds it will send VP/VC-RDI to the far end. (Continuity cells may also be sent continuously every second together with user cells.)

VP/VC performance monitoring is based on monitoring blocks of N user cells. N may have the value 128, 256, 512 or 1024, with a tolerance of $\pm 50\%$ each. Specific performance monitoring/reporting cells are inserted and evaluated, respectively. Errored blocks, loss/misinsertion of cells, cell delay and so on can be detected or measured by means of VP/VC performance monitoring (see Figure 6.7).

Performance monitoring and the continuity check function can be activated either during connection establishment or at any time after the connection has been established via the TMN or by means of so-called activation/deactivation OAM cells.

For on-demand connectivity verification a loopback cell can be sent out. This loopback cell is returned either at a specifically addressed network element, at the end of a segment or at the end of a VPC/VCC. It can be used in-service or as a pre-service connectivity verification tool.

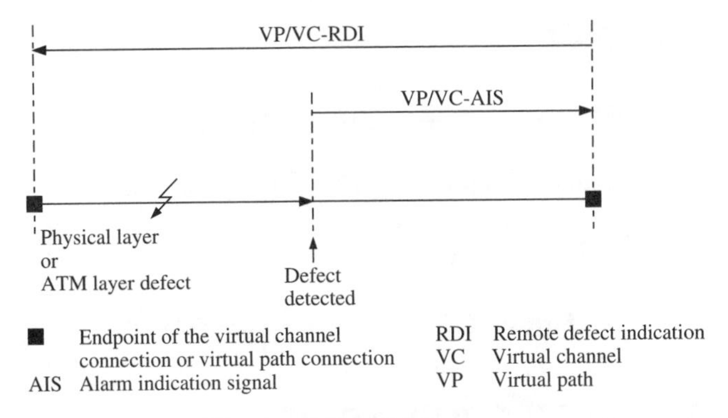

Figure 6.6 VP/VC alarms.

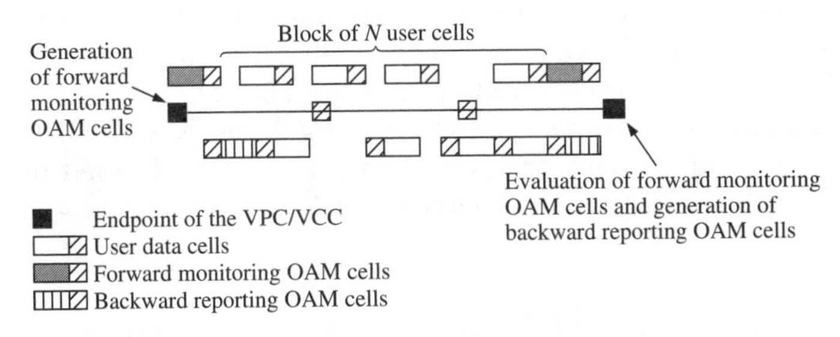

Figure 6.7 Performance monitoring at the ATM layer.

By shortening the loop a failure localization becomes possible, see Figure 6.8. The advantages of the loopback function are its simplicity and the capability of obtaining VPC/VCC status information from a single end of the connection.

Further study is required on how to apply the existing OAM functions on point-to-multipoint connections. Imagine an upstream failure situation for such a connection type. The sender node would be flooded by many RDI cells generated by all the leaf nodes.

Another area currently being discussed within ITU-T is **automatic protection switching** mechanisms (APS) at the ATM layer. One variant is likely to be designed to substitute physical layer protection switching. This is achieved by bundling several VPs into a so-called 'VP group'. A specific OAM channel will be defined which triggers the protection switching of such a VP group. A second variant might only protect single VPs.

Both schemes have in common that a so-called 'protected domain' in which a backup path is established needs to be defined between two network elements .

It should be noted that the ATM layer OAM measures discussed above require the allocation of some extra cell rate to an ATM connection.

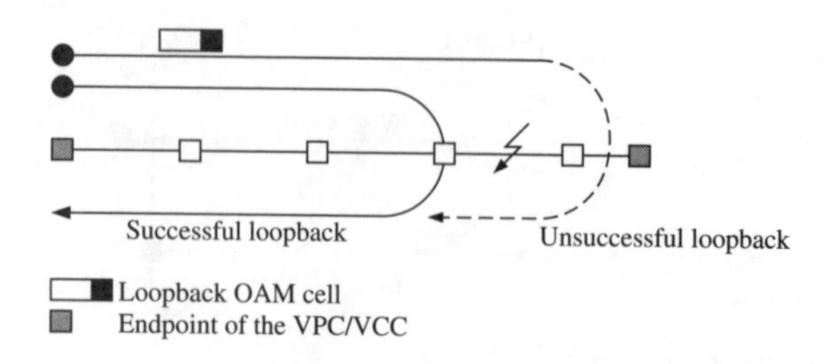

Figure 6.8 The loopback principle.

6.3 Implementation issues

6.3.1 Physical layer

Transmission overhead allocation and coding of the SDH physical layer functions, as defined in Table 6.2, are given in Table 6.5 (details of the SDH frame structure are presented in Section 5.3.1).

Transmission performance monitoring across the B-ISDN UNI in order to detect and report transmission errors is performed at section and path levels, as indicated in Section 6.2.1.

Table 6.5 SDH overhead allocation at the B-ISDN UNI.

Byte	*Function*	*Coding (only OAM-relevant parts)*
STM-1 section overhead:		
A1, A2	Frame alignment	
J0	Regenerator section trace[1]	
B1	Regenerator section error monitoring[2]	BIP-8
B2	Multiplex section error monitoring	BIP-24 / BPI-96[3]
H1, H2	AU AIS/AU-4 pointer	all 1s
H3	Pointer action	
K2 (bits 6–8)	Section AIS/section RDI	111/110
M1	Section error reporting (REI)	B2 error count
VC-4 path overhead:		
J1	Access point ID/verification	
B3	Path error monitoring	BIP-8
C2	Path signal label	ATM cells
G1 (bits 1–4)	Path error reporting (REI)	B3 error count
G1 (bit 5)	Path RDI	1

[1] Need for further study.

[2] The use of B1 for regenerator section error monitoring across the B-ISDN UNI is application dependent and therefore optional.

[3] BIP-24 for the 155 Mbit/s interface, BIP-96 for the 622 Mbit/s interface.

At the SDH section level, an incoming signal is monitored at the section termination point by means of the bit-interleaved parity 24 (BIP-24), which is inserted into the B2 field at the other section termination point.

The **remote error indication** (REI) is used to monitor an outgoing signal. This error count, obtained by comparing the calculated BIP-24 and the B2 value of the incoming signal at the far end, is inserted in the Z2 field bits 18 to 24 and then sent back to the near-end section termination point.

In a similar way, at SDH path level an incoming signal is monitored using BIP-8 of the B3 byte. An outgoing signal is monitored using the path REI of bits 1 to 4 of the G1 byte. This concept is illustrated in Figure 6.9 (only the case 'A sending towards B' is shown for simplicity).

SDH, as described in ITU-T Recommendation G.707 (ITU-T, 1996c), provides additional OAM means that will not necessarily be used in the customer access network. One example is automatic protection switching across the B-ISDN UNI. In the event of failure of the transmission line, the system could automatically switch to a standby line to prevent longer out-of-order periods.

Allocation of the OAM functions (see Table 6.3) to the F1/F3 OAM cells is included in ITU-T Recommendation I.432 (ITU-T, 1996j) for cell-based interfaces.

6.3.2 ATM layer

ATM layer OAM flows F4 and F5 (see Table 6.4) are provided by cells dedicated to ATM layer OAM functions. This type of cell can be identified by specific values of the payload type identifier or VCI. The F4 cells use reserved VCI values while F5 cells use specific payload type identifiers (see Table 6.6). ATM layer OAM cells have a common information field format (see Figure 6.10).

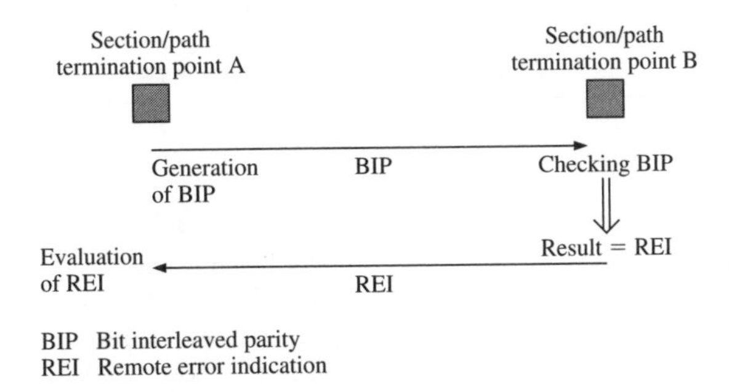

Figure 6.9 Illustration of transmission performance monitoring.

Table 6.6 OAM F4/F5 cell headers.

OAM flow	*Span*	*Header*
F4	End-to-end	VCI = 00000000 00000100
F4	Segment	VCI = 00000000 00000011
F5	End-to-end	PTI = 101
F5	Segment	PTI = 100

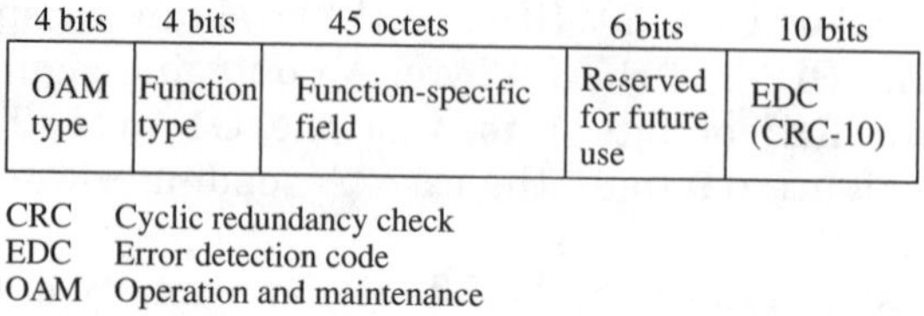

4 bits	4 bits	45 octets	6 bits	10 bits
OAM type	Function type	Function-specific field	Reserved for future use	EDC (CRC-10)

CRC Cyclic redundancy check
EDC Error detection code
OAM Operation and maintenance

Figure 6.10 Information field of the ATM layer OAM cells.

The OAM type and function type codings are given in Table 6.7. Note that unused information field bits (incomplete octets) of the OAM cells are coded all zero while unused (complete) octets are coded 0110 1010.

Figure 6.11(a) shows the function-specific fields for AIS/RDI cells. The defect type field might optionally carry information about the detected defect, for example a physical layer defect or loss of continuity defect. The network element which detected the faulty situation might place a specific identifier into the defect location field which may allow the failure to be located. Currently, neither function seems to be well enough defined to be widely used.

No function-specific fields for the continuity check cells have yet been defined.

Table 6.7 OAM type/function codings.

OAM type	*4 bit*	*Function type*	*4 bit*
Fault management	0001	AIS	0000
		RDI	0001
		Loopback	1000
		Continuity check	0100
Performance management	0010	Forward monitoring	0000
		Backward reporting	0001
Activation/deactivation	1000	Performance monitoring	0000
		Continuity check	0001
System management	1111	For use by end-systems only	

Figure 6.11(b) illustrates the function-specific fields of performance monitoring cells. The **total user cell number** (TUC) indicates how many user cells the sender transmitted within that specific OAM block. The **total received cell count** (TRCC) contains information on the number of user cells received at the other end of the segment/connection. Calculating the difference between the two counts allows the number of lost user cells (differentiated between the CLP = 0 and CLP = 0 + 1 cell flows) to be estimated. The **block error detection code** (BEDC) and **block error result** (BLER) fields are used for estimating and reporting, respectively, by means of a 'bit interleaved parity' check, bit errors occurring in the user cell's payload fields. The **monitoring cell sequence number** (MCSN) is a simple running

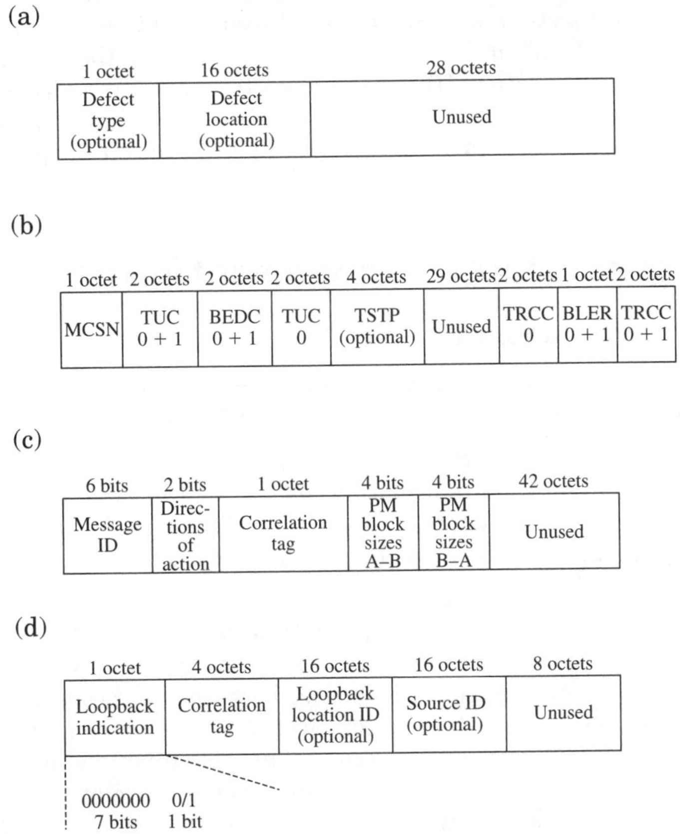

(a)

1 octet	16 octets	28 octets
Defect type (optional)	Defect location (optional)	Unused

(b)

1 octet	2 octets	2 octets	2 octets	4 octets	29 octets	2 octets	1 octet	2 octets
MCSN	TUC 0 + 1	BEDC 0 + 1	TUC 0	TSTP (optional)	Unused	TRCC 0	BLER 0 + 1	TRCC 0 + 1

(c)

6 bits	2 bits	1 octet	4 bits	4 bits	42 octets
Message ID	Directions of action	Correlation tag	PM block sizes A–B	PM block sizes B–A	Unused

(d)

1 octet	4 octets	16 octets	16 octets	8 octets
Loopback indication	Correlation tag	Loopback location ID (optional)	Source ID (optional)	Unused

Loopback indication: 0000000 (7 bits), 0/1 (1 bit)

BEDC Block error detection code
BLER Block error result
ID Identifier
MCSN Monitoring cell sequence number
PM Performance monitoring
TRCC Total received cell count
TSTP Time stamp
TUC Total user cell number

Figure 6.11 Function-specific fields of the ATM layer OAM cells: (a) AIS/RDI cell; (b) performance monitoring cell; (c) activation/deactivation cell; (d) loopback cell.

counter which allows lost performance monitoring cells to be detected. Use of the optional **time stamp** (TSTP) field is not further defined.

In Figure 6.11(c) the content of an activation/deactivation cell is outlined. It contains fields which allow negotiation about the block size used for performance monitoring and the direction in which the performance flow is used. The latter field is also used for activating/deactivating the continuity check.

The function-specific fields of a loopback cell can be found in Figure 6.11(d). The last bit of the loopback indication is used to differentiate between already looped cells and those cells that still have to be looped. The sending node sets this bit to 1. At the loopback point it is reset to 0. A loopback cell with a loopback indication of all 0 is never looped again. This mechanism prevents the occurrence of infinite loops. The correlation tag, together with the optional source ID field, allows a cell to be unanimously identified when received again by the sender. The optional loopback location ID can be used to explicitly address a network element where the cell shall be looped. As an example, this allows the loopback function to be used for fault localization purposes, as described in Section 6.2.2.

6.4 Integrated local management interface

6.4.1 Scope and functionality

The **integrated local management interface** (ILMI) has been specified in the ATM Forum's ILMI specification (ATM Forum, 1996h):

> 'to provide any ATM device (e.g. End-systems, Switches, etc.) with status and configuration information concerning the Virtual Path, Virtual Channel Connections, registered ATM Network Prefixes, registered ATM Addresses, registered services, and capabilities available at its ATM interfaces.'

Until the completion of this specification, the ILMI was part of the ATM Forum's UNI specification (ATM Forum, 1993c). In this context ILMI stood for **interim local management interface** because it was considered to be a preliminary standard which might eventually be replaced with an agreed ITU-T standard offering more comprehensive functionality.

Since, firstly such an ITU-T standard never developed and secondly, the ILMI protocol was widely used, especially in private ATM networks, the ATM Forum decided to use it for an indefinite period. The original specification was considerably enhanced and is now contained in a separate specification (ATM Forum, 1996h).

ILMI refers to the private and public UNI and the private NNI (see Figure 6.12).

ILMI supports the bidirectional exchange of management information between **ATM interface management entities** (IMEs)

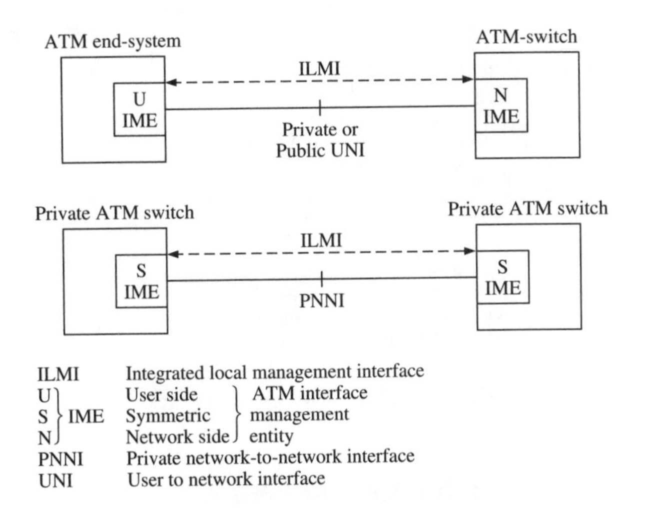

Figure 6.12 Application scope of ILMI.

which are located at both ATM end-systems and ATM switches (see Figure 6.12) and which terminate the ILMI protocol. Such a pair of two IMEs are called adjacent IMEs.

In terms of physical equipment, ILMI will be implemented in workstations and computers with ATM interfaces, in ATM switches, and in higher-layer switches (for example, frame relay switches, LAN bridges/routers) that transfer their frames within ATM cells.

The ILMI functions pertain to the physical layer and the ATM layer of an ATM interface. Furthermore, ILMI provides for an automatic address registration at UNIs and for an auto-configuration for **LAN emulation clients** (LECs), see Section 11.2.2.

The status and configuration of an ATM device is represented in an ATM interface **management information base** (MIB). Every IME contains the same ATM interface MIB and provides an agent application as well as a management application. At the user-side of a UNI the 'user IME' is implemented whereas at the network side of the UNI the 'network IME' can be found. Across a private NNI 'symmetric IMEs' are used.

Objects in the ATM interface MIB are defined according to **Abstract Syntax Notation One** (ASN.1) (ISO, 1987).

6.4.2 ILMI protocol

ILMI employs the protocol stack shown in Figure 6.13.

The ILMI protocol is based on the **Simple Network Management Protocol Version 1** (SNMPv1) as defined by the IETF (1990) without UDP and IP addressing.

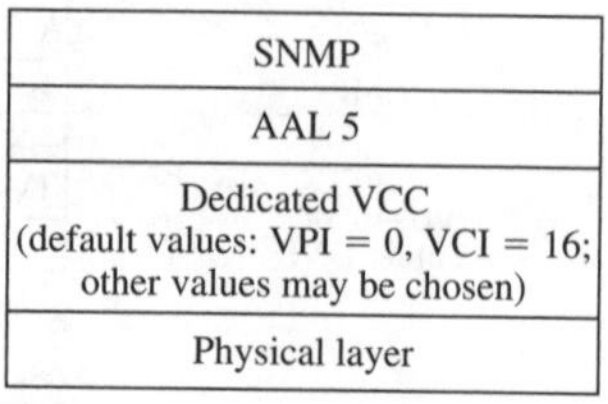

SNMP
AAL 5
Dedicated VCC (default values: VPI = 0, VCI = 16; other values may be chosen)
Physical layer

AAL ATM adaptation layer
SNMP Simple network management protocol
VCC Virtual channel connection
VCI Virtual channel identifier
VPI Virtual path identifier

Figure 6.13 ILMI protocol stack.

Table 6.8 SNMP operations.

Operation	*Function*
Get	Retrieval of specific management information
Get-next	Step-by-step retrieval
Set	Altering management information
Trap	Reporting extraordinary events

A dedicated, permanently available VCC is used for sending AAL-encapsulated SNMP messages between adjacent IMEs. SNMP comprises four types of operation which are used to manipulate management information (see Table 6.8).

SNMP messages of up to and including 484 octets are normally allowed. (The use of larger messages may be negotiated.) The SNMP traffic via the ILMI VCC should comply with a SCR of 1% of the physical line rate and a maximum PCR of 5% of the physical link. The MBS is limited to about 11 cells. (For a definition of SCR, PCR and MBS see Section 4.5.2.)

In 95% of all cases the SNMP message response time should be lower than 1 second. The transmitted data should not be older than 30 seconds, that is no information is older than 30 seconds. The two supported traps (**coldstart** and **enterprisespecific**) should be reported within 2 seconds.

Upon a (re-)start of an IME it first initializes itself. Afterwards, a coldstart trap is sent towards its peer IME and the ILMI connectivity procedure is started. ILMI connectivity is tested periodically by issuing a get-request or get-next-request. A matching ILMI response indicates ILMI connectivity.

As an option a following automatic configuration procedure allows automatic configuration of ATM interface type (public or private), the

IME type (user, network or symmetric), ATM layer attributes and signaling VCC transmission parameters (traffic parameters). After such an auto-configuration address registration may take place.

Any failure or modification of local attributes requires the adjacent IMEs to be re-initialized. For this purpose the coldstart trap is sent to the peer IME.

Using a proxy agent within an ATM device enables an external network management station to access the local as well as the remote ATM interface MIBs.

Furthermore, the ILMI specification also supports virtual UNIs as defined in ATM Forum (1996d). In this case several logical UNIs are concentrated on a single physical link, for example an intermediate VP cross-connect. In this case the cross-connect needs to be transparent for the adjacent IMEs using ILMI.

6.4.3 ILMI services and management information base

ILMI provides management information on the physical layer and ATM layer for monitoring and controlling ATM-based UNIs. This information will be accommodated by the management information base (MIB), which supports the following elements:

- system
- physical layer
- ATM layer
- virtual path/virtual channel connections
- link management traps
- service registry.

System MIB information:

- provides general system information according to RFC 1213 (MIB-II) (IETF, 1991).

It should be noted that the following MIB portions include an interface index attribute as well.

Physical layer MIB information includes (most of the following attributes are only used for being backward compatible with previous specifications):

- interface address
- transmission type
- media type
- operational status: in-service, out-of-service, loopback, other mode
- port specific information
- adjacency information.

ATM layer MIB information comprises:

- maximum number of active VPI/VCI bits
- maximum number of VPCs/VCCs
- number of configured VPCs/VCCs
- maximum switched VPC/VCC VPI
- minimum switched VCC VCI
- ATM interface type (public/private ATM interface)
- ATM device type (user/node)
- ILMI version
- UNI/NNI signaling version.

The VPC/VCC-related MIB information provides information for each VPC/VCC on:

- VPI/VCI value
- operational status
- transmit/receive traffic descriptor
- best effort indicator
- transmit/receive QoS class
- service category
- ABR operational parameters (only if supported at this interface)
- transmit/receive frame discard indicator (for VCCs only).

The management traps are used for indicating:

- a newly configured VPC/VCC
- a modified VPC/VCC
- a deleted VPC/VCC.

The address registration part of the MIB contains information on:

- network prefix
- network prefix status
- ATM address
- ATM address status information
- ATM address organizational scope indication.

The service registry part of the MIB contains information for locating **LAN emulation configuration server** (LECS) and/or **ATM name server** (ANS):

- service identifier
- service address

- service address index
- additional service information (not used by LECS and ANS service registry procedures).

Review questions

1. What are the OAM flows at the physical layer called?
2. What are the OAM flows at the ATM layer called?
3. What are the ATM OAM functions?
4. What are the applications for the loopback function and the continuity check functions, respectively?
5. Where is the ILMI used? What is it used for?

7

Traffic management

As already mentioned in Section 4.5, traffic management within ATM networks has two main goals: to provide specified and guaranteed levels of **quality of service** (QoS) while efficiently using available network resources. Since these two goals are somewhat contradictory it is difficult to find procedures which meet both goals simultaneously. Sophisticated measures needed to be defined which are explained in more detail throughout this chapter.

Traditional networks such as the 'old' Internet can only provide a 'best effort' service; that is, there is almost no guarantee on the time when the information is delivered at the correct destination or that the information is delivered at all.

With ATM, for each virtual connection a specific QoS can be chosen which influences the ATM transfer performance parameters such as cell loss, cell delay and cell delay variation (see Section 4.4).

Through **statistical multiplexing**, traffic management allows the existing transmission capacities to be exploited as far as possible.

7.1 Traffic control procedures and their impact on resource management

The choice of traffic control algorithms directly affects a network's resource allocation strategy. For example, if only the peak cell rate of a connection is considered for **connection admission control** (CAC), then this peak bit rate would have to be allocated to the connection. If this connection had a low average bit rate, then most of the time the network efficiency would be poor. Nevertheless, such simple strategies can assist in quickly introducing ATM-based networks. As long as knowledge about the management of ATM traffic flows is rather limited, because neither the source traffic characteristics nor the actual traffic mixes on a link are sufficiently clear, it might be wise to stay on safe ground even if a considerable amount of network capacity is wasted.

Besides peak bit rate reservation, restricted utilization (for example, 85%) of the cell transport capability of a link helps to avoid congestion.

The basic problem of ATM networks is the statistical behavior of the cell arrival process (for example, at a buffer where cells generated by several sources are multiplexed together).

ATM traffic can be described by a three-level hierarchical model with different time scales, as depicted in Figure 7.1 (Hui, 1988; Schoute, 1988).

The call level has a typical time scale of seconds (for example, a short telephone call) up to hours (for example, a long-lasting video conference), the burst level a millisecond range up to seconds, and the cell level a microsecond range. These levels have different impacts on an ATM network:

Call level: This level is controlled by the **connection admission control** (CAC) algorithms which decide whether a connection can be accepted or not. If too many connections need to be rejected the network may not be very well dimensioned. Additional network resources should be deployed. The time scale at this level is determined by the number of end users who want to transmit information concurrently.

Burst level: This level can be controlled by fast resource reservation mechanisms and adaptive flow control protocols like ABR. It determines the large buffers needed for non-real-time connections. Furthermore, specific CAC algorithms may allow connections to be statistically multiplexed at burst level; for example, the SIGMA Rule (Hauber, Wallmeier, 1991) can be applied on VBR-rt connections. Burst periods are generated by end applications which transmit more or less information during the connection's lifetime.

Cell level: This level can be controlled by mechanisms such as policing, priority control or traffic shaping. At multiplexing points the traffic pattern at cell level determines the buffer size required for real-time connections. Cell arrival variations are caused by the principles of asynchronous multiplexing of cells.

Each of these levels requires a dedicated set of traffic control functions. This leads to a hierarchical traffic management model where functions with a larger time scale rely on those with a lower time scale. As an example, the decisions of the CAC (call level) make sense only if the actual cell flow is controlled by a policing function (burst and cell level) like UPC/NPC (see Section 7.2.3).

All mechanisms need to be aligned carefully for achieving a consistent and efficient traffic management concept.

Based on the traffic model shown in Figure 7.1, mathematical models have been defined with different levels of abstraction (Kuehn, 1989), including:

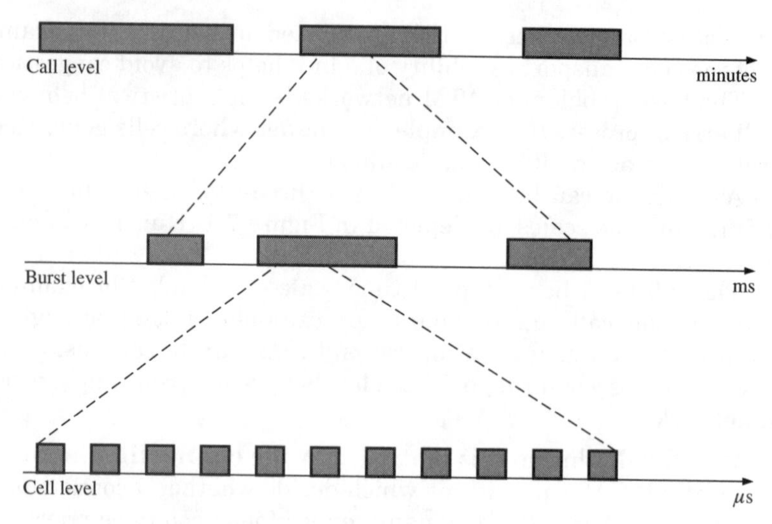

Figure 7.1 Hierarchical modeling of ATM traffic.

- burstiness (peak bit rate/average bit rate)
- geometrically distributed burst lengths
- switched Poisson process
- Markov-modulated Poisson process
- generally modulated deterministic process.

It has been found that the quality of service parameters, such as CDV and loss probability, are very sensitive to the assumed source characteristic. Therefore, the considered traffic management functions need to be robust enough to cope with different traffic characteristics.

7.2 Mechanisms to achieve a specified QoS

In this section the mechanisms that can be used to provide a specified QoS are outlined. Section 7.3 describes mechanisms that allow the network utilization to be increased.

7.2.1 Use of virtual paths

One instrument of network resource management which can be employed for traffic control is the virtual path technique (see Section 5.6.3). By grouping several virtual channels together into a virtual path, connection admission control (Section 7.2.2) and usage/network parameter control (Section 7.2.3) can be simplified as only the aggregated traffic of an entire virtual path has to be handled. Priority control (Section 7.2.4) can be supported by re-aggregating traffic types requiring different qualities of service through virtual paths. Messages for the operation of traffic control (for example, congestion indication) can more easily be distributed: a single message referring to all the virtual channels within a virtual path will suffice.

Virtual paths can play an important role in supporting statistical multiplexing, if, for example, a separation of statistically multiplexed traffic from guaranteed bit rate traffic is desired.

7.2.2 Connection admission control

Connection admission control (CAC) is defined as 'the set of actions taken by the network at the call establishment phase (or during the call renegotiation phase) in order to establish whether a VC/VP connection can be accepted' (ITU-T, 1996i).

A connection can only be accepted if sufficient network resources are available to establish the connection end-to-end with its required quality of service. The agreed quality of service of existing connections in the network must not be affected by the new connection.

During connection establishment CAC is provided with:

- the required ATM layer transfer capability
- the source traffic descriptors
- the required QoS class (including CDVT).

With the CAC process a kind of 'traffic contract' between subscriber and network is established. Depending on this traffic contract different policing configurations (see Table 7.1) will be applied. For each traffic category a distinct CAC algorithm might be used; for example, for CBR the available link capacity needs to be checked, whereas for UBR the available buffer space could be the prime factor to be considered. A promising way to increase network utilization while still guaranteeing QoS is the use of CAC algorithms based on 'measurements' of the actual network load and cell transmission performance.

Table 7.1 Possible policing configurations according to the ATM Forum.

Classification	*PCR policing on cells with a CLP bit setting of*	*SCR policing on cells with a CLP bit setting of*	*Optional cell tagging*
CBR.1	0 + 1	n/a	n/a
VBR.1	0 + 1	0 + 1	n/a
VBR.2	0 + 1	0	No
VBR.3	0 + 1	0	Yes
UBR.1	0 + 1	n/a	n/a
UBR.2	0 + 1	n/a	Yes

n/a Not applicable

7.2.3 Usage parameter control and network parameter control

Usage parameter control (UPC) and **network parameter control** (NPC) perform similar functions at different interfaces. The UPC function is performed at the user–network interface, whereas the NPC function is performed at the internetwork interface.

The use of a UPC function is recommended to control (and limit) the amount of traffic entering a network. Whether or not the operator chooses to use the UPC/NPC function, the network-edge-to-network-edge and user-to-user performance objectives need to be met.

UPC/NPC is defined as:

> 'the set of actions taken by the network to monitor and control that the traffic contract is respected in terms of traffic offered and validity of the ATM connection, at the user access and the network access respectively. Their main purpose is to protect network resources from malicious as well as unintentional misbehavior which can affect the quality of service of other already established connections by detecting violations of negotiated parameters and procedures and taking appropriate actions. Connection monitoring encompasses all connections crossing the user–network or internetwork interface. UPC/NPC apply to user VCCs/VPCs, signaling and meta-signaling virtual channels' (ITU-T, 1996i).

Usage parameter monitoring includes the following functions:

- Checking of the validity of VPI/VCI values.
- Monitoring the traffic volume entering the network from all active VP and VC connections to ensure that the agreed parameters are not violated.

The parameters which may be subject to monitoring and control are those used for source traffic characterization and possibly other dyanamic parameters specific to a given ATM transfer capability. In any case, the peak cell rate should never be exceeded.

Usage/network parameter control can simply **discard** cells that violate the negotiated traffic parameters. In addition, a 'guilty' connection may be released. Another, less rigorous, option would be **tagging** of violating cells. This means that the CLP bit setting is changed from 0 to 1, thus increasing the cell loss probability. These cells can be transferred as long as they do not cause any serious harm to the network. However, in case of network congestion they will be discarded first (see Section 7.2.4). Thus the overall throughput of ATM cells might possibly be raised. At the ATM block level (this is a burst of ATM cells delimited by resource management cells, see Section 7.3.2.), UPC/NPC may discard all the remaining cells in the ATM block.

When, optionally, usage parameter control and traffic shaping (Section 7.2.5) are combined, cells may be rescheduled in order to improve network performance.

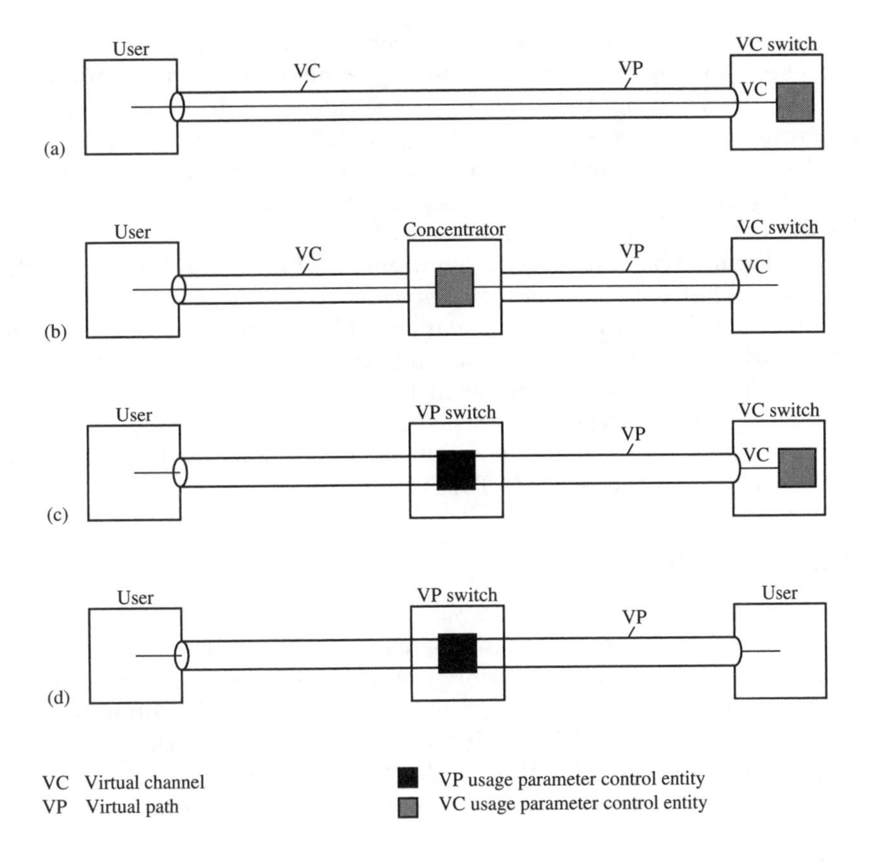

Figure 7.2 Illustration of usage parameter control.

To illustrate the concept of usage parameter control, Figure 7.2 shows different access network arrangements with the appropriate usage parameter control measures applied to VCs or VPs at the point where they are first accessible. This point may be a switch/cross-connect or a concentrator.

In case (a), a user is connected directly to a VC switch. Usage parameter control is performed within the VC switch on VCs before they are switched.

In case (b), a user is connected to a VC switch via a concentrator. Usage parameter control is performed within the concentrator on VCs only.

In case (c), a user is connected to a VC switch via a VP switch. Here usage parameter control is performed within the VP switch on VPs only and within the VC switch on VCs only.

In case (d), a user is connected to another user via a VP switch. Usage parameter control is now performed within the VP switch on VPs only. Users are responsible for controlling the VCs.

Different usage parameter control mechanisms have been proposed, namely leaky bucket, sliding window, jumping window and exponentially weighted moving average. Their modeling and performance are presented in (Rathgeb, 1990). A comparison of these mechanisms has shown that the leaky bucket and the exponentially weighted moving average are the most promising with respect to flexibility and implementation complexity.

It should be noted that the **generic cell rate algorithm** (GCRA) (ITU-T, 1996i) only provides a conformance definition at UNIs/NNIs; that is, a check whether a cell stream complies with the 'traffic contract'. The GCRA as such is not a policing or shaping function. However, policing devices implementing or at least approximating the GCRA are best suited for UPC/NPC.

Table 7.1 shows the possible policing configurations for CBR, VBR and UBR traffic as per (ATM Forum, 1996c).

7.2.4 Priority control

In order to fulfill different ATM transfer performance requirements several priority control mechanisms can be applied. First, within each virtual connection a cell loss priority can be assigned on a per cell basis. Second, different cell delay priorities can be used for the different ATM transfer capabilities. Third, individual connection priority control mechanisms are possible. Modern ATM switch architectures support a sophisticated combination of all three priority control mechanisms.

Cell loss priority

ATM cells have an explicit cell loss priority bit in the header (see Section 5.6.2), so at least two different ATM priority classes can be distinguished. A single ATM connection (on virtual path or channel level) could comprise both priority classes when the information to be transmitted is classified by the user into more and less important parts (one possible application is layered video-coding). In this case the two priority classes are treated separately by connection admission control and usage parameter control.

Selective cell discard, that is, discarding cells with low priority (CLP = 1), either assigned by the user or tagged by the network UPC/NPC, is a possible priority control function currently specified. The applicability of selective cell discard depends on the ATM transfer capability. For CLP = 1 cells no guarantee is given for a maximum cell loss ratio.

Delay priority

During connection establishment each virtual connection is assigned to a specific ATM transfer capability. Among the different transfer capabilities different delay priorities can be used. For real-time and non-real-time connections, VBR-nrt, ABR and UBR for example, within a switch different queues can be provided. Whenever a free cell transmission slot is available one cell out of these three queues can be transmitted. One might want to give VBR-nrt connections priority over ABR connections which in turn might have a higher priority than UBR connections. In this scenario VBR-nrt cells would be transmitted first, and UBR cells would be transmitted only if the other two cell queues were empty. In the case of a high network load more and more cells would be queued. However, with this mechanism it is not possible to guarantee a **minimum cell rate** (MCR) per connection as it might be required for ABR, for example.

Individual connection priority control

The most sophisticated priority control can be achieved if each individual connection is handled by a separate cell buffer queue. Such a scheme allows misbehaving connections to be isolated so that they do not affect others and it can provide fairness among all connections competing for the available transmission capacity. This buffering scheme is also referred to as *per VC queuing*. These connection queues can be selected for cell transmission one after the other. This is called *round robin* scheduling and is the simplest method. Another, more complex, scheduling strategy is known as **weighted fair queuing** (WFQ). With WFQ it is possible to assign different weights for each connection. The available transmission capacity is shared among all connections in proportion to the connections' weights. With WFQ it is also possible to guarantee an MCR and to 'shape' the traffic (see Section 7.2.5).

7.2.5 Traffic shaping

Traffic shaping actively alters the traffic characteristics of a stream of cells on a VPC or VCC in order to reduce the peak cell rate, limit the burst length or reduce the cell delay variation by suitably spacing cells in time. Of course, traffic shaping must maintain the cell sequence integrity of an ATM connection.

Traffic shaping is an option for both network operators and users. Within a customer network it can be used to ensure conformance to the traffic contract of the traffic across the user–network interface. Within a network traffic shaping might be necessary for non-real-time connections if large buffers are used. Queuing a nrt-connection in such a large buffer may alter the traffic pattern in such a manner that a subsequent NPC function might discard cells. Traffic shaping would allow to reconstruct the original traffic pattern. (For real-time connections the situation is believed to be less critical since rt-connections are only 'queued' in small buffers. Hence, the original traffic pattern is only slightly changed.

Shaping of cell bursts or eliminating the CDV can lead to better network utilization but it also increases the cell transfer delay. At points where VPs are resolved, the traffic parameters, for example PCR, SCR and MBS, of the individual VCs can be re-established by traffic shaping. If traffic shaping is not applied the cell flow characteristic might be such that these connections suffer from being policed at the next NPC. Sometimes traffic shaping and policing are combined in a function called 'spacing-policer'.

7.3 Statistical multiplexing in ATM networks

For connections with varying cell rates it is possible to statistically reconcile the required transmission capacities among several connections. (Connections mutually utilize gaps in other connections' cell flows.) In this case the required transmission capacity within a network lies below the sum of the connections' peak cell rates. This is referred to as **statistical multiplexing**. Statistical multiplexing allows to increase network utilization considerably and hence it is a pre-requisite for modern ATM networks. For CBR connections no statistical multiplexing is possible. This means that the PCR for each connection needs to be reserved along the entire path throughout the network.

For VBR connections, under certain conditions a statistical multiplexing gain can be achieved. If the PCR of VBR connections remains far below the available line transmission capacity (at most about 5% of the link rate) then more connections can be allowed compared to a pure PCR reservation. If enough VBR connections simultaneously exist the likelihood that all connections send at their PCR at the same time is very low. Connection acceptance algorithms for this kind of statistical multiplexing are, for example, based on 'effective bandwidth' calculations or the SIGMA rule (Hauber, Wallmeier, 1991).

Fast resource management is a tool that enables the immediate allocation of necessary capacity, such as bit rate or buffer space, to individual burst-type connections for the duration of a cell burst.

Both the ABT and the ABR transfer capabilities make use of fast resource management functions. Resource management cells (see Section 5.6) carry the necessary information to be exchanged between the user and the network, or inside networks.

With these mechanisms the traffic entering a network can be controlled depending on the actual load of the network. This mechanism is explained for ABR and ABT in the following subsections.

Another possibility of increasing network utilization is provided by the UBR traffic class. A UBR cell is transmitted whenever no cell of any other traffic class needs to be transmitted. Thus UBR cells can fill the gaps in the traffic flow. As mentioned earlier, UBR can only be a 'best effort' service since no one can say how often these gaps in cell flow may occur. Furthermore, in case of congestion (see Section 7.4), UBR cells will be discarded first. However:

- UBR is a very simple service – only a PCR needs to specified and no complex source control mechanism exists; and
- UBR can be expected to be a cheap service since no QoS guarantees are provided by the network.

This may lead to a wide use of UBR. Because of this there is now some discussion on offering a guaranteed **minimum cell rate** (MCR) for UBR connections. This would allow at least a basic QoS guarantee to be provided.

7.3.1 The ABR flow control protocol

The ABR (and ABT) mechanism belongs to the class of fast resource allocation procedures which build upon a 'communication' between network and end-equipment about the currently available resources and resource needs. The basic idea behind ABR is that during periods of low network utilization sources can transmit information at higher cell rates than during periods of high network loads. The network informs the source about the current network state and the source in turn is expected to tailor its cell emission rate accordingly. This information is conveyed via specific **resource management** (RM) cells. The source regularly inserts forward RM (FRM) cells into the cell flow, for example after 32 user cells. These cells are turned around at the connection's destination and travel back to the source as backward RM (BRM) cells. On their way through the network the content of the RM cells can be modified by congested network elements. Two basic mechanisms exist:

- a simple binary feedback indication where only a single bit is set during congested periods, either the **explicit forward congestion indication** (EFCI) bit in user cell headers or the **congestion indication** (CI)/**no increase** (NI) bit in RM cells;
- an **explicit rate** (ER) feedback where a network element can explicitly indicate the cell rate at which a source is allowed to transmit cells.

Figure 7.3 shows the fields of RM cells and describes their meaning and use.

All sources are expected to react to both indications according to the rules specified in (ATM Forum, 1996c). The policy for setting the indications within network elements is not standardized but any mechanism is expected to share the available network resources in a fair manner among all competing connections. Figure 7.4(a) shows the RM cell flow.

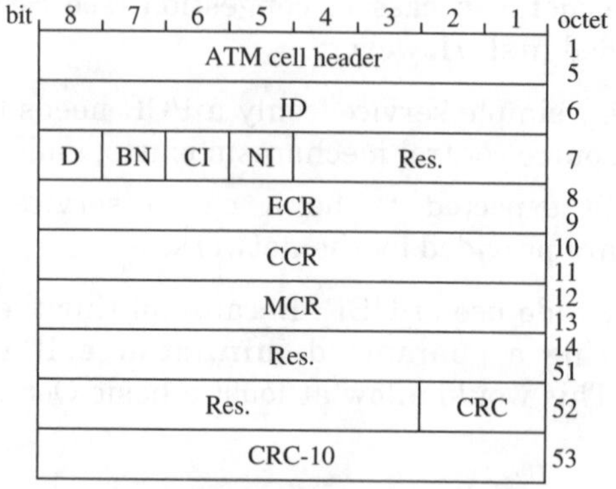

bit 8 7 6 5 4 3 2 1	octet
ATM cell header	1–5
ID	6
D | BN | CI | NI | Res.	7
ECR	8–9
CCR	10–11
MCR	12–13
Res.	14–51
Res. | CRC	52
CRC-10	53

BN Backward explicit congestion notification
CCR Current cell rate, set by the source to its allowed cell rate
CI Congestion indication, source must decrease cell rate
CRC Cyclic redundancy check
D Direction, forward or backward RM cell
ECR Explicit cell rate
ID Protocol identifier
MCR Minimum cell rate
NI No increase, source must not increase cell rate
Res. Reserved

Figure 7.3 RM cell used for ABR.

A number of parameters and detailed behavior descriptions are defined in (ATM Forum, 1996c). Here only the basic principles of the protocol are outlined.

The cell rate at which a source is allowed to transmit cells is called the **allowed cell rate** (ACR). Upon start-up a source begins sending at its **initial cell rate** (ICR). As long as the network is not congested the source can increase its sending rate by a fixed value, called an 'additive increase', with each transmitted RM cell. If BRM cells with the CI bit or EFCI bit set are received the source must reduce its ACR by a certain factor. This results in a 'multiplicative decrease'. If an ER value is received the ACR is immediately set to this value.

(It should be noted that in fact more than an additive increase will occur because the ACR is adopted with each transmitted FRM cell. Since the FRM cells are inserted after 32 user cells their occurrence increases with the ACR. This results in a nonlinear increase of the ACR. With short control loops the corresponding effect influences the 'multiplicative decrease' as well.)

For wide area connections especially, ABR parameters should be carefully selected to achieve good network utilization without increasing cell loss. As long as the source and destination behave according to the defined rules they can expect an almost loss-free cell transmission service.

In wide area networks the *bandwidth * round-trip delay* can achieve considerable values, because when a network becomes congested many cells are already on their way before the source receives an indication to lower its ACR. To guarantee loss-free transmission the buffers within

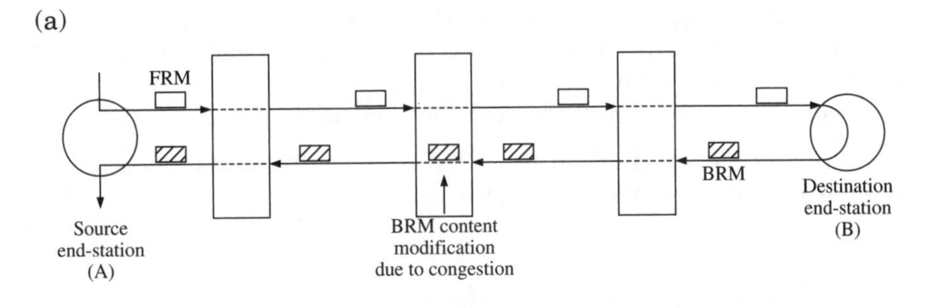

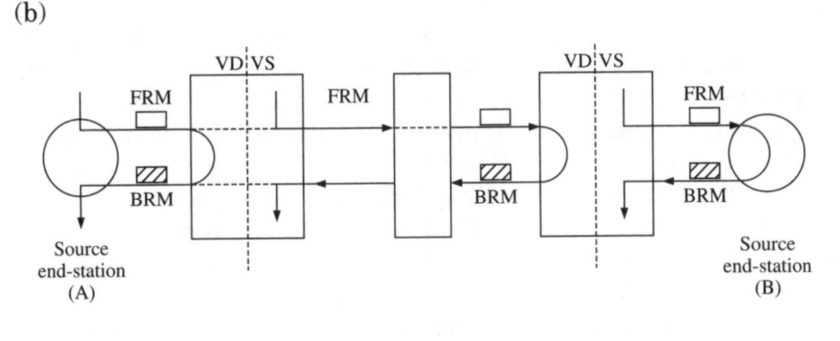

FRM Forward resource management cell VS Virtual source
BRM Backward resource management cell VD Virtual destination

Only a unidirectional transmission from A to B is shown.
For traffic from B to A a second FRM/BRM cell flow is superimposed.

Figure 7.4 The RM cell flow for the ABR service.

the network elements need to be large enough for all these cells. In order to reduce the buffer size required, the ABR control loop can be shortened by introducing **virtual sources** (VSs) and **virtual destinations** (VDs). As shown in Figure 7.4(b) a connection might be split into several ABR control sections, each delimited by a VS/VD combination.

7.3.2 ATM block transfer

The **ATM block transfer** (ABT) capability is a further development of the so-called 'fast reservation protocol'. A burst of ATM cells, called a block, is delimited by an RM cell at the beginning and end of the cell block, respectively. At connection establishment only the maximum PCR will be specified. During the connection's lifetime the network and the end stations may change the actual cell transmission rate from block to block. There are two operation modes:

- **Direct transmission**: in this case the end station sends a block without prior acknowledgment from the network. In case of congestion the entire cell block will be discarded.

- **Delayed transmission**: in this case a cell block is only transmitted after the network has acknowledged the required transmission capacity. For this cell block the QoS provided can be considered as being of the same level as for a CBR connection.

Since ABT is not specified in the ATM Forum's TM4.0 (1996c), it will probably not have the same relevance as ABR.

7.4 Congestion control

Congestion is a state of network elements (for example, switches, concentrators, transmission links) in which the network cannot guarantee the negotiated quality of service to the established connections. Congestion can be caused by unpredictable statistical fluctuations of traffic flows or a network fault. Consider the ABR case where the source might be sending out cells for quite a long time before the network can decrease the cell emission rate. All these cells need to be stored in a congested network element and when the buffer space is exhausted cells will be lost.

Congestion control is a means of minimizing congestion effects and preventing congestion from spreading.

It can employ connection admission, usage/network parameter control procedures or priority control to avoid overload situations.

Another optional congestion control mechanism is **explicit forward congestion indication** (EFCI). A network element in a congested state may set an explicit forward congestion indication in the cell header (see Section 5.6.2). At the receiving end, the customer equipment may use this indication to implement protocols which adaptively lower the cell rate of the connection during congestion. However, it has been shown that EFCI only makes sense in networks with large buffers. If only small buffers are used the end user can detect an overload situation from the increasing cell loss almost as fast as with the EFCI mechanism.

Two very important congestion control procedures are **early packet discard** (EPD) and **partial packet discard** (PPD). Virtual connections using AAL type 5 provide a means to indicate the end of a higher-layer data frame, for example an IP frame, to the ATM layer via the PTI field (see Section 5.7.5). If a network element encounters congestion it may start discarding ATM cells. Since even a single cell loss in most cases corrupts the entire data frame, no cells belonging to the same frame are of any use to the receiver. Hence, all these subsequent cells can be discarded by the network element which had to discard the first cell. Thus no unnecessary information needs to be conveyed and the congestion state can be relieved. This increases the chances of other data frames reaching their destinations uncorrupted.

If this mechanism starts 'in the middle' of a frame it is called PPD. If it is applied from the first cell of a frame onwards it is called EPD. Using EPD and PPD can considerably increase the throughput of usable data frames (also called 'goodput').

Review questions

1. What are the two main tasks for traffic management in ATM-based networks?
2. What is the CAC used for?
3. What are the tasks of UPC/NPC?
4. What is meant by priority control?
5. What is meant by traffic shaping?
6. Which ATM transfer capabilities make use of RM cells?
7. Why is the use of EPD/PPD recommended?

8

Signaling, routing and addressing

Basic aspects of B-ISDN signaling have already been described in Section 4.3. This chapter deals with ITU/ATM Forum-based signaling specifications for both UNI and NNI, public and private networks.

8.1 Introduction

The first set of 13 recommendations adopted by ITU-T in 1990 created a framework for the introduction of ATM-based networks, their subscriber equipment and applications. Only one of these recommendations included some basic aspects of signaling (ITU-T, 1996f).

On the one hand ATM-based networks should be available as early as possible, whereas on the other a variety of services (very simple as well as highly sophisticated) should be integrated in such a network. It is obvious that these two goals could not be met in a single step. Therefore, a phased approach was required for the introduction of ATM-based networks supporting switched services.

ITU-T was aware of this fact and developed a timetable of B-ISDN network and service aspects. This concept comprises three steps which are called capability sets (CS1, CS2, CS3). The main characteristic features influencing signaling are summarized in Table 8.1 (Brill et al., 1993).

In CS1, simple switched services with constant bit rates are provided and basic interworking with the existing 64 kbit/s ISDN is foreseen. (Note that while variable bit rate services can be transported, a peak bit rate will be allocated.) Right from the beginning two signaling access configurations (see Section 8.2) can be used; one of them requires meta-signaling (see Section 8.3).

More sophisticated services with variable bit rates, point-to-multipoint connections and multi-connections will be supported by CS2.

Table 8.1 Capability sets for B-ISDN signaling.

CS1	*CS2*	*CS3*
Constant bit rate Connection-oriented service with end-to-end timing	Variable bit rate Connection-oriented service Quality of service indication by the user	Multimedia and distributive service Quality of service negotiation
Point-to-point connections (uni- and bidirectional, symmetric and asymmetric)	Point-to-multipoint connections (multipoint-to-multipoint in step 2 of CS2)	Broadcast connections Leaf-initiated join
Single connection, simultaneous establishment	Multi-connection, delayed establishment Use of cell loss priority	Third party control Switched VPs
Indication of peak bandwidth	Negotiation and renegotiation of bandwidth	
Peak rate allocation	Bandwidth allocation based on traffic characteristics	
Basic interworking with 64 kbit/s ISDN		Incorporation of IN (see Section 14.4) principles into B-ISDN
Point-to-point or point-to-multipoint signaling access		
Meta-signaling		
Limited set of supplementary services	Supplementary services	Access to the Internet

With CS2, **call** and **connection control** will be separated, that is, connections can be set up and released during a call.

Finally, CS3 provides the full range of services, including multimedia and distributive services.

It is obvious that these CSs have a strong influence on the signaling application protocols which will be described in Sections 8.5 and 8.6. It is assumed that this influence is not so strong on the signaling transfer protocols. Therefore, these protocols should be applicable for all CSs (maybe with some small extensions and/or modifications).

The ATM Forum has been focusing on the protocols necessary for UNI signaling. The specifications delivered by this forum (ATM Forum, 1993c, 1994c, 1996d) are based on the specifications of ITU-T. However, modifications and enhancements have been included. One of the main differences is that right from the beginning the ATM Forum provided switched point-to-multipoint connections. The UNI 4.0 specification (1996d) provides features such as **anycast** (see Section 8.10), **leaf-initiated join** (the ability to join an already established VCC), and **proxy signaling** (the ability of a user to perform signaling for one or more other users).

8.2 Protocol architecture for CS1

ITU-T differentiated between the following signaling access configurations at the UNI (ITU-T, 1995h):

- point-to-point signaling access configuration
- point-to-multipoint signaling access configuration.

In the case of point-to-point signaling access configurations there is only one signaling endpoint on the user side. This may be a single terminal or an intelligent B-NT2 (for example, PBX) depending on the customer network configuration (see Section 4.7). Only a single permanently established point-to-point SVC is required for this signaling access configuration. This channel is used for call offering, call establishment and release.

Several signaling endpoints are located at the user side (for example, a point-to-multipoint terminal configuration) in point-to-multipoint signaling access configurations. In this case meta-signaling (see Section 8.3) is necessary to manage other signaling relations.

The protocol stacks related to these two signaling access configurations are shown on the left-hand side of Figure 8.1.

At the NNI, either the existing STM-based common channel signaling system no. 7 (SS7) or an ATM-based network can be used to transport the signaling messages (ITU-T, 1995h). For both scenarios an appropriate application protocol is required which will be described in Section 8.5.2. The signaling protocol stacks associated with each option are depicted on the right-hand side of Figure 8.1.

It is evident that the reuse of SS7 allows the rapid introduction of B-ISDN. However, a signaling transport network can also take advantage of the ATM technology, so in the future an ATM-based

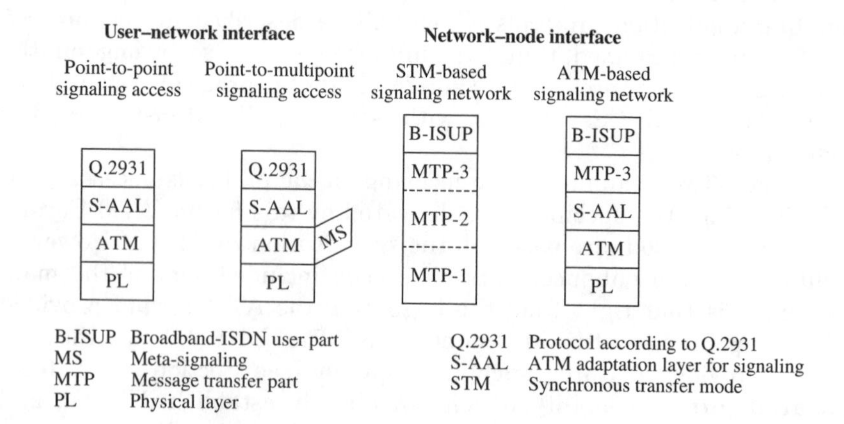

Figure 8.1 Protocol architecture for B-ISDN signaling.

signaling transport network may be the preferred solution. In the case of ATM-based signaling transfer, the SVCs between two switches are managed by OAM protocols and therefore meta-signaling is not required at the NNI.

8.3 Meta-signaling

8.3.1 General

ITU-T Recommendation Q.2120 (ITU-T, 1995i) describes the meta-signaling protocol which is used for establishing, maintaining and removing user–network signaling connections at the UNI.

The meta-signaling protocol only operates over the MSVC. Currently it is the only protocol which uses that channel. The MSVC is defined by VCI = 1 in every VP (see Table 5.8) and its default peak rate is 42 cells/s. In principle, the meta-signaling protocol can operate on each active VP. However, it is sufficient to use it only in one VP between the user and a local exchange (Huber et al., 1992). This reduces the implementation effort.

The meta-signaling protocol is part of the ATM layer. It is located within the layer management plane (see Section 5.1) and is under the control of plane management. Currently, the ATM Forum does not support meta-signaling because the forum only considers point-to-point signaling configurations.

8.3.2 Scope and application

The meta-signaling protocol provides procedures for:

- the assignment and removal of point-to-point SVCs (PSVCs) and their associated BSVCs
- checking the status of these two channel types.

With these procedures it is possible to:

- associate a signaling endpoint with a PSVC and a BSVC
- allocate the cell rate to SVCs
- resolve possible contention problems for SVCs.

The assignment, checking and removal procedures are independent of each other. The necessary relationship between them is performed via plane management.

Normally the meta-signaling protocol is used between the user and the network. However, two users which are connected via a VPC can also use this protocol for managing PSVCs and BSVCs. In that scenario the predefined VCI value for the MSCV should be used (see Table 5.8).

As mentioned in Section 8.2, the meta-signaling protocol is only used in point-to-multipoint signaling access configurations. As a network option it can also be applied in a **dynamic** signaling access configuration. In this case the user can use meta-signaling to inform the network whether the signaling access configuration is point-to-point or point-to-multipoint.

8.3.3 Protocol issues

In this section we will focus on the following details of the meta-signaling protocol:

- message formatting
- procedures.

Message formatting

The message format for meta-signaling messages is depicted in Figure 8.2. It is independent of the message type: each message consists of the same structure. Fields which are not used by a message are coded with a 'null' value.

The first field is called the **protocol discriminator** (PD). It identifies messages on the meta-signaling channel as meta-signaling messages or messages belonging to another protocol. Currently the meta-signaling protocol is the only protocol using the meta-signaling channel.

The **protocol version** (PV) field differentiates between the individual versions of the meta-signaling protocol and identifies the general message format being used. The format described in this section is valid for version 1.

The names of the messages are identified by the **message type** (MT) field. It is also used to determine the exact function and detailed format of each message.

The following six messages are used for the meta-signaling protocol (their application is described in the following subsection):

(1) ASSIGN REQUEST
(2) ASSIGNED
(3) DENIED
(4) CHECK REQUEST
(5) CHECK RESPONSE
(6) REMOVED.

The **reference identifier** (RI) field is used to differentiate between a number of simultaneous assignment procedures. The value is randomly generated for each ASSIGN REQUEST message by the individual

Bit 8 7 6 5 4 3 2 1	Octet number
Protocol discriminator	1
Protocol version	2
Message type	3
Reference identifier	4 … 5
Signaling configuration	6
Signaling virtual channel identifier A	7 … 8
Signaling virtual channel identifier B	9 … 10
Point-to-point SVC cell rate	11
Cause	12
Service profile identifier	13 … 23
00000000	24 … 43
00101000	44
Cyclic redundancy check	45 … 48

SVC Signaling virtual channel

Figure 8.2 Format of meta-signaling messages.

terminal. The associated answer is identified by the same value of the reference identifier.

The **signaling configuration** (SCON) indicates 'point-to-multipoint' or 'point-to-point'.

The contents of **signaling virtual channel identifier A** and **B** (SVCI A and B) fields depend on the procedure in use. The contents may be:

- a point-to-point signaling virtual channel identifier (PSVCI)
- a broadcast signaling virtual channel identifier (BSVCI)
- a global signaling virtual channel identifier (GSVCI).

A PSVC is identified by the PSVCI. The PSVC conveys all point-to-point call and/or bearer control signalling (see Section 8.5.1) for a given signaling endpoint. The BSVCI indicates the VCI value of a BSVC. A BSVC is always unidirectional and used in the direction from the network to the user for call offering. This channel may be the *general* BSVC or the *selective* BSVC. The general BSVC is identified by VCI = 2 (see Table 5.8), whereas the selective BSVC is identified by a VCI value defined during the assignment phase. The selective BSVC has a close relationship to the service profile identifier which will be described later on in this section. The GSVCI indicates all signaling channels except the meta-signaling channel itself. It is a unique identifier and identified by VCI = 1. It is used for a global check (this is only one example); that is, all active signaling endpoints have to deliver a CHECK RESPONSE message after receiving a CHECK REQUEST containing the GSVCI.

The **point-to-point SVC cell rate** (PCR) field indicates the requested/allocated cell rate for the point-to-point signaling virtual channel. Currently only peak cell rate allocation is used. Later on, when it may be desired to take advantage of statistical multiplexing, the coding of this field may be extended to cover this enhanced feature. The following values are defined: 42, 84, 125, 167, 250, 344, 500, 667, 1000, 1334, 2000, 2667 cells/s. (Note: 42 cells/s are approximately 16 kbit/s and 2667 cells/s are approximately 1024 kbit/s.)

The **cause** (CAU) field provides the user with more detailed information; it indicates the reason for sending the message (for example, why the network has sent a DENIED message).

The **service profile identifier** (SPID) field is used by the user to request a basic or specific level of service. The SPID points to the service profile, which is a set of information maintained by the network to provide either the basic or any specific service to a signaling endpoint. When both the network and the user support the service profile concept, calls can be offered via the selective BSVC. In this case only a subset of all signaling endpoints located at the user side receive the offered call and therefore the call offering procedure has been simplified.

The last field in the meta-signaling message is the CRC field, which is used to detect errors in meta-signaling messages.

All meta-signaling messages, their associated parameters and their directions of flow are shown in Table 8.2.

Procedures

The following procedures are used for the meta-signaling protocol:

- assignment
- check
- removal.

The assignment procedure is invoked by the user side sending an ASSIGN REQUEST to the network asking for a PSVCI and BSVCI. The network will – depending on its condition – send either an ASSIGNED message indicating the PSVCI/BSVCI pair or a DENIED message with the appropriate reason. This procedure is supervised by a timer at the user side to cope with loss of messages.

The check procedure is initiated by the network sending a CHECK REQUEST message and waiting for CHECK RESPONSE. The simplest check concerns only a single PSVCI/BSVCI pair, whereas the most general check concerns all signaling channels. This procedure is also supervised by a timer at the network side.

The removal procedure can be initiated either by the network side or by the user side. In contrast to the other two procedures, no

Table 8.2 Meta-signaling messages and associated parameters.

	Messages					
	ASSIGN REQUEST	*ASSIGNED*	*DENIED*	*CHECK REQUEST*	*CHECK RESPONSE*	*REMOVED*
PD	M	M	M	M	M	M
PV	M	M	M	M	M	M
MT	M	M	M	M	M	M
RI	M	M	M	–	–	–
SCON	M	–	–	–	–	–
SVCI A	–	M	–	M	M	M
SVCI B	–	M	–	–	M	–
PCR	M	–	–	–	–	–
CAU	–	M	M	–	–	M
SPID	M	–	–	–	M	–
CRC	M	M	M	M	M	M
Direction	U→N	N→U	N→U	N→U	U→N	U→N or N→U

M Mandatory valid value must be present
– Coded with null value
N→U Network to user direction
U→N User to network direction

handshake procedure is used. The initiating side sends only a REMOVED message followed, after a random interval, by a second REMOVED message. This mechanism is very simple and prevents errors occurring if one REMOVED message is lost.

At the user side each message is delayed for some random time before it is delivered to the network. This is done for all messages even if it is not always necessary. This simple mechanism prevents certain network overload conditions which might otherwise occur (for example, when the network starts a global check all terminals would answer almost simultaneously).

8.4 ATM adaptation layer for signaling

A suitable signaling AAL (S-AAL) is required in order to adapt the signaling application protocols for the UNI and the NNI (see Section 8.5) to the services provided by the underlying ATM layer (see Section 5.6). The S-AAL is subdivided into the following two parts (see Figure 8.3):

- common part (CP)
- service-specific part (SSP).

Section 8.4.1 discusses some aspects of the common part, while Section 8.4.2 deals with the service-specific part of the S-AAL.

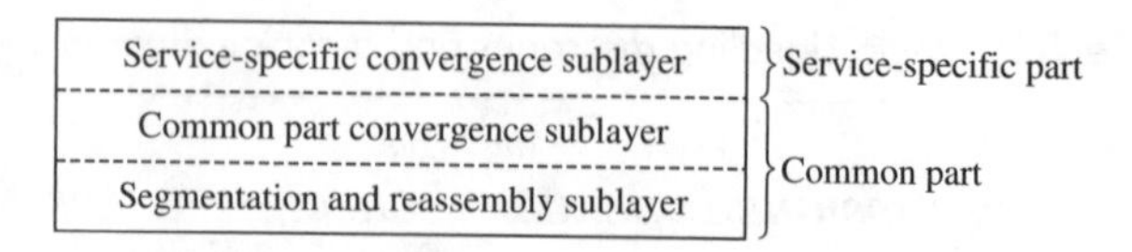

Figure 8.3 ATM adaptation layer for signaling.

8.4.1 Common part

In principle both AAL type 3/4 (see Section 5.7.4) and AAL type 5 (see Section 5.7.5) are suitable for the CP. ITU-T decided to use AAL type 5 (ITU-T, 1996h) because it is a simple protocol with only a small overhead.

AAL type 5 has a better performance (for example, higher throughput, lower mean transfer time) for long messages. However, in the case of small messages (1 to 3 cells) and a high cell loss probability, AAL type 3/4 has the better performance (Renger and Briem, private communication). This is because if the last cell of an AAL-PDU is lost, the error cannot be detected immediately by AAL type 5. The error will be detected after receiving the adjacent PDU. However, the receiver cannot indicate which AAL-PDU has been errored and therefore discards both PDUs.

8.4.2 Service-specific part

ITU-T decided to use a common protocol for the UNI and the NNI, which is described in ITU-T Recommendations Q.2100 (ITU-T, 1994a), Q.2110 (ITU-T, 1994b), Q.2130 (ITU-T, 1994c) and Q.2140 ITU-T, 1995j), for the SSP. The commonality between UNI and NNI provides the following benefits:

- reduced complexity at network nodes
- short specification time
- flexibility in network operation and configuration
- efficient operation and maintenance of the network
- reduced operating and manufacturing costs.

It should be noted that full commonality is not possible because of the different protocols located above the S-AAL at the UNI and the NNI. To meet the different requirements of the protocols located above the S-AAL, the architecture shown in Figure 8.4 has been chosen.

The **service-specific connection-oriented protocol** (SSCOP) provides mechanisms for the establishment and release of connections and the reliable exchange of signaling information between signaling entities. The **service-specific coordination functions** (SSCFs) map the requirements of the layer above to the requirements of the next lower layer.

UNI	NNI
SSCFs for UNI	SSCFs for NNI
Service-specific connection-oriented protocol	

NNI Network–node interface
SSCF Service-specific coordination function
UNI User–network interface

Figure 8.4 Service-specific part of S-AAL.

In the case of the SSCOP it would have been possible to use an existing data link layer protocol, such as the data link layer protocol for frame mode bearer services (ITU-T, 1992c), with some modifications and enhancements (Frantzen et al., 1992). Reuse of an existing protocol would have been the same approach as for the signaling application protocols for CS1, providing the same advantages (for example, short specification process). However, ITU-T decided to specify a new protocol for SSCOP, which has to be simple and efficient.

The new SSCOP performs the following functions:

Sequence integrity: Preserves the order of the SSCOP-SDUs.

Error correction by retransmission: Missing SSCOP-SDUs are detected by the receiver using a sequencing mechanism. SSCOP then corrects the error by selective retransmission.

Flow control: The receiver controls the rate of messages which its associated peer transmitter sends by using a dynamic window mechanism.

Error reporting to layer management: The occurrence of errors is indicated to layer management.

Keep alive: Two peer entities will remain in connection even in the event of a prolonged absence of data transfer.

Local data retrieval: SDUs can be retrieved which have not yet been delivered or acknowledged.

Link management: This function establishes, releases and resets SSCOP connections.

Transfer of data: This function performs the transfer of messages between two peer-to-peer entities. SSCOP provides the **assured** and **unassured** data transfer mode.

PCI error detection: Errors within the PCI are detected.

Status reporting: The SSCOP transmitter and receiver can exchange status information using this function.

In order to achieve a high-speed protocol, the transmitter and receiver state machines are decoupled. This decoupling, the use of selective retransmission and the dynamic window mechanism for flow control are the major feature changes compared with existing data link layer protocols. A more detailed list of protocol features, the messages used and the protocol operation are described in ITU-T Recommendation Q.2110 (ITU-T, 1994b).

As shown in Figure 8.4, two different SSCFs are needed, one for the UNI and another for the NNI. SSCFs (ITU-T, 1994c) map the service primitives between SSCOP and the layer 3 protocol Q.2931 (ITU-T, 1995o) at the UNI. At the NNI, the SSCFs (ITU-T, 1995j) perform mapping of primitives between SSCOP and message transfer part level 3 (MTP-3) (ITU-T, 1996l) and some local functions such as local retrieval.

8.5 Signaling protocols for CS1

The approach chosen for CS1 was the reuse of existing protocols with some modifications. The existing layer 3 protocol for 64 kbit/s ISDN according to ITU-T Recommendation Q.931 (ITU-T, 1993v) served as a basis for the UNI, whereas the protocol at the NNI was based on the ISDN user part (ISUP) (ITU-T, 1993r, 1993s, 1993t, 1993u).

The objective was to keep the modifications as simple as possible. However, to ensure a smooth transition from CS1 implementations to further solutions, measures were taken in the CS1 protocols to prepare for the separation of call and bearer control. This extension of the basic approach guarantees compatibility and allows the future reuse of CS1 protocols thereby preserving investment.

In the following two subsections we will describe only those protocol issues which are related to the ATM-specific features of the UNI and NNI signaling protocols.

8.5.1 User–network interface signaling

The layer 3 signaling protocol for B-ISDN is described in ITU-T Recommendation Q.2931 (ITU-T, 1995o). It resides directly above the S-AAL (see Section 8.4). Q.2931 includes the specification of the signaling messages, information elements and communication procedures between signaling endpoints for the B-ISDN UNI.

Because Q.2931 is based on the existing layer 3 signaling specification for 64 kbit/s it provides the same set of functions for B-ISDN as are available for 64 kbit/s ISDN. However, Q.2931 is independent of the Q.931 protocol: this is achieved by the use of a new protocol discriminator for broadband signaling. The structure of Q.2931 directly corresponds to the well-known structure of Q.931. Most of the procedures of Q.931 have been transferred to Q.2931 with merely editorial modifications, in particular to the call establishment and release procedures.

There are two reasons why modifications were made to Q.931:

- adaptation to the new transfer mode ATM
- additional modifications which allow a smooth transition from CS1 signaling protocol to the CS2 and CS3 protocols.

Adaptation to the new B-ISDN transmission system mainly refers to the description of the B-ISDN bearer services. A new broadband bearer capability has been developed which is used for the description of these services. In addition, the ATM cell rate describes the throughput of an ATM connection. For CS1 the value of the ATM cell rate is selectable, but only as a peak cell rate resulting in a constant throughput.

As described in Section 5.1.5, users can select between different AAL classes and the associated protocols for their communication needs. A new information element was developed describing the attributes of the user plane AAL connection which are chosen by the user.

The old channel identifier of Q.931 became obsolete and was replaced by the new connection identifier information element. This information element consists of the virtual path connection identifier (VPCI) and the VCI already introduced in Section 5.6.2. The VPCI identifies a VPC, whereas a VPI only identifies a VP link. The use of the VPCI in the connection identifier information element is necessary because the user may be connected via a VP cross-connect network to the local exchange (see Figure 8.5). The VP network is controlled by network management and not by signaling. (A signaling endpoint only knows about its own VPI and not the VPI of the associated signaling endpoint.) Therefore, it became necessary to introduce the unique identifier VPCI between the user and the local exchange.

The modifications described above are the main ones. However, some parts of the messages and procedures also need to be modified slightly. Most of them remain the same as in Q.931. In addition to the existing procedures, the interworking between ISDN and B-ISDN services is supported by Q.2931.

Besides the modifications which result from the use of ATM for broadband communications, Q.2931 also provides means to support a smooth transition between B-ISDN CS1 and future releases. In addition to some improvements and simplifications of the coding rules (see Table 8.3), the inclusion of compatibility instructors into messages and information elements of Q.2931 is discussed with regard to this objective.

8.5.2 Network–node interface signaling

The signaling application protocol for the NNI is described in ITU-T Recommendations Q.2761 to Q.2764 (ITU-T, 1995k, 1995l, 1995m,

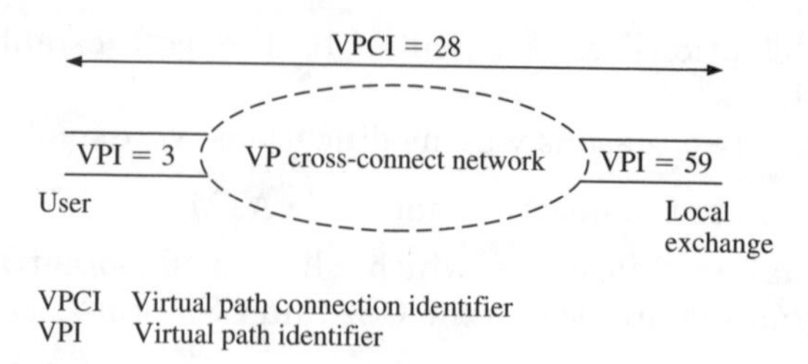

Figure 8.5 Example of the use of the virtual path connection identifier at the B-ISDN UNI.

Table 8.3 Overview of major changes in ITU-T Recommendation Q.2931 compared to ITU-T Recommendation Q.931.

Protocol changes	
New/modified information elements	Broadband bearer capability ATM cell rate AAL parameter information element Connection identifier
New/modified coding rules	Indication of the message length Common variable format for information elements Free ordering of information elements within a signaling message except for the message header Inclusion of compatibility information
Modified procedures	Called-side compatibility checking VPCI/VCI allocation/selection

1995n). These recommendations are based on the description and definitions of the existing monolithic version for ISUP of 64 kbit/s ISDN (ITU-T, 1993r, 1993s, 1993t, 1993u).

The signaling application protocol for the NNI is called broadband-ISUP (B-ISUP). To allow independence from existing protocols, it was agreed that the new protocol is a new ISUP. B-ISUP uses a new code point for the service information octet which differentiates it from the existing ISUP.

In order to meet the time schedule for B-ISUP the same approach is used as for the Q.2931 protocol. Modifications for B-ISDN can be categorized into two parts:

- modifications required to provide signaling for B-ISDN
- preparations to allow a smooth transition towards future releases.

Mandatory modifications for B-ISDN

The most important modification with respect to the ISUP for 64 kbit/s is the substitution of the **circuit identification code** (CIC) and its functions. This also influenced the procedures that are related to the CIC. In ISUP the CIC identifies the following:

- transmission channel (connection element)
- bearer control association
- call control association.

Transmission channel: The 12-bit value of the CIC is strictly related to a circuit, that is, a 64 kbit/s channel. In ATM, the transmission channel has to be identified by a new identifier relating to the VP and the VC. The VC is identified by the VCI, whereas the VPCI is required to identify the VP. The reasons why the VPI is not useful as a unique identifier at the UNI are also valid at the NNI. The VPCI is semi-permanent and consists of 16 bits. Both the VPCI and VCI are encoded in the connection element identifier (CEI) parameter. The CEI will be assigned and used bidirectionally, that is, the same value will be used for both directions.

Bearer control association: The dynamic behavior of the CEI, which consists of 32 bits, is very disadvantageous for the identification of the logical bearer control association. Therefore, the control plane association is identified by a separate bearer control identifier (BCI). A transaction mechanism (that is, using origination and destination BCI) will be used which is similar to the transaction capabilities application part (TCAP).

Call control association: In CS1 there is no distinction between call and connection. Therefore, identification of the call control association is implicitly provided by the BCI.

Bandwidth handling: In contrast to 64 kbit/s ISDN, the bandwidth of a connection has to be indicated explicitly. The new information transfer rate parameter will be used for this purpose. This parameter consists of the forward and backward peak cell rate.

OAM procedures: In ISUP, OAM procedures like *reset*, *blocking* and *testing* are strictly related to the CIC. Because CIC was removed, these procedures cannot be used for B-ISDN. It is very likely that these procedures will be separated from the bearer control procedures.

Bothway operation: Bothway operation of channels is

applicable to VPs and VCs. In B-ISDN two types of dual seizure can be identified:

- dual seizure of the CEI
- dual seizure of cell rate, that is, simultaneous request for the last available cell rate in a VPC.

As the procedures for prevention and resolution of dual seizure in ISUP are not applicable, new ones have to be defined.

B-ISDN routing: Routing in ISUP is based on the indication of the required information transfer capability and signaling capability; both are single values. Routing in B-ISDN will become more complex because more information (cell rate, quality of service, broadband connection-oriented service subcategory, symmetry, and so on) has to be taken into account.

Parameter format: All B-ISUP parameters will have a common format: parameter name followed by a length indicator and the parameter content. This format is identical to the one used at the UNI.

Message format: Simplification of the parameter format (only one for all parameters), means that the message format will be simplified to include the routing label followed by the message type, a length indicator and the B-ISUP parameters. The fixed mandatory part and the variable mandatory part are no longer necessary. The ordering of parameters can be chosen arbitrarily. This format is also very similar to that used at the UNI.

Compatibility mechanism: To allow for forward compatibility, the ISUP mechanism will also be applied to B-ISUP. All messages and parameters will contain compatibility instructions for handling unknown (new) signaling information.

Protocol evolution

For CS2 and CS3, call control and connection/bearer control are to be separated. The application layer structure for OSI (ITU-T, 1993w) will be applied to protocols. B-ISUP will mainly be used for bearer control. Therefore, reuse of the B-ISUP – with some modifications and enhancements – as a bearer control protocol in CS2 and CS3 should be possible (see Figure 8.6). Some provisions, for example on the syntax level, are required to allow for simple protocol enhancements and evolution. The use of modern description techniques like **abstract syntax notation no. 1** (ASN.1) (ITU-T, 1988g), should be possible. Furthermore, harmonization between the UNI and NNI protocol elements will simplify interworking between them.

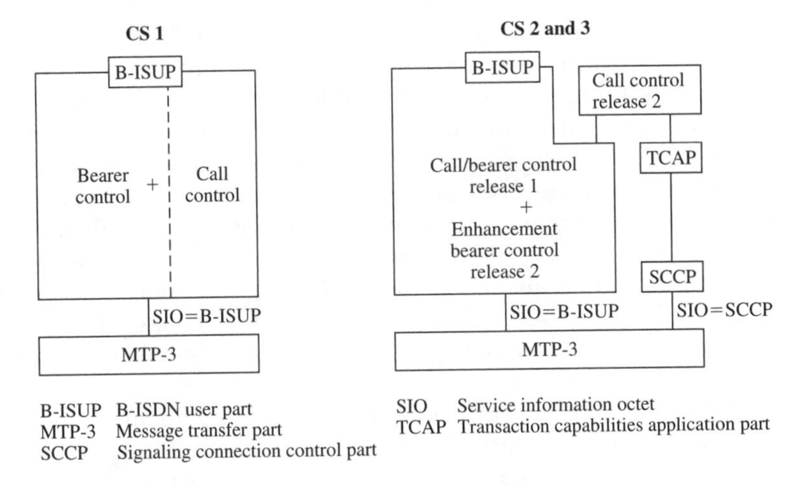

Figure 8.6 Evolutionary path for B-ISDN NNI signaling protocol.

8.6 Requirements for CS2 and CS3 signaling protocols

It has been agreed that these specifications should be based on the separation of call control and bearer/connection control. Using this approach it will be possible to have calls with several connections (multi-connection calls), which is one of the powerful features required for the provision of multimedia services. In addition, separation between call and bearer/connection control will allow:

- Establishment and release of a call without any connection. Such a call might be useful for the negotiation of end-to-end service compatibility prior to reserving resources (look-ahead function), and is also useful for some supplementary services.
- Dynamic allocation/deallocation of connections to/from a call.

8.7 Private network–network interface (PNNI)

The private network–network interface (P-NNI) (ATM Forum, 1996e, 1996f, 1997f) is a **trunking, routing and signaling protocol** specified by the ATM Forum. It is an inter-switch protocol which supports SVCs between switches of multiple vendors. Whereas the predecessor of the P-NNI, which was called **Interim Interswitch Signaling Protocol** (ATM Forum, 1994d) only provided for signaling and some table-based rerouting and therefore could only be applied to small ATM networks, the P-NNI supports topology discovery by distribution of reachability information, hierarchical routing and addressing – both indispensable in the case of large networks – and quality of service. P-NNI covers intra-domain as well as inter-domain applications, in contrast to traditional data network protocols which are tailored for one application only.

In the hierarchical network model upon which the P-NNI protocol is based, several entities – at the lowest level these are physical devices – form a **peer group** (where one element is elected as the **peer group leader**) acting as a single **logical group node** at the next higher level. So it is sufficient that the P-NNI protocol provides switches with comprehensive information about their local topology, but only summarized information about distant parts of the ATM network. The P-NNI protocol uses NSAP addresses (see Section 8.10). Members of a peer group can be identified through a common address prefix.

ATM nodes exchange all necessary information about the network structure when the network is being established. Later on, either regularly or initiated by changes of topology or capacity in the network, the same procedure is applied.

An efficient routing algorithm has to meet the following requirements:

- To determine a set of end-to-end connections complying with the requested quality of service.
- To choose the preferred one (and provide further solutions in case the latter fails) so as to optimize the use of the network (for example, distribute the network load as equally as possible).
- To react dynamically to network failures or shortcomings.
- To avoid too complex route calculations and set-up procedures (as there are stringent requirements on connection set-up times).

P-NNI was conceived to support networks with different levels of sophistication employing different routing schemes. So the protocol cannot rely on nodes always having sufficient knowledge to support the 'optimal' route for a desired connection. As a consequence, the P-NNI, for example, replaces connection admission control (CAC, see Section 7.2.2) actually performed at an ATM switching node with a **Generic CAC** (GCAC), a model assumption of how the switches will react.

Note that, if no suitable path can be found at all, alternative paths can be selected as a fall-back solution whereby some of the connection requests may have to be relaxed.

The P-NNI also supports **soft permanent virtual connections** (this is a technique where an SVC-capable network accomplishes a connection between two endpoints when both have configured a permanent connection to their associated node) and **multicasting** (see Section 8.10) via point-to-multipoint connections. However, the present P-NNI (ATM Forum, 1996e, 1996f, 1997f) does not have all the features of the present UNI signaling 4.0 (ATM Forum, 1996d); for example, it does not support multipoint-to-multipoint connections.

8.8 Broadband inter carrier interface (BICI)

The **Broadband inter carrier interface** (B-ICI) (ATM Forum, 1995e, 1996g) was specified by the ATM Forum for the connection of ATM networks operated by different service providers with the aim of providing the user with end-to-end services across network boundaries. The B-ICI protocol therefore includes service definitions, signaling, traffic management, operation and maintenance, and accounting. The following services are supported: frame relaying, cell relaying, SMDS/CBDS and circuit emulation.

B-ICI has taken into account the existing ITU-T NNI recommendations. The physical layer of the B-ICI may be SONET/SDH-based (at 155 and 622 Mbit/s) or PDH-based (E3/T3 at 34 and 45 Mbit/s, respectively). ITU-T's B-ISUP signaling protocol and the MTP Level 3 routing protocol are used. Note that the MTP protocol also supports signaling network management functions.

Whereas the first B-ICI release (ATM Forum, 1993d) only supported permanent virtual connections and SMDS, version 2.0 (ATM Forum, 1995e, 1996g), which is based on the ATM Forum's UNI 3.1 (ATM Forum, 1994c) and on ITU-T's B-ISUP Release 1 with Capability Set 2 extensions (ITU-T, 1996s), in addition facilitates, for instance, switched virtual connections, variable bit rate service, multipoint connections, congestion management and OAM functions.

B-ICI permits both virtual path connections and virtual channel connections.

8.9 ATM inter-network interface (AINI)

The **ATM inter-network interface** (AINI) is a network-to-network interface for the interconnection of ATM networks which may internally use either PNNI or BISUP or even other protocols. The AINI specification drafted by the ATM Forum (1997g) is based on a symmetric peer-to-peer signaling protocol like the PNNI signaling protocol. As an option, exchange of routing and topology information and further capabilities to support billing can be added. Thus a set of capabilities is provided which satisfies the needs of operators in different interworking environments.

In order to avoid interoperability problems and to allow for fast development and implementation, only features which are supported by both PNNI and BISUP and by BICI (version 2.1) are supported for the time being (for example, anycasting – see Section 8.10 – is not supported as it is not supported by BISUP).

Figure 8.7 gives an example where AINI my be deployed. Note that the ATM networks A, B and C in this figure may be either public (carrier) or private (enterprise) networks and may use different internal protocols.

AINI can also be employed for the connection of ATM access networks with ATM core networks.

As to AINI routing, (ATM Forum, 1997g) states: 'Topology and reachability information about the network on the other side of a network boundary is in general static in nature and not likely to change very often (most likely it is a matter of contract between the networks what information is given to the partner). For a PNNI network the AINI, as an interface between different networks, is regarded to be an exterior link. ... Therefore, as a default, no routing and topology information is exchanged over the AINI. The necessary information concerning e.g. address reachability across the AINI is configured via Network Management in both networks. However, a PNNI node at the AINI may advertise information within the PNNI network.'

8.10 Addressing issues

With regard to addressing, there are two totally different schemes employed by the telecommunication world (at least where based on telephone networks) and the computer network community, respectively. Whereas in telecommunications the *global ISDN numbering plan* detailed in ITU-T Recommendation E.164 (ITU-T, 1991a) is used, computer networks mostly employ the OSI *network service access point* (NSAP) addressing mechanism. The actual assignment of addresses to end-systems or subscribers is done by means of addressing plans.

E.164 addresses comprise 16 digits (eight bytes each). This is the conventional numbering scheme well known to everyone who has ever dialed a phone number. E.164 addresses contain a country code, area or city code, and the subscriber number.

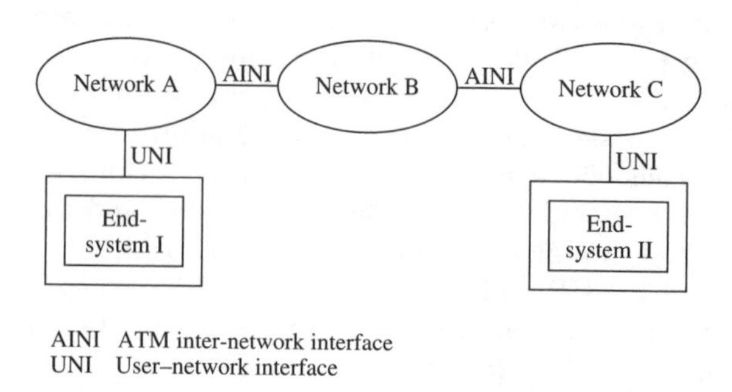

Figure 8.7 Example configuration for AINI usage.

NSAP addresses are defined in ISO 8348 and ITU-T Recommendation X.213 (ITU-T, 1995p). The ATM Forum has chosen this 20-octet NSAP address format and encoding for the addressing of ATM systems connected to a private network; systems connected to an ATM public network can, according to the ATM Forum specifications (1994c, 1996d), use either NSAP or E.164 addresses. Interworking between ATM networks which employ different addressing schemes (for example, between public and private ATM networks) requires translation of one address type into the other at the border of the two networks.

Special addresses can be used to support *broadcasting*, *multicasting* and *anycasting*.

Broadcasting means the sending of a message to *all* end-systems connected to a network; multicasting is sending to any subset of end-systems. For a large ATM network deploying physical point-to-point connections, broadcasting and multicasting are not trivial issues. They can be accomplished, for example, by means of a server in the ATM network, which upon a broadcast/multicast request establishes multipoint connections on which broadcast/multicast cells can be delivered to their destinations. The ATM Forum's LAN emulation standard (see Section 11.2) introduces such a server (called *Broadcast and Unknown Server*), which is identified by a special address.

Anycasting is a service specified by the ATM Forum in UNI Signaling 4.0 (ATM Forum, 1996d): a user of a specific service need not know which entity in the network actually performs the service, and instead can use a published group address assigned to this service. The network can automatically distribute service requests to the service-providing group members.

Review questions

1. What ATM features do the ITU-T signaling capability sets support? What additional signaling functions do the ATM Forum specifications provide?
2. What does the protocol architecture for B-ISDN signaling look like at the UNI and NNI?
3. Which AAL type has been chosen for signaling?
4. Where are the following protocols employed and what are their main functionalities?
   - PNNI
   - BICI
   - AINI
5. Where are E.164 and NSAP addressing used and how are they characterized?

9

The Internet and ATM

The **transmission control protocol** and **Internet protocol** (TCP/IP) are defined in (IETF 1981). They constitute one of the most widely used protocol suites in the computer communication area. A very large number of workstations, hosts and even PCs are interconnected using TCP/IP. The Internet and especially the **world wide web** (WWW) which is also based on IP are very popular today. Many subscribers throughout the world 'surf' the Internet which has exponential growth rates. Hence ATM-based networks must also be capable of carrying and supporting these protocols. In ATM networks TCP/IP protocols will be placed on top of ATM.

If native ATM applications using the UBR or ABR transfer capabilities are developed they might build the Internet of the future. However, this will definitely take some more time and thus interworking with TCP/IP-based networks will be essential for the next decade.

IP is a connectionless protocol designed for data applications. It can be run over various link-layer transmission systems like Ethernet, token ring and leased lines. The IP addresses are subdivided into a network identifier (ID) and a host ID. IP datagrams contain the source and destination addresses. Sometimes, for an IP address the corresponding HW address (this typically is a **medium access control** (MAC) address or in the context of IP and ATM the destination's ATM address) is not known. In this case the **address resolution protocol** (ARP) will be used: the sender broadcasts an ARP request message to all stations. A station recognizing its own IP address within this ARP request will send back an ARP reply containing its HW address (and IP address). This enables the source to send data to the designated destination.

IP packets are transferred in a best-effort manner. This means that no error correction capabilities, no QoS guarantees and no information about successful delivery are provided. In the case of an unknown destination address the **internet control message protocol** (ICMP) can be used to inform the sender. ICMP messages are encapsulated and conveyed in IP packets.

IP data packets are transported in an unassured manner without any guarantee that they will arrive in the same order as sent out. Therefore, higher-layer protocols, such as TCP, have to ensure data integrity. TCP is a connection-oriented reliable transport protocol, whereas the **user datagram protocol** (UDP) is a connectionless unreliable transport protocol. On top of these protocols, applications like **simple network management protocol** (SNMP), **file transfer protocol** (FTP), **simple mail transfer protocol** (SMTP), and **remote login** (TELNET) can be found.

Although the layered approach of the Internet and ATM allows the reuse of higher-layer protocols without changes, some problems have to be resolved at the boundary between IP and AAL. Some of these problems and possible solutions are discussed in this chapter. In addition consideration should be given to the following issues:

- **TCP aspects**: In principle it is possible to use the higher-layer TCP without any changes. For protocols implementing flow control or acknowledgment mechanisms the *bandwidth * round trip-delay* product has to be taken into account. For an ATM-based WAN this product might be high. Extensions of TCP are available which can be used on paths with a large bandwidth delay product (IETF, 1992). Since TPP/IP constitutes a flow control mechanism above IP the question of how to interwork it with ATM-based flow control, for example, ABR, arises. It can be assumed it is not a trivial task to align both protocols parameter settings for an optimal end-to-end throughput.

- **Security**: The transfer of data through a public network results in higher security requirements compared with the (physically) limited extent of, for example, LANs, where normally all attached stations belong to the same domain. This issue has to be resolved before any end user will send sensitive data through a public network. This has already been identified and taken into account by IP sec (IPv4) and IP version 6 (IPv6).

9.1 IP over ATM

When discussing IP over ATM three main aspects need to be considered: encapsulation, routing and addressing/address resolution. In the following the basic principles of carrying IP traffic over an ATM-based network are described. These mechanisms are based on the following **Internet Engineering Task Force** (IETF) specifications, called **requests for comment** (RFCs):

RFC 1483 'Multiprotocol encapsulation over ATM adaptation layer 5' (IETF, 1993)

RFC 1577 'Classical IP and ARP over ATM' (IETF, 1994)

RFC 1626 'Default IP MTU for use over ATM AAL' 5 (IETF, 1994)

RFC 1755 'ATM signaling support for IP over ATM' (IETF, 1995)

RFC 2020 'Support for multicast over UNI 3.0/3.1-based ATM networks' (IETF, 1996) and several Internet drafts such as 'Next hop resolution protocol' (NHRP) (IETF, 1996k), and 'IP broadcast over ATM' (IETF, 1996b).

The most mature mechanism is based on RFC 1577, its draft update (IETF, 1996c) and other associated RFCs. This mechanism is also known as *classical IP and ARP over ATM* and further elaborated in the following.

This model considers ATM to be a 'wire' for the transport of higher-layer (that is, from layer 2b of the OSI reference model onwards) protocol information. The concept is similar to LANE (see Section 11.2) but uses a different addressing level, IP, instead of the **medium access control** (MAC) level. It describes the transmission of IP datagrams, their encapsulation into ATM cells and the ATM address resolution protocol (ATM-ARP).

In order to reflect the fact that the IP network on top of ATM builds a logical association of network elements, such as hosts or routers, which is independent of the underlying ATM network, the notion **logical IP subnetwork** (LIS) was introduced; 'an LIS consists of a group of IP nodes (hosts or routers) that connect to a single ATM network and belong to the same IP subnetwork'.

The model is called 'classical' because the classical concept of subnetworks is retained. Each subnetwork can be of any type, for example frame relay, Ethernet or ATM. Within a subnetwork, namely an LIS, direct connections between stations are possible and connections towards a station outside an LIS must be conveyed via a default router. This router would constitute an ATM endpoint and can belong to several LISs via a single physical interface.

However, such a router could easily become a bottleneck and end-to-end connections with a guaranteed QoS cannot be established. One attempt to solve this problem is illustrated in Section 9.2 which describes the **next hop resolution protocol** (NHRP).

Within an LIS all ATM connections used for information transfers are VCCs which can be either permanent or switched connections.

The **maximum transmission unit** (MTU) size within an LIS is the same for all VCs. The default value has been specified as being 9180 bytes as in the 'IP over SMDS' case. Together with the 8 octets LLC/SNAP header (see Section 9.1.1) this results in an AAL 5 PDU size of 9188 octets. However, since the AAL 5 protocol allows for much larger values, up to 64 kbyte MTU sizes might optionally be used. In general a higher MTU value will result in a better performance although larger data frames are more susceptible to be 'hit' by a transmission error. An optimum size has to be chosen carefully. In order to retain the advantages of the maximum MTU size, routers and hosts must implement the 'IP path MTU discovery' mechanism RFC 1191 (IETF, 1990a) plus RFC 1435 (IETF, 1993a). With this mechanism IP fragmentation in the Internet can be reduced. The ATM signaling procedure allows for negotiation of the MTU size (backward/forward maximum CPCS-SDU size for use with AAL 5).

9.1.1 Encapsulation

In traditional LAN conventions 'encapsulation' denotes the **logical link control** (LLC) multiplexing and addressing function of layer 2b. Higher-layer protocol packets (such as IP, IPX, AppleTalk, and so on) are encapsulated with the LLC information. This LLC information can be considered as being an 'in-band' signaling message which addresses the correct higher-layer protocol at the destination.

In ATM networks 'out-of-band' signaling is used; that is, the signaling information is conveyed via a separate virtual connection and is not bound to the actual user-data transfer (see Section 4.1).

With IP over ATM these two paradigms come together and hence several proposals exist on how to 'marry' the two worlds.

Three main encapsulation schemes can be differentiated: LLC/SNAP encapsulation, VC multiplexing and TULIP/TUNIC. Using the network layer model Figure 9.1 illustrates these three mechanisms and shows from which point onwards the ATM/AAL 5 connection is used.

There is a general trade-off between these three models: the greater the amount of the multiplexing and addressing function realized in the end-station, the less information needs to be transported via the ATM signaling protocol (Table 9.1).

Common to all schemes is the use of AAL type 5, or more specifically, the data frames are directly carried over the CPCS of AAL 5 with an empty SSCS (see Section 5.7.5 for more AAL5 details).

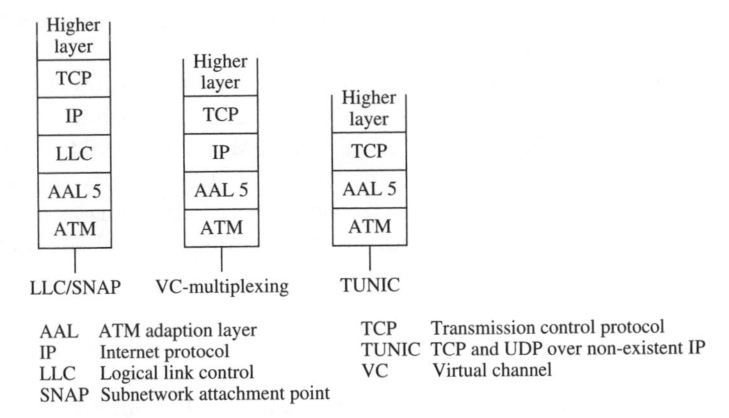

Figure 9.1 IP over ATM encapsulation schemes.

Table 9.1 IP encapsulation overhead.

Encapsulation method	*Demultiplexing in end-system*	*In signaling message*
LLC/SNAP	Source address Destination address Protocol family Protocol Port	Nothing
VC multiplexing	Source address Destination address Protocol Port	Protocol familiy
TUNIC (TCP and UDP over a non-existent IP connection)	Nothing	Source address Destination address Protocol family Protocol Port

Logical link control (LLC)/subnetwork attachment point (SNAP)

This scheme identifies the protocol by an IEEE 802.2 (IEEE, 1989) LLC type 1 (unacknowledged, connectionless mode) header (3 octets) which is followed by a subnetwork attachment point (SNAP) header (5 octets). Figure 9.2 shows the entire encapsulation for a routed IP frame.

Most of the multiplexing function needs to exist in the end-station where the LLC/SNAP header has to be generated or evaluated, respectively.

This encapsulation method is best suited for PVC-based networks, or networks not capable of carrying many parallel VCs, where several higher-layer protocol connections can easily share a single VCC. All traffic between adjacent routers, for example, could go over a single VCC. Although the LLC/SNAP encapsulation involves more processing and transmission overhead it represents the default method for many networking scenarios, including, for example, LANEv2 or MPOA (see Sections 11.2.4 and 11.3).

However, switched VC connections for connecting two end-systems can be established according to RFC 1755 which is based on the ATM Forum's UNI 3.1 specification (ATM Forum, 1994c). The B-LLI of the SETUP message indicates that an LLC entity constitutes the endpoint for that ATM/AAL 5 connection; no information about the sender's higher-layer protocol (for example, IP, IPX, AppleTalk) is transmitted via the signaling channel.

VC multiplexing

This is also known as 'null encapsulation' since an LLC function does not exist. Layer 3 packets, such as IP packets, are directly mapped onto

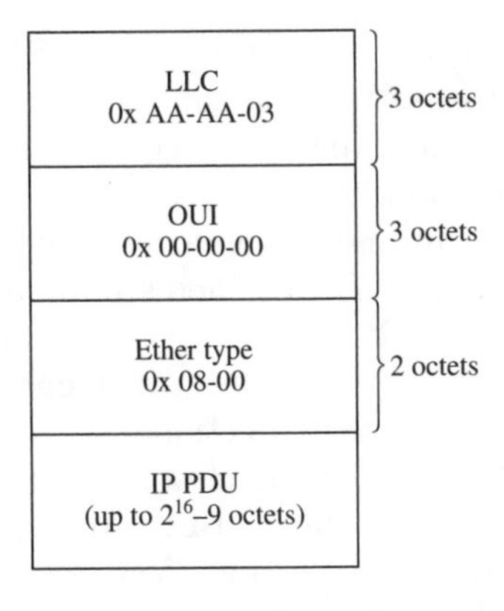

Ether type	Indicating an P PDU
IP	Internet protocol
LLC	Logical link control, indicating the presence of a SNAP header
OUI	Organizationally unique identifier, indicating that the following two octets are an ethertype
PDU	Protocol data unit
SNAP	Submetoarly attachment point

Figure 9.2 LLC/SNAP encapsulation format for routed IP PDUs.

an ATM/AAL 5 connection. The B-LLI specifies the layer 3 entity as the destination for that connection.

Although VC multiplexing incurs less overhead than LLC/SNAP encapsulation it requires many VCCs, a separate one for each individual layer 3 protocol pair between two ATM endpoints, and generates potentially high amounts of signaling traffic. LANEv1 is one example that makes use of VC multiplexing (see Section 11.2).

TCP and UDP over lightweight IP (TULIP)/TCP and UDP over a non-existent IP connection (TUNIC)

TULIP and TUNIC were presented as proposals for a further reduction of overhead but are not part of the 'classical IP and ARP over ATM' model. TUNIC, for example, not only eliminates the LLC layer but even omits the entire IP network layer. The ATM/AAL 5 connection exists directly between transport layer entities such as TCP or UDP. However, use of this method is not widespread and has almost no practical relevance; addressing transport layer entities is not possible with the existing signaling protocol mechanisms and would require certain extensions.

One major restriction with TUNIC is that it requires single hop reachability between endpoints. An intermediate router would not know the destination IP address and could not forward the packet.

The two main encapsulation methods, LLC/SNAP and VC multiplexing, relate to the connectionless versus connection-oriented paradigms of LANs and ATM networks, respectively. LLC/SNAP encapsulation for TCP, for example, appears to be a somewhat cumbersome approach: mapping a connection-oriented protocol, TCP, on top of a connectionless protocol, IP, which is finally conveyed via a connection-oriented network ATM. However, in most scenarios the LLC/SNAP encapsulation method will be used since it provides simple and smooth interworking with existing procedures. It is used as the default method for 'classical IP and ARP over ATM', for the **multicast address resolution server** (MARS) as described in Section 9.4, and has also been adopted for use by LANEv2 and MPOA. However, the signaling protocols provide a mechanism, using the B-LLI, to negotiate the preferred encapsulation method on an SVC by SVC basis. As more future networks allow for single hop connections and as higher-layer protocols demand support for different QoS levels, the need for and use of 'single protocol connections' will increase.

9.1.2 Addressing and routing

Today, the most commonly used addressing model is also referred to as the **overlay model**. In this model both IP and ATM addresses are used. To resolve an IP to an ATM address the classical ARP mechanism was adopted for use in ATM based networks (see Section 9.1.3).

Apart from this, there are three other main addressing models: the **peer model**, the **integrated model** and **integrated routing**. The peer model assumes that network layer addresses (here IP addresses) are algorithmically mapped into ATM addresses (NSAP-based). This would allow routing information to be easily exchanged among IP hosts or routers and ATM network entities, while each network retains its classical routing mechanisms. A connection set-up request containing an IP address, for example, could be routed using the PNNI protocol. The main advantage of the peer model is that it makes a separate ARP mechanism obsolete. However, it was felt that using different routing protocols in the IP and ATM networks, respectively, may lead to less optimal end-to-end routes and that other technical issues need to be resolved. Therefore, the IETF and ATM Forum decided not to investigate the peer model further but to concentrate on other models first.

Based on the PNNI, the **integrated model** is currently being considered by the MPOA group of the ATM Forum. Route calculations will take into account network topologies and the link states of both the IP and the ATM network. This approach will result in an optimal route for IP packets in mixed IP/ATM environments.

A further extension of the integrated model is the **integrated routing model**. This model, also known as integrated PNNI, proposes using the PNNI protocol for routing in ATM switches and router-based IP networks. Since PNNI is considered to be one of the most powerful routing protocols ever specified, it may lead to optimal end-to-end routes through switched and routed networks as well. It is inherently well suited for the integrated services internet as it supports QoS routing. However, many technical problems still need to be resolved, such as how to use PNNI in a broadcast media environment, and a migration path from existing routing protocols to integrated PNNI must be defined. One can assume that it will be some time before integrated PNNI is widely used; if it is ever used.

9.1.3 Address resolution

For the classical IP over ATM approach a mechanism is required to resolve target IP addresses to target ATM addresses and vice versa. ATM-ARP, which is based on 'normal' ARP and **inverse ARP** (InvARP) procedures with some extensions, are used for this purpose, but in a slightly different manner for PVCs and SVCs.

Since ATM networks do not allow the use of an ARP mechanism being based on broadcasts as, for example on Ethernets, a specific ATM-ARP server needs to be deployed.

Although the classical model is based on a single ATM-ARP server within a LIS, future standards may allow for multiple servers providing increased reliability. All ATM-ARPs stay within a LIS.

When a host station is switched on, it automatically establishes a connection to the ATM-ARP server. The server is found via a configured, unique ATM address. After this connection has been established the host must register its addresses at the ATM-ARP server. Depending on the ATM network, an ATM client needs to register its IP address, its ATM-NSAP or E.164 or both addresses at the server. These address mappings are stored in the server's database.

Clients may now send ATM-ARP requests to the server for IP addresses to be resolved. With the returned ATM address a VCC to the desired destination can be established.

Any VCC to another station is invalidated after 15 minutes but not immediately released. Prior to any subsequent data transmission the client must revalidate its own address mapping with the ATM-ARP server by issuing an ATM-ARP request. In some failure situations the server might not be able to respond. In this case an InvARP request can be sent directly to the destination client via the still existing VCC. If no InvARP reply is received, this destination can no longer be reached via this connection and the SVC is released.

Address entries within the ATM-ARP server are timed out after 20 minutes. However, normally an ATM client will refresh its own address mappings before the end of that time period, namely every 15 minutes. Hence server entries are normally revalidated by this procedure before they time out.

ATM-ARP and InvARP packet formats are based on the classical ARP protocol with some extensions. ATM-ARP is differentiated from the classical ARP by a unique **hardware type value** see (IETF, 1982) for details. ATM-ARP packets are LLC/SNAP encapsulated and transported via an ATM/AAL 5 VCC.

Since ATM does not directly support broadcast/multicast communications, the corresponding mapping of an IP broadcast/multicast address to a broadcast/multicast ATM address does not exist. However, clients should reply with their ATM address if they receive an IP broadcast ATM-ARP request within their LIS.

9.2 Next hop resolution protocol

As described earlier, the *classical IP and ARP over ATM* approach requires that connections between hosts in different LISs are established via (several) router hops. Since every router constitutes a potential bottleneck and prevents the use of a transparent ATM end-to-end connection, the **next hop resolution protocol** (NHRP) allows a *cut-through* connection through several LISs to be established. Only the establishment of a direct and transparent ATM connection end-to-end allows, for example, ATM's particular QoS capabilities to be exploited.

The **next hop server** (NHS) performs the next hop resolution protocol and realizes an inter-LIS ARP mechanism which can be used by a source host to establish such a 'cut-through' connection. If the destination host is not part of the ATM network the NHS provides the address of the egress router through which this destination can best be reached. The NHRP can be used by any host or router attached to an ATM network. Coexistence with the existing, classical ARP mechanisms is possible.

Although NHRP is described here in the context of IP and ATM networks, generally it can be used for other networks as well. Therefore, it has been adopted by the ATM Forum as the basis for carrying multiple protocols over ATM, that is, MPOA. The use of NHRP between two routers is proposed in (IETF, 1996i). Furthermore, it is likely that this protocol will be adopted for use in classical IP networks (or other broadcast, multiple access networks) as well.

The NHRP as such is not a routing protocol but a specific ARP mechanism that avoids extra IP hops. Its address resolution is based on routing tables created either manually or by a dynamic routing protocol. The manually created, static routing tables, of course, are only applicable to small network configurations. The dynamic protocols are assumed to create a loop-free routing table.

Figure 9.3 shows how a host issues an NHRP resolution request which is passed along several other NHSs until the request can be answered. When it receives the NHRP reply the host knows the destination's address and can thus establish a direct connection.

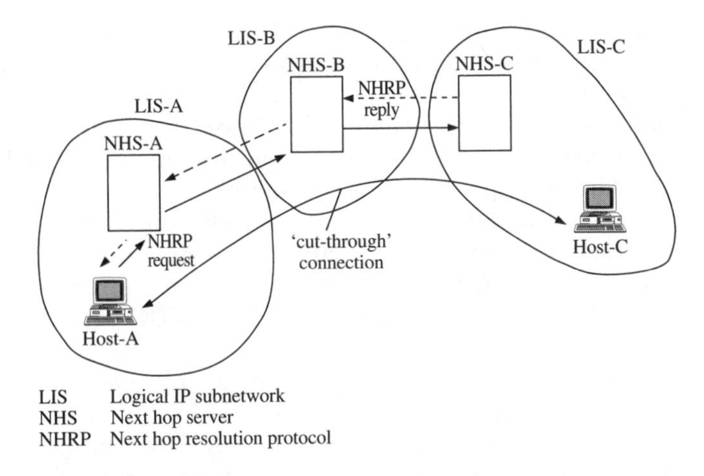

Figure 9.3 Operation of the next hop resolution protocol.

Using the **server cache synchronization protocol** (SCSP) (IETF, 1996j) several NHSs can be interconnected. This allows NHRP to be scaled for use within large networks.

Even within networks with NHRP capabilities the hosts or routers still have the freedom to decide whether a cut-through connection should be established or whether a packet should be routed hop-by-hop. This is a local decision which has to be taken by the application. Since an end-to-end connection needs to be established, cut-through connections might be more expensive or introduce an additional latency. Criteria such as QoS requirements, the amount of data to be transferred or the connection's duration (long-lived versus short-lived traffic) might be evaluated before taking the final decision. Hence one can assume that routers will still be used to provide a connectionless service. Both NHRP with RFC 1937 'Local/Remote' Forwarding Decision in Switched Data Link Subnetworks (IETF, 1996d) and MPOA explicitly suggest the use of both connection types. This is specifically important during the migration period towards an all-NHRP-based network. During that period a host might not be able to set up a cut-through connection and the only way to reach a destination might be the routed path to decide whether a connection is 'cut-through' or routed.

A connection's duration, for example, is the parameter evaluated by a mechanism called IP switching (see Section 9.6) in a pure IP network.

9.2.1 How does the next hop resolution protocol work?

This subsection summarizes the basic operations of the NHRP as currently proposed by (IETF, 1996k). At the time of writing this was an IETF draft only and some modifications are still possible. Here the term NHRP 'client' denotes either an attached end-station which does not forward data packets to other stations, that is, a host, or an NHRP function being implemented and used by a router which forwards data packets to other routers or hosts.

NHRP is based on the deployment of **next hop servers** (NHSs). In essence, an NHS is an ARP server which contains cache tables with IP to ATM address mappings. These mappings can be constructed by several means, such as NHRP registration packets or NHRP resolution requests/replies. After power-up a client is supposed to register its addresses at its NHS. A client itself may also maintain an address mapping table. There should always be a direct VCC between a client and its NHS.

As usual, all cached address mappings are aged out by timer supervision. In addition, a purge message is sent out by an NHS whenever it detects a topological change. This purge message invalidates all cached address information from that NHS in its neighbor stations. A purge might also be sent out by a client to its NHS to de-register itself. In this case the NHS has to forward this purge message to all stations that might have cached the client's address information.

The normal operation is similar to the classical ARP mechanisms. When a source station wants to set up a direct connection to a destination it may find the appropriate address in its local cache table. If the destination address is not known, the source station issues an NHRP resolution request packet with its own IP and ATM address and the (requested) destination's IP address to its NHS. Typically this will be a pre-configured default server which might also act as the client's default router. If this server knows the requested address mapping it will answer the NHRP resolution request. If it cannot resolve the address it forwards it to the next NHS along the path towards the destination. Normally, several other NHSs can be reached and hence a routing table needs to be maintained to store information about the NHS 'closest' to the destination. At the next NHS the same procedure takes place, until an NHS which can answer the request is found.

This answer is sent back to the source station along the same path, thus traversing all NHSs which forwarded the request. This mechanism enables intermediate NHSs to learn the resolved address mapping. Thus, they can directly answer the next NHRP resolution request for that destination IP address without involving other NHSs.

Two exceptions to this procedure should be mentioned:

- A source may decide to send out an **authoritative** request. In this case intermediate NHSs with indirectly learned and cached address mappings are not allowed to answer this request. Only the final NHS serving the destination station can answer an authoritative request.
- Under certain circumstances (see IETF, (1996k) for details) the final NHS might directly answer the requesting station, for

example via an existing VCC and without traversing any other NHS. This feature allows response times to be reduced.

If an address cannot be resolved a negative NHRP resolution reply (NAK) is returned to the requester. In case of multicast addresses more than one NHS might return a part of the requested multicast address.

While the source station is waiting for the NHRP reply it may drop, store or forward the data packet along the routed path towards the destination. Although the last of these options is the preferred behavior since it reduces latency, no mechanism is yet described for how to maintain the frame sequence order after a direct VCC could be established. (LAN emulation (LANE) (see Section 11.2) uses a specific flush message to overcome this problem.)

Whenever an NHRP request (or reply) crosses the border of a LIS, the next ingress NHS can be configured not to forward or answer all messages. Thus firewall policies for certain addresses can be realized. Since only the address resolution requests need to be screened, an NHS is not burdened with those tasks for the actual data transfer.

However, NHRP firewall mechanisms need to be complemented by an ATM address filtering mechanism within the ATM network since the source node might have obtained the destination's ATM address through other means than the NHRP.

The NHRP specification allows for additional options such as:

- **address aggregation**, based on IP address prefixes (subnetwork address masks);
- **route record**, to trace the routed path for maintenance purposes, for example loop detection.

NHRP packets are encapsulated using either the LLC/SNAP or the VC multiplexing technique as described above (Section 9.1.1). Since NHRP is not only used in the IP and ATM context, every NHRP packet contains information specific to the actual internetworking case.

Although NHRP constitutes a powerful protocol some shortcomings need to be addressed. The router to router operation which may result in stable loops needs to be resolved. More work in the area of multicast/broadcast traffic seems to be necessary. Auto- and self-configuration mechanisms which would allow for 'plug and play' operations are an important feature in the ATM world; these mechanisms are not currently provided by the NHRP specification. However, despite these limitations NHRP will play a key role, especially in the MPOA context as currently defined by the ATM Forum (see Section 11.3).

9.3 SVC establishment

Although all of the IP over ATM issues addressed so far can be realized in a pure PVC-based ATM network, it has already been mentioned that

SVCs are considered to provide a much more flexible and powerful means of carrying IP traffic over ATM connections. Therefore, RFC 1755 (IETF, 1995) and (IETF, 1996e) describe in more detail how UNI 3.1 and UNI 4.0 procedures, respectively, are to be applied for IP over ATM.

RFC 1755 provides guidelines on how SVCs between ATM stations should be used, based on UNI 3.1. For example, it suggests default times for holding a connection or inactivity timers for clearing an SVC. Table 9.2 shows the **information elements** (IE) and their default coding as they are used for IP over ATM. Calling/called party subaddress and transit network selection IEs might optionally be used in addition to these mandatory IEs.

Compared to LANE the requirements for connection establishment delays are considered to be more stringent since no **broadcast and unknown server** (BUS) can be used for an immediate data transfer (see Section 11.2).

The draft IETF (1996e) suggests some enhancements to RFC 1755 for systems using UNI 4.0 (ATM Forum, 1996d). The most relevant UNI 4.0 features considered are ABR support, traffic parameter negotiation and the frame discard capability. For networks supporting the **resource reservation protocol** (RSVP), (IETF, 1996e) proposes that an SVC should not be discarded after a predefined idle period but that the connection should be released when the receiver no longer refreshes the RSVP reservation.

Although the ABR traffic class is now also available for SVCs, it is still perceived that UBR constitutes the best match for IP over ATM traffic. For traffic parameter negotiation purposes, the PCR must be requested to be equal to the link rate and the 'minimum acceptable ATM traffic descriptor' IE must allow for a PCR of zero. Use of the frame discard capability should be made as it has the potential to increase the overall throughput in a congested network.

Beside these relatively minor enhancements, the exploration and impact of leaf-initiated joins, ATM anycast capabilities and switched VP

Table 9.2 Signaling information elements for IP over ATM SVCs.

Information element (IE)	*Default coding*
AAL parameters	AAL 5, SSCS = 0, MTU size = 65 535 (although the default size is only 9180 bytes)
ATM traffic descriptor	UBR, peak cell rate = line rate
Broadband bearer capability	BCOB = X
Broadband low layer information	LLC/SNAP encapsulation
QoS parameter	QoS class = 0
Called party number*	E.164/AESA
Calling party number*	
Transit network selection*	

* Use is optional for IP over ATM

services leave enough room for further developments. The leaf-initiated join capability especially is presumed to impact the multicast/broadcast realm of functions.

9.4 Multicast and broadcast support for IP over ATM

Up to now the 'classical' model has not tackled the problem of how to support the connectionless IP multicast service via a connection-oriented ATM network. Before describing the multicast/broadcast support mechanisms the following terms should be introduced:

- **unicast traffic**: an information flow destined for a single, specific receiver;
- **multicast traffic**: an information flow sent to a specific group of receivers;
- **broadcast traffic**: information sent to all destinations within a specific networking environment.

With IP over a shared medium like Ethernet a source can easily send information to a multicast group address (that is, a class D address). The source 'broadcasts' its message via the shared medium to all attached stations. Each individual station then decides for itself if it wishes to join this multicast group. The different groups are identified by means of the multicast group address contained in the source's message. Joining or leaving a group is performed by individual hosts, their decision has no impact on any other hosts attached to the same network. (However, **internet group management protocol** (IGMP) messages are used by hosts to inform IP multicast routers when they join or leave a multicast group. See RFC 1112 (IETF, 1989) for more details.)

Although UNI 3.1 specifies the use of point-to-multipoint connections, such a connection type cannot be considered to be equivalent to a multicast group address since a sender node needs to be aware of all stations that want to be members of that group. An explicit (leaf) connection to each recipient needs to be established. These multipoint connections are unidirectional only and the source determines the group membership.

RFC 2022, 'Support for Multicast over UNI 3.0/3.1 based ATM Networks' (IETF, 1996a) describes how UNI 3.1-based ATM transport capabilities in conjunction with a **multicast address resolution server** (MARS) can be used for IP multicast applications.

A group of endpoints, called a 'cluster', is connected to a MARS in order to obtain information about other nodes which are part of that multicast group. This allows endpoints to distribute information either via meshes of VCs or by using ATM multicast servers.

Each LIS is served by a separate multicast address resolution server (MARS) and IP multicast routers need to be deployed for interconnecting several LISs. Within a LIS all ATM endpoints or only a subset might want to exchange multicast information among themselves. Those endpoints actually participating in a multicast communication join the MARS and are called cluster members.

9.4.1 Multicast address resolution server operations

As an NHS a MARS can be considered to be an ARP server dedicated to the resolution of multicast addresses. Instead of providing a one-to-one mapping of one IP to one ATM address, multiple ATM addresses are associated with a single (multicast) IP address. The returned ATM address format is either E.164, NSAP or based on both addressing schemes.

Upon registration a MARS client establishes a point-to-point connection with its MARS which is used for registration and request/reply message exchange purposes. A client registers its ATM address for all the multicast groups it wants to join (or leave). In turn the MARS adds this client as a leaf to its **cluster control VC** and assigns a **cluster member identifier** (CMI).

The ATM address of at least one active MARS needs to be configured within all endpoints that may want to participate in a multicast communication. Several MARSs within a LIS might be used for redundancy reasons.

Whenever a client station needs to resolve an IP multicast address it sends the appropriate request message via the point-to-point VCC towards its MARS. The MARS in turn replies with all ATM addresses which are currently members of that group. With this information a host or router may establish a point-to-multipoint connection or disseminate the multicast information through a **multicast server** (MCS).

By using multipoint connections any ATM station might be the source (or root) of one or more point-to-multipoint VCC trees and simultaneously the destination (or leaf) of other multipoint VCCs. This overlaying of several multipoint connections is called a **VC mesh**. All group members must support the UNI 3.1 multicast procedures.

It should be noted that, like every address resolution server, the MARS does *not* participate in the actual datagram multicasting.

Alternatively, a client establishes a point-to-point connection with an MCS which in turn forwards the multicast traffic via a point-to-multipoint connection to all group members. Thus, each station uses *one* outgoing VCC to send information to the group and *one* incoming VCC to receive multicast information. A more detailed design and implementation description of MCSs is currently under development by the IETF draft 'Multicast Server Architectures for MARS-based ATM multicasting' (IETF, 1997). The solid lines in Figure 9.4 illustrate the use of a VC mesh while the dotted lines denote point-to-point VCCs handled by an MCS.

Both models have advantages and drawbacks (see RFC 2022 (IETF 1996a)); for example, use of an MCS might result in reflected packets. This means that a source receives its own packet. However, those packets can be filtered using the CMI. The MARS architecture allows VC meshes and MCSs to be used.

In order to be able to inform clients about changes in the group membership, the MARS itself uses a point-to-multipoint connection to each endpoint, called a **cluster control VC**. Upon reception of a **group membership change** notification via this VCC the client adds (or drops) the appropriate leaves from its point-to-multipoint VCCs. MCSs are informed via a point-to-point server control VC. The default encapsulation method for any MARS control traffic is LLC/SNAP.

In order to save connection resources all VCCs, between clients and MARS or among clients, are supervised by idle detection mechanisms and discarded after a certain period of inactivity.

Broadcast traffic can be seen as a special case of multicast in the sense that all members of a LIS are members of a multicast group. For example, broadcast traffic plays a role in diskless boot scenarios where a host must first obtain its IP addresses from a server via the 'all 1's' broadcast address. Details on using a MARS to support IP broadcasts over ATM networks are currently discussed within the IETF's draft 'IP Broadcast over ATM Networks' (IETF, 1996b).

It should be mentioned that with UNI 4.0 (ATMF 1996d) a leaf-initiated join procedure has been specified. This capability will certainly influence the 'multicast over ATM' scenario as described here. However, more details on the best way of exploiting this feature still need to be settled.

As with NHRP, the MARS model might be adopted for MPOA as well after adding some more sophisticated auto configuration capabilities.

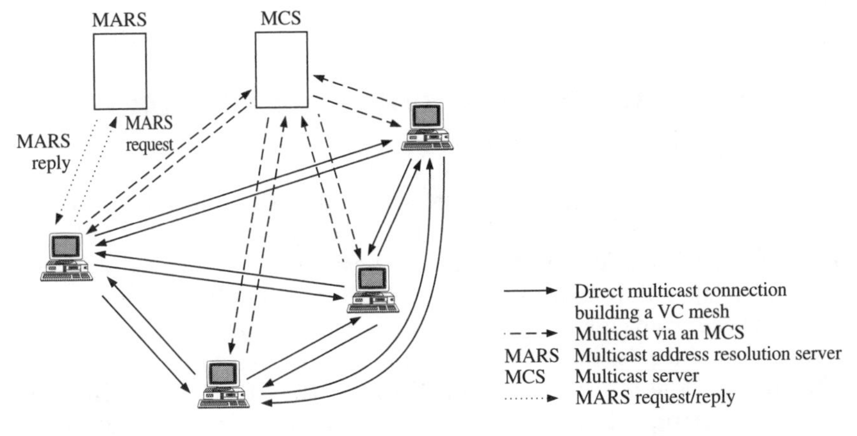

Figure 9.4 Multicast support for IP over ATM.

9.5 IP Version 6 over ATM

At the time of writing, mechanisms for using IP Version 6 (IPv6) (IETF, 1996g) over ATM were still under discussion. However, basically the same mechanisms, for example RFC 1483-based encapsulation and the NHRP, as for IPv4 are used for IPv6 as well. This allows a smooth upgrade of existing network implementations, thus saving most of the existing investment.

Draft 'IPv6 over NBMA Networks' (IETF, 1996f) describes the use of the NHRP for IPv6 neighbor discovery and proposes to use an **interface token** for IPv6 address auto configuration support.

Establishment of the first VCC to the NHS needs to be manually configured or can be based on a media-specific group address. This VCC is used to register the host's link-local IPv6 address and its ATM address at the NHS. Afterwards the host requests the IPv6 addresses of the routers in its (sub-) network by issuing an address resolution request for the link-local all-routers multicast address.

Consider IETF (1996f) and the ongoing discussions within the IETF for more details on aspects such as router bootstrap processing, Internet control message protocol for IPv6 redirect messages (ICMPv6) and neighbor unreachability detection used on ATM-based networks.

IETF (1996h) proposes to establish a cut-through connection whenever an IP traffic flow is identified. Initially, a default router is used by its hosts for destinations which are not considered to be neighbors. Whenever a router detects an IP flow it resolves the destination's IP address via the NHRP. By sending a redirect message to the flow's source the destination is identified as a transient neighbor. Thus a cut-through connection is established.

Furthermore, it is proposed to base the distribution of neighbor discovery messages on the MARS model. The neighbor/router discovery protocols can remain unchanged except for the cut-through connection establishment procedures.

9.6 IP switching, tag switching and carrier scale internetworking

The need for supporting IP in large-scale, high throughput and reliable transport networks has recently led to a variety of proposals on how best to deal with IP traffic. Common to all these ideas is to speed up layer-3 routing by layer-2 switching ('layer' refers to the OSI reference model). Most of these proposals originate from company-specific solutions. Some of them are largely based on existing standards, while attempts are being made to standardize others through the IETF's **multiprotocol label switching** (MPLS) working group; a first draft of RFC can be found in (IETF, 1997a).

IP solutions considered in this section are:

- **IP switching:** a proposal from Ipsilon, described in several information RFCs (RFC 1953, RFC 1954, RFC 1987);
- **tag switching:** a Cisco solution which is currently studied in the MPLS working group;
- **carrier scale internetworking:** the Siemens/Newbridge solution for IP traffic, largely based on existing standards.

Others which are currently being discussed within the IETF include:

- Aggregate Routing based on IP-Switching (ARIS), by IBM
- Fast-IP, by 3COM
- IP-Navigator, by Cascade
- Secure Fast Virtual Networking (SFVN), by Cabletron
- Flow Attribute Notification Protocol (FANP), by Toshiba.

9.6.1 IP switching

IP switching combines 'the intelligence of routing with the performance of switching'. This method was first introduced by Ipsilon in early 1996. The basic idea is to use ATM hardware for IP flow forwarding and routing software for short-lived IP frames. The network is built upon ATM switches running a specific software package called **general switch management protocol** (GSMP).

An IP frame is forwarded on a default VPI/VCI (0/15) and routed from switch to switch as in the case of a classical router. As soon as several IP frames with the same destination address are detected, that is, an IP flow exists, this flow is shifted from the default connection onto a dedicated ATM connection. This decision is local to each network element. It informs its previous neighbor node about the ATM connection used by assigning a specific flow label via a proprietary protocol. Flows carried by a dedicated ATM connection do not further burden the routing machine. Furthermore, they benefit from the high-speed ATM connection. In addition the RSVP might be evaluated in order to support a specific QoS via the ATM connections.

As soon as an IP flow is carried over an ATM connection end-to-end this is referred to as **cut-through**. Since such a connection is not explicitly established prior to the data transfer it is also called a **soft-state connection**.

9.6.2 Tag switching

Tag switching 'combines the performance and traffic management capabilities of layer 2 (data link layer) switching with the proven scalability of layer 3 (network layer) routing'. The basic building blocks for a tag switching internet are **tag edge routers**, **tag switches** and the **tag distribution protocol**.

Tag edge routers are located at the entrance and exit points of a tag switching network, respectively. The network itself is built upon tag switches.

The basic idea behind tag switching is to allocate a tag for a given IP destination address prefix. The tag edge routers evalute the IP destination address and obtain a tag for that IP frame from the routing table. This tag is then placed in the ATM cell's VCI field. Subsequent tag switches can forward these cells according to the tag value at the same speed as they could forward an ATM cell according to the VCI field. Re-creation of the original IP frame takes place at the destination tag edge router by removing the tag.

Tag edge routers and tag switches run any of the existing routing protocols (OSPF, BGP, and so on) to populate their routing tables which also contain the tag values. These tag values are distributed via the **tag distribution protocol** (TDP). As in the case of IP switching no ATM connection set-up procedures are used. However, in contrast to IP switching, tag switching is applied on short-lived packets as well as on IP flows. With IP switching a flow label is assigned for each individual flow whereas with tag switching several IP flows (all having the same destination) share the same tag.

9.6.3 Carrier scale internetworking

Carrier scale internetworking (CSI) is the Siemens/Newbridge solution for a large-scale internet. It is based on **edge forwarders**, **default forwarders**, **configuration servers** and **route servers**.

All internet traffic is first handled by the edge forwarding function which resides in an ATM capable device connected via a UNI. Access to the edge forwarder can be provided using a dedicated ATM PVC or alternative access technologies such as frame relay. This edge forwarder fulfills three main tasks:

- handling of all remote access functions, such as user identification, password verification, billing (if applied);
- default forwarding of IP frames;
- establishment of cut-through ATM connections.

By default the IP traffic is forwarded via ATM (long-lived) SVCs to the next **default forwarder**. Thus the packet is routed through the network hop-by-hop. Upon detection of an IP flow, or by handling RSVP requests or by evaluation of higher-layer protocol information, the edge forwarder can decide to establish a *cut-through* ATM connection. These IP connections then benefit from ATM's QoS and high throughput capabilities. The basis for the establishment of such a *cut-through* is the knowledge of the destination edge-forwarder's ATM address. This address resolution takes place via the route server using the NHRP.

Instead of running route protocols between and among access and default forwarders, the route server, as a centralized router, performs all necessary calculations and transmits its results to the network elements.

The route server will also handle ICMP and other IP-specific tasks. Furthermore, initially it also contains a multicast server for IP multi-/broadcasts. The main benefits of this centralized router approach are:

- complex (and hence expensive) route calculations are concentrated in centralized devices;
- these (few) devices are easier to manage;
- for reliability reasons these devices can easily be built redundantly.

The main application for CSI is the building of **virtual private networks** (VPN), offering value-added IP services for corporates, for example, teleworking, mobile access. As part of such a VPN edge forwarders may be interconnected via existing (long-lived SVC) connections. Even if such a connection does not exist, the edge forwarder will at least know most of the destination addresses which it stores locally in an address cache table. For IP broadcasts within a VPN the edge forwarders establish ATM point-to-multipoint connections. Therefore, use of the default forwarder and/or the route server will be minimized.

Whereas all IP traffic normally uses a UBR ATM connection, within a VPN connections offering higher QoS levels can be used.

For an easy installation of CSI within a network the following auto-configuration mechanisms are used. Within a PNNI peer group the anycast address of the next configuration server is distributed to all local exchange switches. Via ILMI the edge forwarders get knowledge of this address. Subsequently the edge devices download all the information that they need from the configuration server, for example addresses of the next default forwarder and route server. This auto-configuration mechanism is derived from the LANE principles (see Section 11.2).

The entire CSI approach is based on existing standards as ILMI, PNNI, LANE, NHRP and MPOA as far as possible thus providing an open, multivendor environment.

Review questions

1. What are the main differences between IP and ATM networking techniques?
2. What is meant by classical IP over ATM?
3. Which are the two basic encapsulation mechanisms used for IP over ATM traffic?
4. What is the NHRP used for?
5. How can multicast/broadcast communication flows be supported in an ATM environment?

10

Interworking with other networks and services

10.1 Interworking principles

For many years the need to interconnect individual computing systems and terminals has been increasing. Today, numerous networks coexist and increasing communication requirements demand that they be interconnected. Interworking units (IWUs) are necessary to achieve this.

Two networks located close to each other can be coupled directly via an IWU. However, networks far from each other can only be interconnected via intermediate subnetworks.

In the following description it is assumed that different protocols are used at layer $(N-1)$. The protocols of layer N and above are identical in the interconnected networks. Networks can be coupled using various approaches (Biersack, 1989; Sunshine, 1990):

(1) Interconnection of the heterogeneous systems is achieved by protocol conversion at layer $(N-1)$. Often protocol transformation is difficult and not all functionalities can be retained after the conversion if the two networks do not provide the same functions.

(2) In another approach, a common global protocol sublayer is placed on top of the different network protocols. This requires an additional adaptation sublayer between all layer $(N-1)$ protocols and the common global protocol.

(3) Coupling of two networks can also be performed at the first common layer. This approach avoids difficult protocol conversion but increases the transfer delay because the IWU has to process an additional layer.

The IWU is involved in important issues other than the selection of a protocol level at which different networks will be interconnected. **Naming**, **addressing** and **routing** are all necessary for the correct delivery of data to its appropriate destination. **Congestion control** has to be applied if the speeds of the networks are not matched. When different maximum packet sizes are defined for the two networks, the IWU has to perform **segmentation** and **reassembly** functions.

The following interconnection approaches are commonly used for today's networks:

- A **repeater** interconnects two networks at layer 1. Its main purpose is the enlargement of small networks.
- Interconnection at layer 2 is performed by **bridges**.
- A **router** interconnects two networks at layer 3.
- An IWU coupling networks at higher layers is called a **gateway**. Normally, only layer 4 gateways and application layer gateways (layer 7) are used.

ATM interworking can be realized as network and service interworking. Whereas network interworking provides for the transparent transport of non-ATM services through the ATM core network, service interworking maps all non-ATM services into a native ATM environment (see Figure 10.1). Service interworking is much more effective (and much more difficult to achieve) than network interworking, as, in contrast to network interworking, it allows connections between terminals designed for different services.

10.2 Circuit emulation service (CES)

In the following, ATM network interworking ('ATM trunking') for narrowband services will be described in more detail. ATM trunking comes in two flavors: as static and switched trunking, where both categories can be subdivided in different subcategories. ATM trunking provides circuit emulation to narrowband services.

Static trunking is a CES which supports AAL 1 (and AAL 2 in the near future). It can provide either unstructured T1/E1 and T3/E3 service or structured (channelized) T1/E1 service including fractional T1/E1 (that is, n times 64 kbit/s) service. In the case of unstructured CES, the total CBR stream, including framing bits, is segmented and packetized into an AAL 1 entity associated with a single VCC. In the case of structured CES, T1/E1 frames are terminated at the IWU which maps n times 64 kbit/s circuits into trunk groups according to their destination. Each n times 64 kbit/s circuit is emulated as a different AAl 1 entity associated with a different VCC. If *channel associated signaling* is present in a structured link, its data is added to the AAL1 payload (for details see ATM Forum, 1997h).

Currently under consideration by the ATM Forum is a mechanism for detection and suppression of inactive 64 kbit/s voice

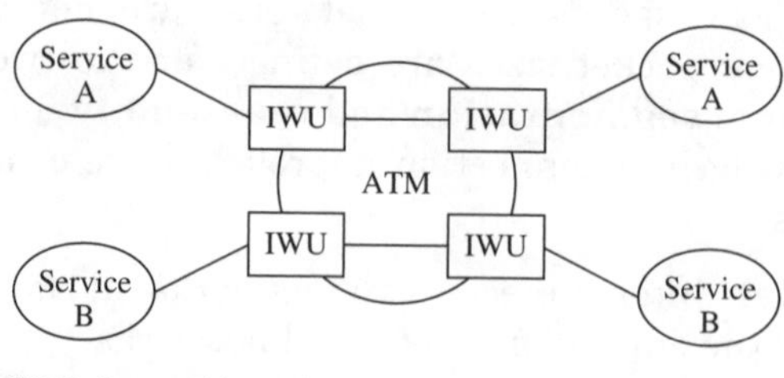

(b)

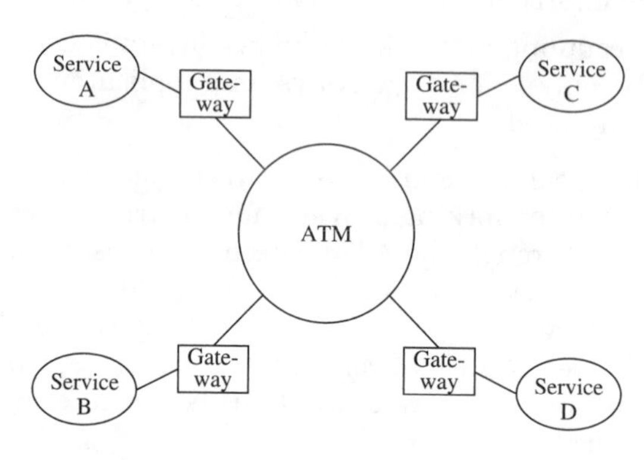

Figure 10.1 ATM interworking classification. (a) Network interworking; (b) service interworking.

channels within a VC ('dynamic bandwidth utilization') which would allow unused bandwidth to be offered to UBR traffic.

Switched trunking is an enhanced CES where the IWU is capable of dynamically switching 64 kbit/s circuits by terminating and processing narrowband signaling (Figure 10.2). Switched trunking can be realized as *one-to-one mapping* (a separate VCC is set up between two peer IWUs across the ATM network for the transport of single 64 kbit/s channels) or *many-to-one mapping* (a VCC is established for the transport of multiple 64 kbit/s channels).

10.3 Connectionless service in B-ISDN

Besides **connection-oriented** (CO) communication, B-ISDN also supports **connectionless** (CL) communication. The provision of CL services in the B-ISDN may be important because one of the first applications in B-ISDN will be the interconnection of LANs/MANs which at present use predominantly CL protocols. As only the CO technique exists at the ATM layer, the CL services have to be realized

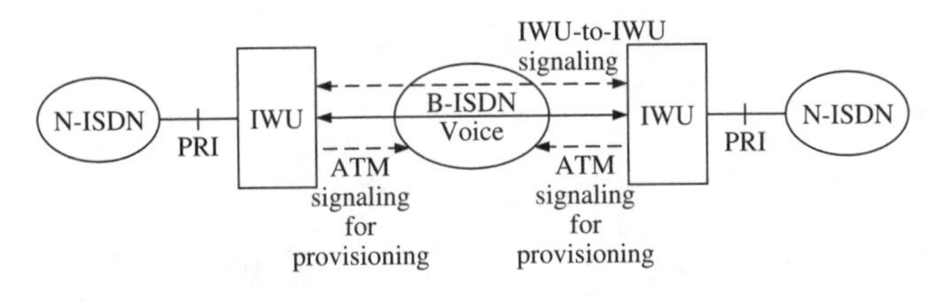

Figure 10.2 Switched trunking.

on top of ATM. ITU-T Recommendation I.211 (ITU-T, 1993g) describes two mechanisms for supporting CL services in the B-ISDN. ITU-T Recommendation I.327 (ITU-T, 1993i) specifies the concept of CL capabilities within the B-ISDN functional architecture.

10.3.1 Connection-oriented and connectionless communication

Many applications, like constant bit rate services or the X.25 data service (ITU-T, 1996n), are best handled by CO communications. A connection has to be established before any information transfer can start. This can be either a physical or a virtual connection. A separate procedure is necessary for establishing such a connection. During this phase the path for the succeeding information transfer will be determined and the necessary resources will be reserved.

However, applications like mail services and other data services are characterized by sporadic behavior and a small amount of data. For reasons of time and expense, no connection is established. Instead, user information is delivered in a message which includes all the necessary addressing and routing information. Each message is handled separately and therefore message sequence integrity cannot be guaranteed. There is also no guarantee of delivery and no acknowledgment of delivery within that layer.

10.3.2 Indirect provision of connectionless service

In the first approach the CL service is provided *indirectly* via the B-ISDN CO service (see Figure 10.3). Transparent ATM layer connections are used between the B-ISDN interfaces. These connections may be either permanent (PVC) or on demand (SVC). The CL protocols are transparent for the B-ISDN because all CL services and AAL functions are implemented outside the B-ISDN. CL services are independent of the protocols within B-ISDN. Support for CL services based on this approach is always possible.

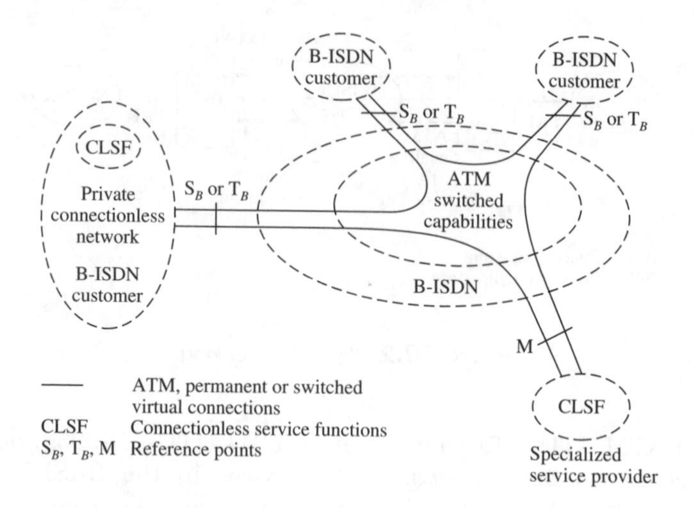

Figure 10.3 Indirect provision of connectionless service.

The full intermeshing of the B-ISDN interfaces requires a lot of ATM layer connections. The use of permanent connections results in an inefficient deployment of these connections, whereas switched connections (establishment at the start of the CL service session) lead to increased signaling traffic for call/connection control as well as long call/connection set-up times. One of the characteristic features of the CL services used in LANs or MANs is the short transfer delay. In order to retain this feature, the connection set-up time should also be very short in ATM.

The indirect provision of CL services is only applicable in the case of a few CL service users which are attached to the B-ISDN. This approach can be considered as an interim solution.

10.3.3 Direct provision of connectionless service

The second approach *directly* supports the CL service via the B-ISDN (see Figure 10.4). **Connectionless service functions** (CLSFs) may be located within or outside the B-ISDN; they terminate the CL protocols and route cells to their destination according to the routing information included in the cells.

The CL service is again on top of ATM. This requires an ATM connection between each user and the CLSFs which are implemented in the CL server. Such a connection will be either permanent or switched. Permanent connections can be VPCs or VCCs, whereas switched connections will typically be realized by VCCs. In the case of permanent VPCs, all VCCs within the VPC could be used by CL services.

In the case of switched connections it may happen that a message cannot be delivered from the destination CLSF to the destination user

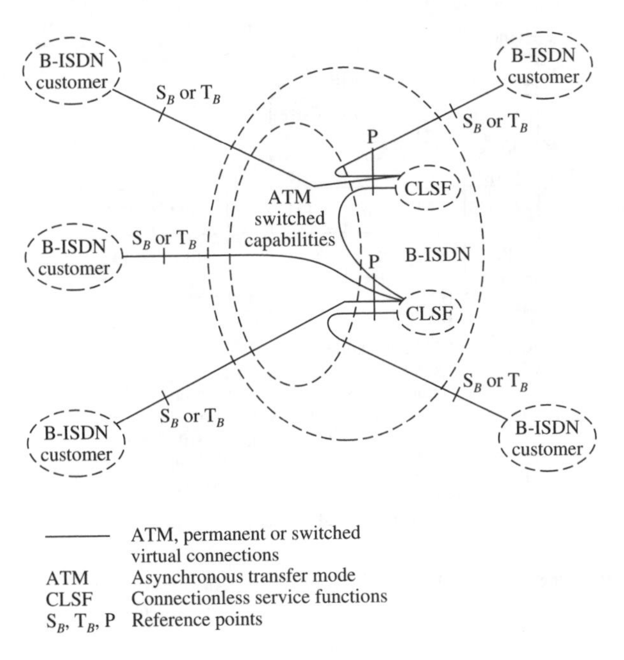

Figure 10.4 Direct provision of connectionless service.

because no connection exists. The destination CLSF will then reject the message.

The indirect provision of CL services for n users requires $n \times (n - 1)/2$ connections for full intermeshing. In a system with direct provision of CL services and only one CL server, n connections are required for n users. The number of CL servers depends on the size of the B-ISDN and the volume of CL traffic to be handled. CLSFs can be implemented in a specific service node or in a separate part of an ATM switching node.

10.3.4 CBDS and SMDS

Connectionless broadband data service (CBDS) and **switched multi-megabit data service** (SMDS) are well-defined services which are both used for the direct provision of a connectionless service within the B-ISDN. CBDS (ETSI, 1993) is the European version of SMDS which was developed by Bellcore (Bellcore, 1989). Since only minor differences between the protocols exist, only CBDS is described in the following.

The ATM network transfers connectionless data units between specific servers which can handle the connectionless protocols. The CL protocol includes such functions as routing, addressing and QoS selection. The adaptation between the CL layer and the ATM layer is

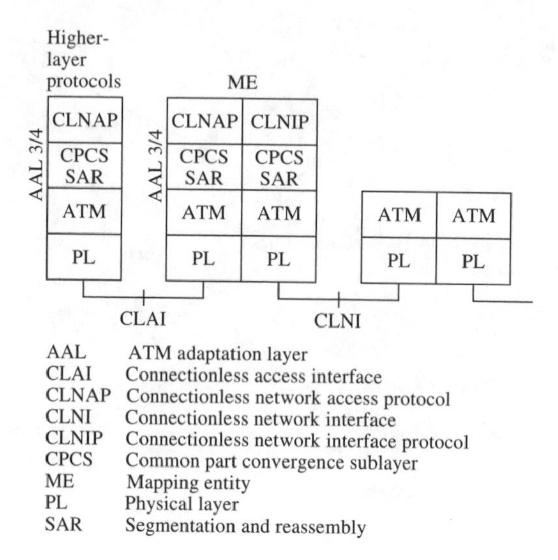

AAL ATM adaptation layer
CLAI Connectionless access interface
CLNAP Connectionless network access protocol
CLNI Connectionless network interface
CLNIP Connectionless network interface protocol
CPCS Common part convergence sublayer
ME Mapping entity
PL Physical layer
SAR Segmentation and reassembly

Figure 10.5 Protocol structure for connectionless service.

performed by AAL type 3/4 (see Section 5.7.4). The general protocol structure for connectionless services in B-ISDN is shown in Figure 10.5.

A distinction has to be made between the **connectionless network access protocol** (CLNAP) and the **connectionless network interface protocol** (CLNIP) which are applied at UNIs and NNIs, respectively. Fortunately, both protocols are similar so the description in the following paragraphs can be applied to either of them unless otherwise stated.

Each CLNAP/CLNIP-PDU contains a source and a destination address. These addresses are modeled according to ITU-T Recommendation E.164 (ITU-T, 1991a). Information transfer at the connectionless layer can be point-to-point or point-to-multipoint. In the latter case a **group address** is used which specifies a set of geographically distinct interfaces. The CLNAP/CLNIP-PDUs are copied inside the network and delivered according to these group addresses. For point-to-point connections, the destination address specifies an individual interface.

The size of the user data fields carrying the CLNAP/CLNIP-SDUs is variable, with an upper limit of 9188 octets for CLNAP-SDUs and 9236 octets for CLNIP-SDUs. The CLNAP-PDU has a maximum length of 9232 octets (maximum 9188 octets user data plus 40 octets CLNAP-PDU header plus 4 octets CRC). With the 4-octet alignment header, the CLNIP-SDU can be as long as 9236 octets. SDUs are transparently transferred through the connectionless layer in a 'best effort' manner; that is, lost or corrupted data units are not retransmitted. Figure 10.6

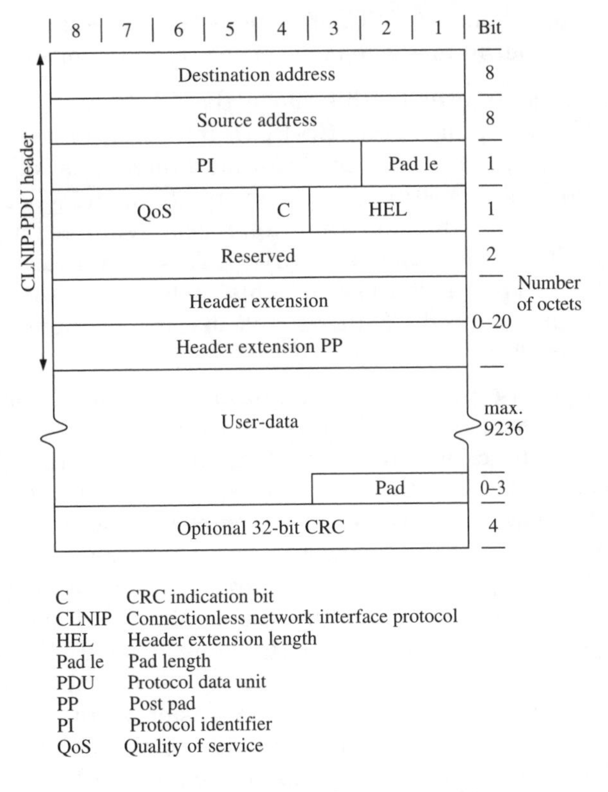

C	CRC indication bit
CLNIP	Connectionless network interface protocol
HEL	Header extension length
Pad le	Pad length
PDU	Protocol data unit
PP	Post pad
PI	Protocol identifier
QoS	Quality of service

Figure 10.6 Structure of the CLNIP-PDU.

shows the structure of the CLNIP-PDU. The structure of the CLNAP-PDU is the same, with three exceptions:

(1) The protocol identifier field, which is called the **higher-layer protocol identifier** (HLPI), is transparently carried end-to-end by the network (except CLNAP-PDUs with HLPI values from 44 to 47, which are discarded).

(2) The header extension post pad field does not exist, so up to 20 octets can be used for the header extension field.

(3) The maximum length of the user data field is 9188 octets.

The functions performed by AAL type 3/4 are identical to those for CLNAP and CLNIP. AAL type 3/4 provides a transparent and sequential transfer of CLNAP/CLNIP-PDUs between two layer entities in an unassured manner. This means that lost or corrupted data units are not retransmitted. More than one AAL 3/4 connection on a single ATM connection is possible by using different MID values (see Section 5.7.4). CLNAP/CLNIP-PDU sequence integrity is preserved for any individual

AAL 3/4 connection. No SSCS functions are necessary; information is transferred either in message mode or in streaming mode:

- **Message mode:** In this mode the CL server collects all SAR-PDUs belonging to one CS-PDU and reconstructs this PDU (see Section 5.7.4). Then the routing function is performed, after which the message is segmented again. Because of the reassembly function, this approach requires a large buffer capacity within each CL server. This may cause high transfer delays. The advantage of this method is the feasibility of detecting errored CS-PDUs and discarding them (unburden the network from worthless PDUs).
- **Streaming mode:** This approach takes advantage of the fact that as soon as a BOM has been received in the CL server the routing function can be performed because the BOM payload includes the destination address. The routing table will be updated with the correspondence of the incoming VPI/VCI/MID combination and the appropriate outgoing one. All COMs and the EOM belonging to a message are routed by a simple look-up in the routing table. After processing the EOM the corresponding registration in the routing table is deleted. When using this method, only a small buffer capacity per CL message is necessary in each CL server, enabling the transfer delay of a message to be kept low. This approach is similar to the CO technique of the ATM layer. A BOM is processed like a connection set-up message and the routing of the COMs and EOMs can be performed by the same mechanism used for cells.

The **mapping entity** (shown in Figure 10.5) is responsible for the proper encapsulation/decapsulation of CLNAP-PDUs into/from CLNIP-SDUs. After adding an alignment header to the CLNAP-PDU it is encapsulated into the CLNIP-PDU. The entire process, including the necessary AAL type 3/4 procedures, is shown in Figure 10.7. (The encapsulation technique need not be applied for NNIs within a single network operator's domain.)

Several 'access classes' are defined for CBDS. These classes constitute the maximum allowed rate with which a user can transfer information into the network. An access class enforcement mechanism is applied at the network entry point, limiting the actual transferred information rate. An access class is defined by three parameters (ETSI, 1993):

(1) **Maximum information rate** (MIR): the maximum instantaneous value of the information rate during transmission of a burst.

(2) **Sustained information rate** (SIR): the long-term average of the information rate for bursty traffic.

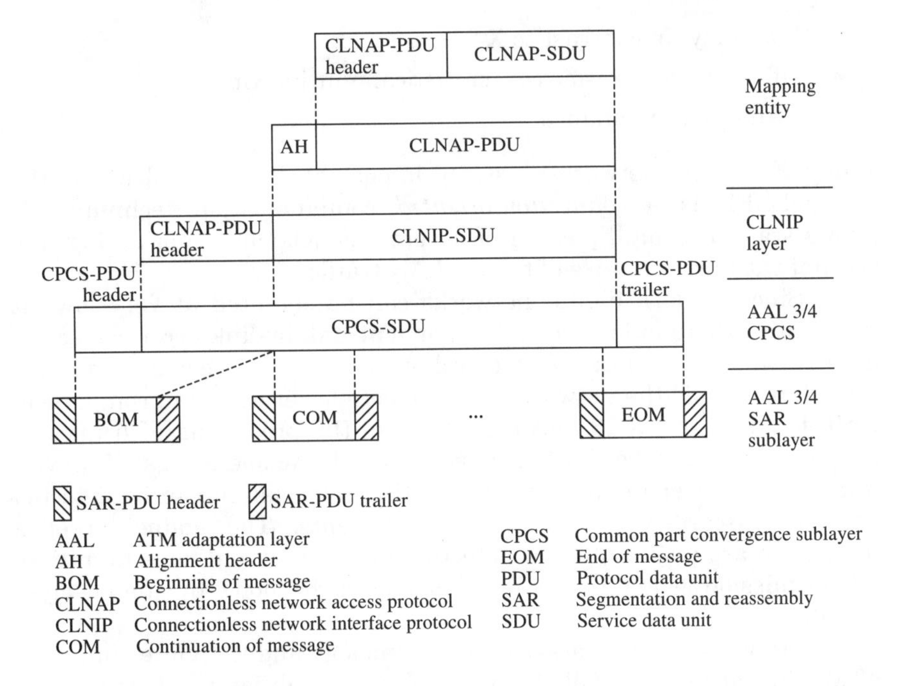

Figure 10.7 CLNAP-PDU encapsulation.

(3) **PDUs per time unit** (PPTU): the long-term average PDU rate for bursty traffic.

At the entry point to an ATM network the UPC function enforces the peak cell rate, which can easily be deduced from the MIR by using the following formula:

$$\textit{ATM peak cell rate} = \textit{MIR} \times (9188 + 44 + 8)/(9188 \times 44 \times 8)$$

How the UPC will enforce the SIR and PPTU is for further study. In DQDB networks these parameters are controlled by a mechanism called the 'credit manager' (IEEE, 1991).

10.4 Frame relay via ATM

10.4.1 What is frame relay good for?

The need for fast and effective interconnection of LANs was the main incentive for the development of **frame relay** (FR). FR is based on the ISDN; compared with the available alternatives, such as leased-line services and X.25 (ITU-T, 1996n) networks, it has some remarkable advantages:

- high throughput

- low delay (in contrast to X.25)
- efficient use of resources (statistical multiplexing)
- standards compliance.

Today FR-based networks are an important building block of the Internet. FR is a *connection-oriented* communication technique. It allows statistical multiplexing of several connections, thus taking into account the bursty nature of typical LAN traffic.

Since modern digital networks can be operated at very low bit error rates, there is less need for stringent link-by-link error correction mechanisms. Therefore, frame relay provides no error correction procedures inside the network. This leads to lower delays in comparison with, for instance, X.25 networks because the processing of frames at network nodes can be limited to some very basic operations. However, frames with bit errors and invalid routing information can be detected and are discarded at FR network elements. The endpoints of a connection are responsible for detecting lost frames and have to initiate retransmission when required. This has to be done by higher-layer protocols.

Frame relay is a *packet-based* technology and therefore similar to ATM. FR packets are called frames. The main difference between the two technologies is the variable length of the frames in FR as compared with the fixed-sized cells in ATM. Although the transfer delay in FR is lower than in X.25, real-time services such as voice and video communications are restricted to small, geographically limited networks.

FR frames carry routing information, called the **data link connection identifier** (DLCI), which has only local (link-by-link) significance. A **discard eligibility** (DE) bit is defined which is similar to the CLP bit, and two bits are used for congestion notification. Frames are relayed according to the address information along a virtual connection. The FR virtual connections can be either permanent (established by subscription) or switched (established and released on demand via ISDN signaling procedures). Several virtual connections can share the same physical access line. The available transmission capacity can be used by each virtual connection up to the physical limits (bandwidth on demand). Bandwidth not being used by one virtual connection can be utilized by others (statistical multiplexing). For management purposes, special management frames are defined which can be recognized through a predefined DLCI value (similar to OAM cells). These management frames are used for the supervision of PVCs. They provide a means of determining the statuses of the physical and logical links between network and user equipment.

ITU-T Recommendation Q.922 (ITU-T, 1992c) describes the core functions used to support the frame relaying bearer service:

- frame delineation and alignment;

- frame multiplexing/demultiplexing using the address field;
- inspection of the frame to ensure that it consists of an integral number of octets;
- inspection of the frame to ensure that it is neither too long nor too short;
- detection of (but not recovery from) transmission errors;
- congestion control functions.

Frame relay operates at OSI layer 2. One great advantage is the possible use of existing hardware and software. Frames carry both FR routing information and user data. The payload is transparently transported, allowing the use of existing higher-layer protocols. A set of standards exists describing the encapsulation of such protocols in FR frames.

Another advantage of FR is that it can convey information of variable size from 1 to 260 octets within a single frame (or more, if agreed at connection establishment), similar to the information generated by data systems (such as LANs). Future standards may specify the maximum payload size up to approximately 8000 octets; this will be the upper margin because of the limited error detection capabilities of the **frame check sequence** (FCS). The default length of the frame header is two octets. However, 3- or 4-octet DLCI fields are also possible.

Figure 10.8 shows the FR frame.

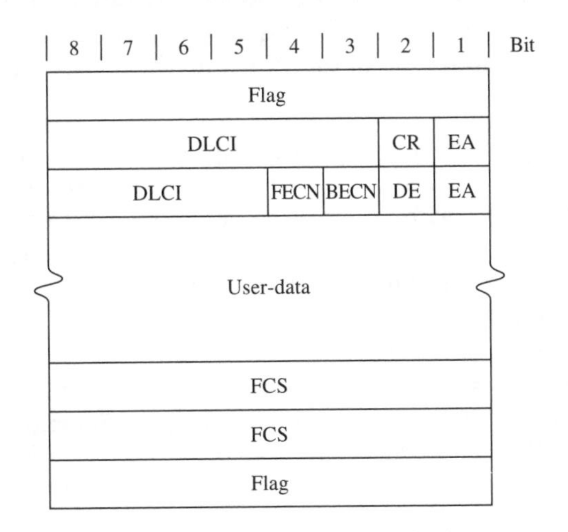

BECN Backward explicit congestion notification
CR Command response bit
DE Discard eligiblity
DLCI Data link connection identifier
EA Address extension bit
FCS Frame check sequence
FECN Forward explicit congestion notification

Figure 10.8 Frame format for frame relay.

Two different bits in the FR frame header are available for traffic management: **forward explicit congestion notification** (FECN) and **backward explicit congestion notification** (BECN). Both bits can be set by network elements when they experience congestion. The FECN might be used by the receiver for throttling down the transmitter via higher-layer protocols. With BECN the transmitting terminal can be informed directly that congestion has occurred. All these mechanisms require that terminals react accordingly, meaning that they reduce the traffic sent into the network. Since user frames are in danger of being discarded at congested network nodes, terminals are assumed to support some kind of reactive congestion control mechanism.

Four parameters are defined in ITU-T Recommendation I.370 (ITU-T, 1991f) for the resource management of individual virtual connections:

(1) **Committed information rate** (CIR): The information rate which the network is committed to transfer under normal conditions.
(2) **Committed burst size** (Bc): The maximum committed amount of data a user may offer to the network during a time interval Tc.
(3) **Excess burst size** (Be): The maximum allowed amount of data by which a user can exceed Bc during time interval Tc. This data (Be) is delivered in general with a lower probability than Bc.
(4) **Committed rate measurement interval** (Tc): The time interval during which the user is allowed to send only the committed amount of data (Bc) and the excess amount of data (Be).

These parameters are defined by subscription or via signaling and may be enforced at the network's entry point. For excess frames (those belonging to Be), the DE bit has to be set indicating that these frames should be discarded in preference to other frames, if necessary. As in ATM, setting of the DE bit is called **tagging**. Frames exceeding the agreed Be can be discarded.

A user can optimize each virtual connection for the traffic to be carried by choosing suitable values for CIR and Be. Additionally the user can set the DE bit according to the relative importance of the frame compared with others.

10.4.2 Frame relay and ATM

The Frame Relay Forum differentiates between two basic modes of operation:

- network interworking, FRF.5 (Frame Relay Forum, 1994), as shown in Figure 10.9(a), and

- service interworking, FRF.8 (Frame Relay Forum, 1995), as shown in Figure 10.9(b).

Both interworking scenarios are based on the same principles, for example for mapping FR header bits to ATM header bits.

Networking interworking uses ATM as a pure transport mechanism between FR networks or FR CPE. If an ATM station implements a complete FR protocol stack, as shown on the right-hand side of Figure 10.10, it can communicate with FR equipment over ATM. However, in both cases the use of ATM is not visible to FR users.

Service interworking is accomplished by employing a specific **interworking function** (IWF). Via this IWF a native FR CPE can communicate with a native ATM CPE without the need for any specific adaptation functionality within the end-equipment; that is the FR-SSCS in Figure 10.10 is a 'null' function. For service interworking a mapping of the encapsulation schemes of higher-layer protocols is defined in Frame Relay Forum (1995). Since there is no equivalent to the command response bit of the FR header in the ATM cell header, this bit is mapped

(a)

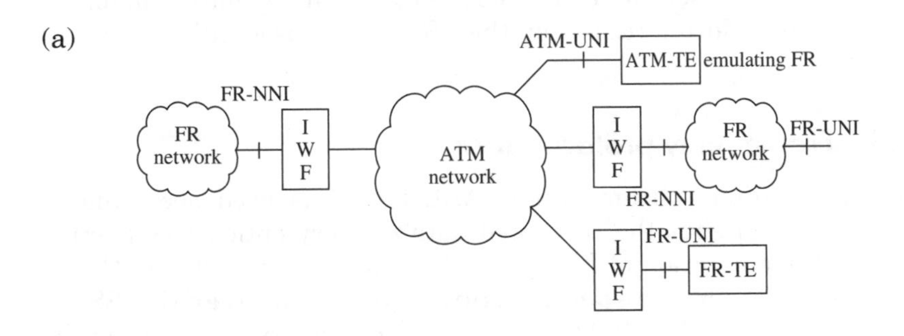

(b)

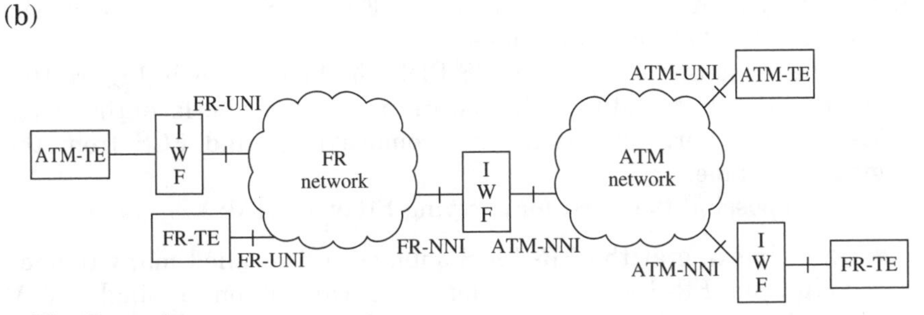

FR Frame relay
IWF Interworking function
NNI Network-to-network interface
TE Terminal equipment
UNI User-to-network interface

Figure 10.9 (a) Frame relay to ATM network interworking; (b) frame relay to ATM service interworking.

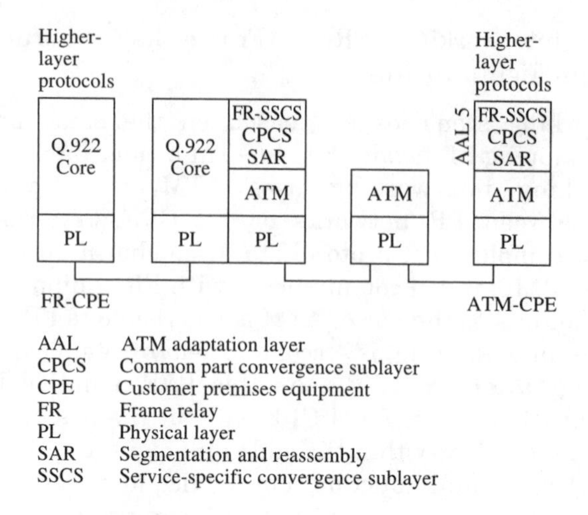

Figure 10.10 Protocol stacks for network interworking between FR and ATM equipment.

onto the least significant bit of the AAL 5 CPCS-UU field (see Section 5.7.5). If SVCs are used the IWF needs to maintain an address mapping cache table. For address resolution the ARP will be used; it is translated by the IWF.

10.4.3 Frame relay protocol stack

For the adaptation of frame relay, AAL type 5 is used operating in message mode without the corrupted data delivery option (see Section 5.7.5). For the emulation of FR by and for interworking with ATM, the **frame relaying service-specific convergence sublayer** (FR-SSCS) has been defined (ITU-T, 1993p). This sublayer preserves the FR-SSCS-SDU sequence integrity. Functions of the FR-SSCS are the same as the Q.922 (ITU-T, 1992c) core functions.

The structure of the FR-SSCS-PDUs is the same as in Figure 10.8 apart from the flags and FCS. Interworking functions perform the Q.922 (ITU-T, 1992c) core operations and remove flags and FCS from the original FR frame.

Two possibilities exist for carrying FR over ATM:

- Multiplexing at the FR-SSCS sublayer (also called many-to-one). Multiple FR-SSCS connections are carried on a single ATM connection. The DLCI allows unambiguous identification of the individual frame relay connection. (Network interworking only)
- Multiplexing at the ATM layer (also called one-to-one). Each FR virtual connection is mapped onto a single ATM connection. The well-known ATM multiplexing mechanisms using VCIs/VPIs are applied.

The mapping of DE to CLP and FECN/BECN to the congestion indication in the ATM cell header is explained in ITU-T Recommendation I.555 (ITU-T, 1993p), and mapping of FR to ATM traffic parameters is defined in the ATM Forum's B-ICI specification, Appendix A (ATM Forum, 1994f).

Review questions

1. What is meant by service/network interworking via an ATM network?
2. Which ATM circuit emulation services are currently defined?
3. What are the ATM connectionless services called in Europe and North America, respectively?
4. Which AAL type is used for the connectionless service?
5. Which AAL type is used for frame relay interworking with ATM?

11

Local area networks and metropolitan area networks

11.1 ATM local area networks

Initially CSMA/CD and token ring/bus LANs were deployed to interconnect workstations and PCs. After a short period these LANs were interconnected and extended to high-speed local area networks, such as FDDI or DQDB. The next generation of LANs should provide:

- real-time information transfer;
- scalable throughput;
- high transfer capacity;
- easy interworking between the LAN and **wide area networks** (WANs), such as the public network;
- standards-based solution.

ATM is deemed to be the appropriate transport technique to fulfill these requirements. Besides traditional data communications, upcoming applications like multimedia or videotelephony demand a high-speed integrated network to carry all types of multiservice traffic. Thus, a mixture of different, specialized networks operating in parallel can be avoided.

ATM allows real-time transfer of information as a result of its short cells. Scalable throughput is automatically achieved by sending as many cells as required. The interfaces currently defined offer transmission capacities up to 622 Mbit/s, which is considered to be enough for most LAN applications for the next decade. Since ATM LANs make use of the same technique as used in public networks, simple interconnection between LAN and WAN can be guaranteed. For the public network several ITU-T standards are available which are taken as a basis for LAN implementations as well. These standards guarantee future-proof solutions. In addition, the use of similar technologies on both public and private networks activates possible synergies and will lead to reduced costs for hardware and software.

ATM is usually introduced initially as a high-capacity campus **backbone network** for transmission and switching. Such a backbone network has to emulate existing LAN concepts. However, since CN equipment can be more rapidly amortized than public networks, instead of an evolutionary approach a sort of revolution may, in some cases, lead to immediate ATM implementation up to the desktop.

Another important aspect is the bandwidth available for each end user. Traditional LANs offer up to 16 Mbit/s which have to be shared between several participants (shared medium configurations). ATM LANs are expected to be mainly **star configured**, so that the individual terminal is connected with a **hub** via a single (physical) line, thus providing the full (physically) possible transfer capacity to each terminal (Newman, 1992). Moving away from ring or bus configurations has further advantages, such as simpler management functions.

Figure 11.1 shows a possible ATM-based customer network (Denzel et al., 1993).

As an important step towards ATM to the desktop, the ATM Forum has specified a so-called **low-cost interface** at 52 Mbit/s and 25.6 Mbit/s (see Section 5.4) which allows the use of copper cables already installed (ATM Forum, 1993a). This is crucial since the laying of new cables consumes large amounts of money.

Despite the high degree of commonality between CN equipment and public network equipment, some differences exist to allow simpler and cheaper solutions for the CN. For corporate networks, smaller switches can be used with somewhat less stringent reliability requirements for the individual end user, thus avoiding complex and expensive redundancy concepts. Simpler solutions for OAM, signaling and management can be adopted, such as use of the **simple network management protocol** (SNMP) (see Section 6.4.2) instead of **common management information service/common management information protocol** (CMIS/CMIP). Tariffing and accounting mechanisms can be reduced in CNs. Traffic management processes can be simplified because terminal compliance with the network rules can be expected. For private interfaces, UPC is not needed. Simplified reactive congestion control schemes can be used because of the smaller extent of LANs, and the network can rely on the behavior of the sources; that is, they will really throttle back when required (Newman, 1992).

The most crucial point that will decide the success of ATM-based LANs is the cost per port. A modular concept has to be provided to allow later growth of the LAN combined with low first-installation costs.

Today the costs for Ethernet-based LANs are lower than for ATM. This has led to the predominance of Ethernet-based LANs over ATM.

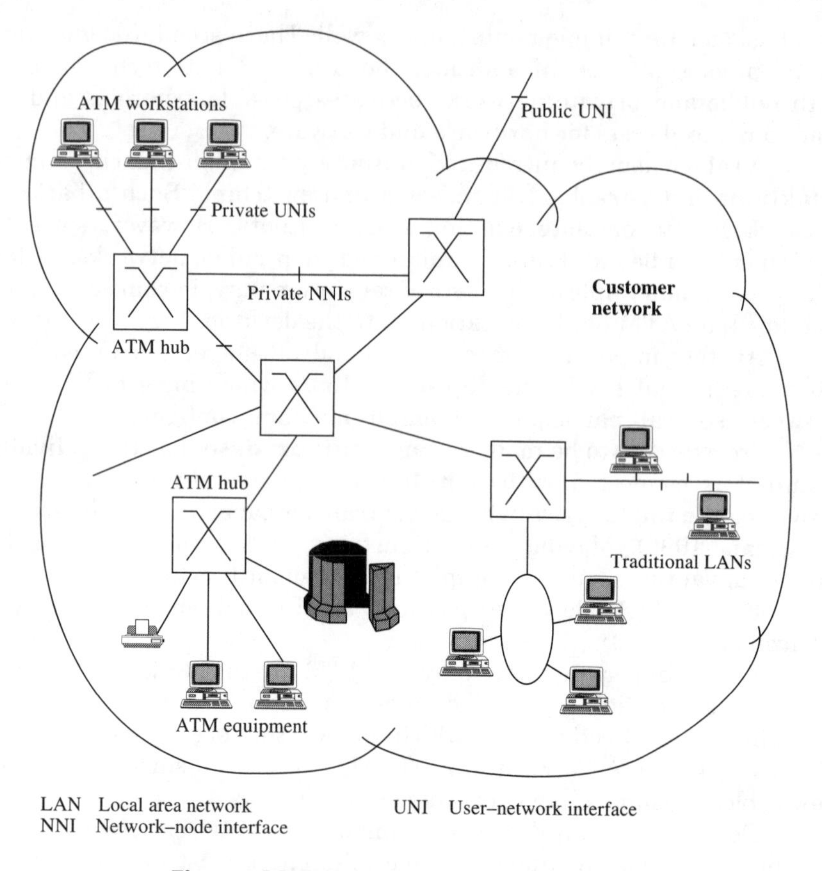

Figure 11.1 ATM-based customer network.

11.2 Local area network emulation

Local area network (LAN) technologies such as Ethernet or token ring/bus are widely employed nowadays and there are many applications which are used via these legacy networks. However, new applications like videoconferencing or multimedia sessions require higher bandwidths, QoS support and easy means of network growth. Among other features, ATM is the technology which can fulfill all these requirements. Hence, it is an attractive solution if used in a LAN. On the other hand, existing devices like PCs, servers and workstations with, for example, Ethernet **network interface cards** (NICs) and many applications, constitute an investment which cannot be replaced overnight. Therefore, the ATM Forum has described a way which allows ATM to be gradually introduced into existing legacy LANs. This method is called **LAN emulation** (LANE) and is specified in ATM Forum (1995b). LANE allows coexistence and interoperability among traditional LAN stations and ATM stations. All existing applications and protocol stacks can be reused while offering a migration path towards ATM-based LANs.

LANE is a function which can reside in edge devices such as LAN switches, bridges, routers, in legacy LAN to ATM conversion equipment or in native ATM hosts. In most cases these ATM hosts (with ATM NICs) will first be application servers such as email or file servers, since they most benefit from the higher bandwidth available with ATM, thus avoiding a potential bottleneck.

The **LAN emulation user network interface** (LUNI) is the defined interface between those devices and the ATM network. The **LAN emulation network-to-network interface** (LNNI) is specified in LANE version 2 (LANEv2) (ATM Forum, 1997c) and describes the interfaces between LANE servers. Those servers are the basic building blocks for an emulated LAN and are described in Section 11.2.2.

Considering the two main differences between ATM and legacy LAN technologies, *connection-oriented communication* with local connection identifiers (VCI/VPI) versus *connectionless communications* with global addressing (MAC), the question of how to combine these contradicting networking paradigms arises.

11.2.1 The basic LANE principles

LANE is used instead of the MAC layer, namely Ethernet and token ring, as shown in Figure 11.2. Instead of a shared medium 'broadcast and filter' transfer technique, point-to-point ATM connections are established between two endpoints. LANE provides the same service primitives to the higher layer as 'standardized' MAC device drivers like the **network driver interface specification** (NDIS), **open data-link interface** (ODI) or **data link provider interface** (DLPI). LANE makes the point-to-point virtual ATM connections look and behave like a shared medium thus achieving transparency for all higher-layer network protocols such as IP, IPX or AppleTalk.

Another considerable advantage of LANE is that it does not impact ATM switches. All LANE procedures are based on standardized ATM protocols, for example UNI 3.0/Q.2931 for setting up an SVC. This decoupling of internetworking protocols and ATM switching is also known as the 'overlay model'.

The basic operation of LANE is to set up a point-to-point ATM connection between two LANE end-systems instead of broadcasting (and filtering) data frames. In order to enable a LANE end-system to set up an ATM connection, a MAC to ATM address resolution needs to take place beforehand.

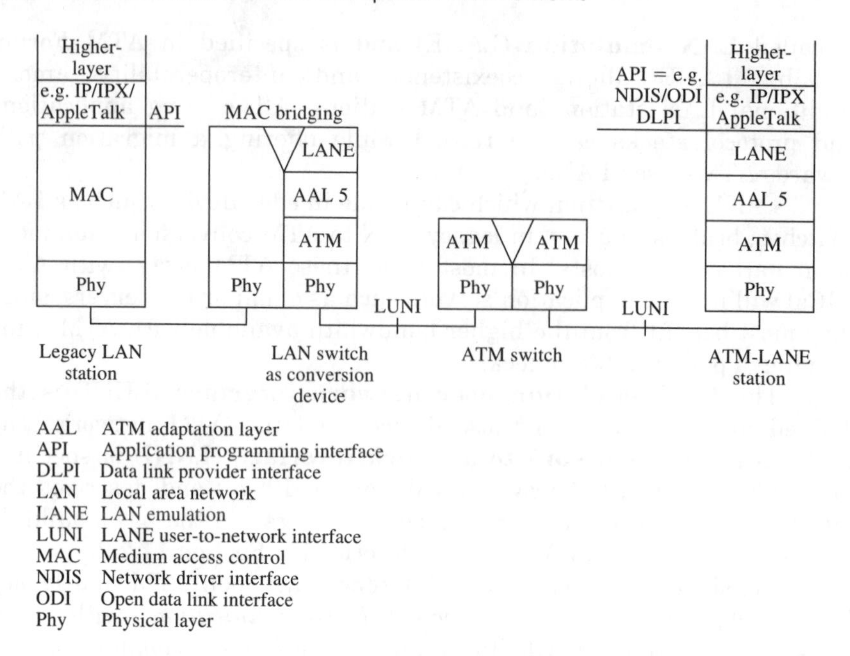

Figure 11.2 The LANE principle of communication between legacy LAN station and ATM LANE station.

11.2.2 The basic building blocks

LAN emulation clients (LECs) are the end-systems of a single **emulated LAN** (ELAN) which communicate with each other via the LUNI. A LEC handles the data forwarding/reception, address resolution and control functions. LECs can be found, for example, in ATM hosts on ATM NICs or in LAN switches. Several LECs may exist in those devices simultaneously if they are connected to several ELANs. Each LEC is identified by a unique ATM address but can be associated with several MAC addresses as in the case of a LAN switch.

Note that here the term **LAN switch** comprises devices like an ATM bridge, ATM router and legacy LAN to ATM conversion equipment. Normally, more than one device is connected to a LAN switch. Therefore, a LAN switch represents more than one MAC address and the LEC in a LAN switch is associated with each of these MAC addresses. This results in a one-to-many relationship for the ATM to MAC addresses. Such a LEC is also known as a **proxy LEC**.

For each ELAN one (logical) **LAN emulation server** (LES) exists. The LES provides the necessary control functions and its main task is to register and to resolve MAC and ATM addresses.

As mentioned earlier, several LANs may be emulated on a single physical ATM network. The **LAN emulation configuration server**

(LECS) has the responsibility of assigning a LEC to a particular ELAN. This is implicitly achieved by transmitting the ATM address of the LES to the LEC. Furthermore, it exchanges information on addresses, LAN type and maximum allowed frame size with the LECs.

The **broadcast and unknown server** (BUS) is used for 'flooding' unknown destination address traffic and multicast/broadcast traffic to all LECs within an ELAN. Each LEC is associated with a single BUS whose unique ATM address can be obtained from the LES. Additionally, the BUS will be used for data transfers between LECs as long as no direct point-to-point VCC could be established.

All these server functions can be physically distributed or co-located. For example, BUS and LES are ideally co-located for sharing a common database, possibly resulting in an optimized BUS forwarding policy.

As mentioned in Section 11.2.1, LANE is defined for Ethernet and token ring-based LANs. However, a LEC can only belong to one type of ELAN and a direct connection between 'Ethernet LEC' and 'token ring LEC' is not possible. At present, linking Ethernet with token ring LANs still requires a router.

Figure 11.3 shows the LANE building blocks and connection types.

Initial configuration

The main goal of the initialization procedure as defined by the ATM Forum is to reduce the required manual configuration of a LEC when it first joins an ELAN. Ideally all configuration and initialization is done automatically, without human interaction. This is achieved by using LECS. The LECS contains all the information required by a LEC such as its own ATM address, the ATM address of its LES, the type and name of the ELAN and the maximum frame size supported. All this information of course needs to be manually configured by a network operator, but only once for the LECS and not for each LEC joining the ELAN.

How does a LEC connect to a LECS after power-up without having any knowledge about its environment? Several possibilities exist to allow a LEC to locate its LECS:

- Use a well-known, permanent VCC (VPI = 0, VCI = 17). This VCC must be pre-configured throughout the entire LAN and its network components and will only be used for finding a LECS.
- Obtain the LECS address from its ATM switch via the ILMI procedure (see Section 6.4). This requires that the LECS address is configured in the ATM device terminating the ILMI request.
- Use a well-known LECS address (but this may lead to interoperability problems because of different addressing formats).

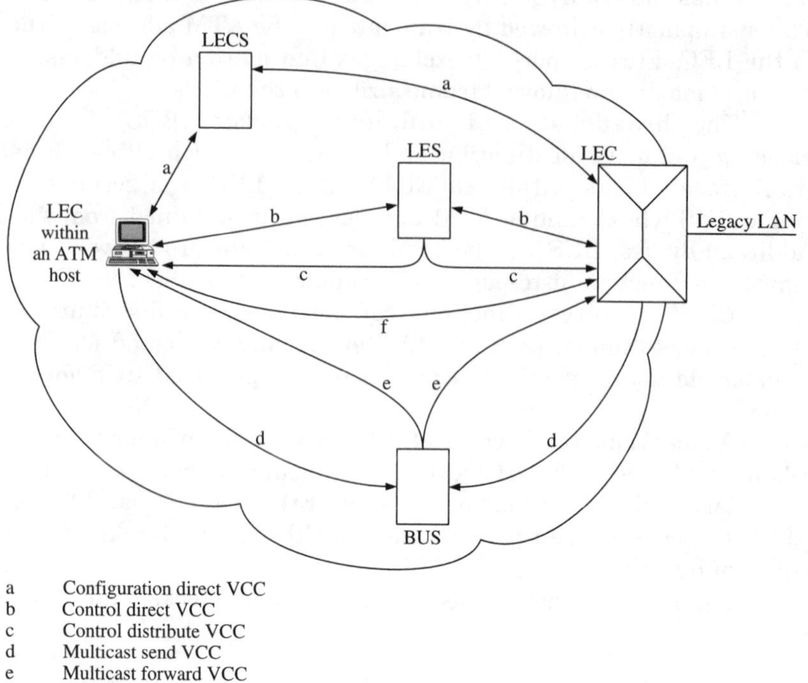

a Configuration direct VCC
b Control direct VCC
c Control distribute VCC
d Multicast send VCC
e Multicast forward VCC
f Data direct VCC
BUS Broadcast and unknown server
LAN Local area network
LEC LANE client
LECS LANE configuration server
LES LANE server

Figure 11.3 LANE elements and connection types.

After finding its LECS the LEC establishes a *configuration-direct VCC* and obtains all the information mentioned above. The LES address provided by the LECS implicitly determines to which ELAN the LEC belongs.

Alternatively, the LECS procedures can be completely omitted if the LES address is configured in the LEC.

Joining an ELAN and registration

After obtaining the necessary configuration data the LEC is able to establish a connection to its LES. This VCC is called a **control-direct VCC**. In the following procedure the LEC registers its addresses (MAC and ATM address) at the LES. (It may register more than one address-set in case of a proxy LEC which represents more than one device. A bridge is a typical example of such a proxy LEC.) In turn the LES assigns a unique LEC-ID. Subsequently the LES establishes a **control-distribute VCC** towards the LEC. In most cases this VCC will be a leaf of a multicast tree.

The control-direct VCC is used by the LEC to issue a **LAN emulation address resolution protocol** (LE-ARP) message for obtaining any unknown ATM address. If the LES does not find the requested address in its cached address table it forwards the LE-ARP request to all attached LECs via the control-distribute VCC. At this point it becomes clear why the control-distribute VCC best constitutes a multicast tree.

The first ATM address a LEC tries to obtain is that of the BUS. This is done by issuing an LE-ARP for the all-1s MAC broadcast address. Now the LEC establishes a point-to-point *multicast send VCC* to the BUS and the BUS in turn adds the LEC as a leaf to the point-to-multipoint *multicast forward VCC*.

After all these initializations the LEC is ready for data transfers.

(As mentioned above, an ATM switch working as a bridge can be a LEC as well. Since this bridge represents more than its own MAC address it is also called a proxy LEC.

A LES normally knows all the addresses of its ATM hosts. It may not know all the MAC addresses associated with a bridge since this could be a considerable number of addresses. In order to relieve the non-proxy LECs from dealing with too many LE-ARPs the LES may establish a second multicast tree to only those LECs that registered as a proxy.)

The data transfer phase

Data is transferred in two possible ways: for broadcast or multicast traffic the LEC transmits the packet to the BUS for further distribution. For unicast traffic (one destination only) the LEC uses a *data direct VCC*.

Using the BUS for multicast traffic involves two major problems. First, the sending LEC will receive its own packet via this distribution mechanism and some LAN protocols cannot tolerate such behavior. To cope with this problem the source's LEC-ID needs to be contained in the protocol information header of each packet. LECs will check the LEC-ID and discard those frames carrying their own LEC-ID.

Secondly, the amount of unnecessary traffic can reach considerable levels if, for example in case of a live video distribution, only a few LECs want to receive the program. This leads to a waste of bandwidth. LANEv2 will specify mechanisms to improve the distribution of vast amounts of traffic to only a subset of all attached LECs (see Section 11.2.4).

In case of unicast traffic the LEC may find an already existing data-direct VCC or at least it knows the destination ATM address. In this case it can directly establish a VCC. However, established connections will be released after a certain idle period and address information will be aged-out after some time. Therefore, the normal case will be that the LEC neither finds a VCC for the data transfer nor does it know the destination ATM address. For those cases it first issues an LE-ARP request to the LES to obtain the correct ATM address. The LES may either know the requested address or forward the LE-ARP request via the control-distribute channel. The LEC that knows the correct address mapping will answer via the control-direct VCC back to the LES. Now the LES may forward this answer either to the requesting LEC only or to all LECs. The latter option enables all LECs to learn this address mapping which might reduce the number of LE-ARP requests.

During the waiting period until an ARP answer is received, the sending LEC could store the packet or, if no buffer space is available, discard it. Neither alternative is ideal. Hence, normally the LEC will send the data packet immediately to the BUS which will flood it to all LECs. As soon as the source LEC receives the expected LE-ARP response it knows the destination ATM address and establishes a data-direct VCC. However, this VCC cannot be used immediately since some cells may still be on their way via the BUS and the cell sequence integrity cannot be guaranteed if cells are sent via the data-direct VCC at this time. To assure that no more cells get 'stuck' on the way to their destination a *flush message* is sent via the first transmission path, that is, via the BUS. This flush message is a specific control cell which needs to be acknowledged by the destination LEC before the data-direct VCC can be used.

It should be noted that, by using the flush message, frame ordering for unicast traffic can be guaranteed but frame sequence integrity for a mixture of unicast and multicast frames originated by a single source cannot always be assured. However, most legacy LANs exhibit the same behavior.

Use of the spanning tree protocol

As in traditional bridged LANs the spanning tree protocol see (IEEE, 1988) for more details) will also be used in ELANs to avoid the occurrence of loops in a network. Spanning tree bridge packets will be exchanged between the LECs of all LAN switches by means of the BUS. As a reaction to the spanning tree protocol information, a LAN switch will 'turn off' one of its ports to break the loop. Those changes may occur relatively frequently and cause an undesirable side-effect: the destination MAC address can no longer be reached via the same LEC address. Since other LECs may use a data-direct VCC towards this LEC or may have stored its address in a local cache, the data flow will be interrupted and never reach the designated destination. To avoid this situation the LEC which caused this topology change will distribute an **LE-topology-request message** to all other LECs via the LES. This causes the invalidation of all cached address information and a subsequent update via ARPs. However, existing data-direct VCCs are not discarded immediately. These VCCs will be released later through the idle detection mechanism since the new address mapping causes all data to be sent via the new connection.

11.2.3 Advantages and disadvantages of LANE

One of the most important advantages of ATM-based LANE is the speed of ATM, for example being capable to operate via 155 Mbit/s pipes. Introducing ATM in the corporate backbone paves the way for a gradual migration towards an all ATM-based environment where each network device makes use of ATM.

Another great advantage of LANE is that it allows network administrators to build **virtual LANs** (VLANs). VLANs allow end-systems as PCs or workstations and servers to be assigned to a logical group. As an example, one could consider the marketing group and the sales department of a company as each being one such logical group. Normally, communication inside a group is not restricted but some 'firewalling' between groups might be desired. In former days all members of a group needed to be attached to the same physical LAN segment. With VLANs logical groups can be built all over the enterprise network independent of their physical location. This enables network administrators to reconfigure working groups, to move, add or remove members thus allowing, for example, an ad hoc cross-functional project team to be established very rapidly.

VLANs are built by allocating a LEC to a specific ELAN but are still being operated via the same physical ATM backbone network. In fact each ELAN constitutes a VLAN and the terms can be used synonymously.

For communications between different ELANs, bridges and routers need to be employed.

As mentioned above, LANE hides the properties of ATM from higher layers. On the one hand this allows good reuse of existing applications and hence saves the investment; on the other hand it does not allow the benefits of ATM such as a guaranteed QoS to be exploited. So far LANE can only be run over UBR connections since these services resemble the nature of MAC traffic.

Since LANE is a bridging technique it is, like every bridged network, susceptible to large amounts of broadcast traffic which might easily flood the entire network. This may cause congestion even on a high-speed ATM network. The servers (LES and BUS) constitute potential bottlenecks as well. Therefore, LANE is limited to 'small' networks. For interconnecting these small ELANs ATM routers need to be used. Such an ATM router implements LANE itself and it typically supports several LECs on a single, physical ATM port (one LEC for each ELAN).

11.2.4 Overview of LANE version 2

With LANE version 2 (ATM Forum, 1997a, 1997c) several considerable enhancements to LANEv1 will be provided:

- use of LLC/SNAP (see Section 9.1.1) multiplexed VCCs;
- support of (up to eight) different QoS levels;
- enhanced multicast support;
- definition of the LAN emulation network–network interface (LNNI), which describes interfaces between the different LANE servers.

LLC/SNAP multiplexing

With LANEv2 the possibility of using a single VCC for different information 'flows' between LANE components will be introduced. For this multiplexing the same **logical link control** (LLC) method is used as described in Section 9.1.1. This method not only allows different flows to be conveyed via a single VCC but also different protocols via the same VCC. LLC/SNAP multiplexing is allowed for all traffic in LANEv2 except for the **configure-direct VCC** which must be a non-multiplexed VCC.

Within a network, LLC/SNAP multiplexed and non-multiplexed VCCs may exist simultaneously and interworking with LANEv1 components must be possible. In order to differentiate multiplexed from non-multiplexed VCCs, different **broadband low-layer information** (B-LLI) values are used in the set-up message for an SVC (see also Section 9.3). Since the only difference between a multiplexed and a non-multiplexed VCC lies in the 8-byte LLC/SNAP header and an additional 4-byte ELAN-ID field (see Figure 11.4), interoperability can be achieved by adding or deleting these two fields. Thus, existing LANEv1 components can be used unchanged in a LANEv2 environment. However, the handling and conversion of multiplexed and non-multiplexed VCCs adds additional complexity (and thus cost) to LANEv2 components.

Quality of service support

With LANEv1 all connections were best-effort VCCs, that is, essentially UBR traffic. This prevented the use of one of the most important ATM capabilities, support of different QoS levels. With LANEv2 a LEC can provide eight levels of QoS to the higher layers, each being associated with a specific UNI call set-up parameter set. Thus, each **data-direct VCC** offers one of the eight QoS levels for data transmissions. A possible mapping of QoS level to call set-up parameters will be recommended by the IETF.

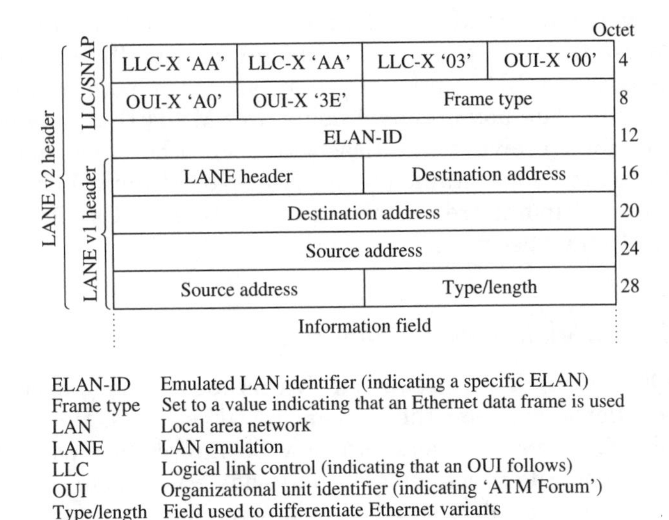

Figure 11.4 LANE frame formats for Ethernet frames.

Enhanced multicast capabilities of LANEv2

With the basic method of LANEv1 a LEC receives all flooded unicast frames and all broadcast/multicast traffic. LANEv2 enables a LEC to indicate for which particular MAC multicast addresses it wants to receive traffic. This is done during registration with the LES which distributes this information to all other LECs and/or the **special multicast server** (SMS).

An SMS can be used to overcome the problem of high amounts of broadcast traffic which may exist in bridged LANs. Compared with the BUS which broadcasts traffic to all attached clients, the SMS forwards traffic only to members of a specific multicast group. This may lead to a considerable reduction in network load, and LECs that are not part of that multicast group are not burdened with unnecessary traffic.

An SMS joins an ELAN following a similar procedure as a LEC and registers a set of multicast MAC addresses with its LES.

LECs which intend to send out traffic to a specific multicast group address will obtain the ATM address of the nearest SMS via a LE-ARP for that group address. After obtaining the SMS's address the LEC establishes a *special multicast send VCC* to the SMS and in turn receives a *special multicast forward VCC* from that SMS. The special multicast send VCC is used by the LEC for forwarding data to a particular multicast group MAC address and the special multicast forward VCC is used by the SMS to distribute the multicast traffic to all members of that multicast group.

However, since a BUS still exists in an ELAN the LEC decides whether to use the SMS or the BUS for multicast traffic. Alternatively, if no SMS exists a LEC optionally may set up a point-to-multipoint connection itself. The advantage of using an SMS is that only one VCC is required for receiving multicast traffic, otherwise many VCC endpoints need to be maintained by a LEC. Each of these VCCs is a leaf of a point-to-multipoint tree established by other LECs. On the other hand, the SMS may become a bottleneck.

The LANE network to network interface

The **LANE network to network interface** (LNNI) specifies the 'information flows' between the various LANE servers, that is, LECS, LES and BUS. Several servers can now be interconnected to provide redundancy among its functions and to provide a means to scale ELANs to larger networks by distributing tasks over several machines, thus avoiding processing bottlenecks.

The LNNI protocol (ATM Forum, 1997c) describes how synchronization between server pairs such as LES–LES, BUS–BUS and LECS–LES is achieved. It is based on four layers: the VCC layer, the flow layer, the spanning tree layer and the **server cache synchronization protocol** (SCSP) layer (IETF, 1996j).

At LNNIs the LLC/SNAP multiplexing method for sharing a single VCC among different flows is used.

The VCC layer is the basic ATM transport layer using AAL 5. Since the VCC layer may consist of several point-to-point or multipoint connections, or even provide several connections between adjacent entities, the flow layer was introduced. This layer abstracts the actual VCC topology and provides reliable, bidirectional connectivity for the next higher layer, the spanning tree layer.

Within a network server-nodes may be arbitrarily interconnected and a specific server might be reached via several different ways. In order to avoid loops, not all links can be used for information exchanges simultaneously. The spanning tree layer is used to obtain spanning tree state information for the SCSP, solving that problem.

Cache information contained in different servers is synchronized via the SCSP, which is build of three protocols:

- the **hello protocol**, which is used by a server when it first joins (or re-joins) a server tree;
- the **cache alignment protocol**, to exchange pertinent cache entries;
- the **client state update protocol**, which is subsequently used for propagating any further changes.

For a more detailed description (ATM Forum, 1997c) can be considered.

11.3 Multiprotocol over ATM

The main goal of the **multiprotocol over ATM** (MPOA) working group of the ATM Forum is to define a service which provides a seamless end-to-end connection at the networking layer, that is, layer 3 of the OSI reference model. Under the classical networking paradigm end-systems within the same (sub)network could reach each other directly whereas devices in other networks can only be reached via a router. As mentioned in Sections 9.1 and 11.2, using LANE or 'classical IP over ATM' respectively, incurs some scaling and performance limitations, as is the case for any bridged network. Communication over several subnetworks requires routers which are potential bottlenecks. Even with high performance engines such effects as latency still exist since first a connection needs to be established to a router which then has to set up a second connection to the destination. This processs may consume even more time if more than a single router gets involved in the connection establishment (that is, multiple hops are necessary).

Another problem of bridged networks is the high amount of broadcast traffic which burdens every attached device.

With LANEv1, use of ATM's property of supporting different QoS levels cannot be exploited and each network layer 'flow' requires the establishment of a separate VCC. ATM also allows much larger **maximum transmission units** (MTUs) to be used compared to, for example, Ethernet. Although some of these points are resolved with LANEv2, and the IETF has defined solutions for some of these problems (for example, RFC 1577, 1483 NHRP (IETF, 1994a, 1993, 1996k)), MPOA attempts to combine those solutions to provide an integrated solution. Its primary goal is to allow optimal, direct end-to-end communication over several subnetworks without crossing any router.

Whereas the IP protocol was designed to be able to run over almost any network, ATM being only one possibility among others, MPOA as defined by the ATM Forum is designed to run any layer 3 protocol over ATM, but over ATM only.

Although this MPOA framework allows any existing network layer protocol to be used, such as IPX, AppleTalk or DECnet, the ATM Forum focused its initial work on supporting TCP/IP networks.

In the context of MPOA and throughout the remainder of this section the notion of an ELAN is used instead of 'layer 3 subnetwork'. The mechanisms described are based on MPOA Version 1.0 (ATM Forum, 1997k).

11.3.1 The basic principles

MPOA makes use of existing networking solutions as far as possible. It provides an evolutionary path for LANE and uses the models defined by the IETF. Data frames are LLC/SNAP encapsulated and conveyed via (AAL 5) VCCs, which are established using UNI 3.0/3.1/4.0 signaling as described in Chapter 8. MPOA is an example of the so-called 'overlay model' where layer 3 addresses are assigned independently of ATM addresses and where routing protocols such as OSPF or BGP run separately from ATM routing such as PNNI.

For resolving layer 3 to ATM addresses MPOA makes use of the NHRP and its NHSs. Furthermore, for multicast traffic the MARS protocol with its MCSs might be used.

MPOA reduces overhead and broadcast traffic while supporting larger MTU sizes and making full use of ATM's QoS capabilities. In addition to these remarkable assets MPOA builds upon a **distributed router** or **virtual router** model.

What does this mean? Looking at a traditional router two distinctive main tasks can be identified:

- Route processing, that is, running routing protocols like OSPF or BGP to obtain topology information and calculate the next hop route. This process is normally performed by dedicated software running on high performance processors.
- Data forwarding, that is, transferring incoming packets to the right output port based on the address information contained in the data frame and the routing information obtained through the route processing function. Typically, specialized hardware is used for data forwarding.

MPOA makes an attempt to distribute these two main router functions on discrete devices, namely **MPOA clients** (MPCs) for switching and forwarding, and **MPOA servers** (MPSs) for building routing tables.

An MPOA client communicates with its MPOA server via defined protocols. MPCs build the endpoints of a communication association. They may reside in MPOA-capable ATM hosts or in devices connecting legacy LAN equipment with an ATM network, the latter also being known as **edge devices**. An MPC can be considered to be an intelligent multilayer switch since it evaluates either the destination MAC address or the destination network layer address to forward data frames. However, MPCs do not run routing protocols.

An MPS is a route server comprising the set of functions to build routing tables and to be able to answer address queries. It maintains MAC-layer, network-layer and ATM-layer address mappings and runs a routing protocol such as OSPF or BGP or a **routing information protocol** (RIP) to communicate with traditional routers. An MPS can be considered to be an enhanced NHS. The routing information calculated by an MPS is then forwarded to an MPC.

The MPS can be realized in a dedicated physical device or be integrated into existing routers or switches.

The following analogies with traditional networking techniques can be made: an MPC within an edge device can be considered to be a router's interface card connecting legacy LAN stations. The ATM switching fabric constitutes a router's backplane and the MPS route server is the router's control processor.

11.3.2 Principles of data flow

As long as a data flow stays within a single ELAN, communication is based on LANE principles as described in Section 11.2. Whenever an inter-ELAN communication is necessary two possibilities exist to forward data: either on a hop-by-hop basis from router to router or using a short-cut connection which directly connects two MPCs. By default, any data frame is bridged via LANE to the default MPS which acts as a 'traditional' router. However, as soon as an MPS detects a 'flow' of packets, that is, several subsequent packets with the same layer 3 address within a given time interval, it attempts to establish a cut-through connection. If the destination ATM address for this layer 3 address is not known, an MPOA resolution request needs to be issued first. This resolution request is resolved according to the principles of the NHRP as outlined in Section 9.2. Finally, the source MPC establishes a cut-through VCC to the destination MPC by means of signaling procedures.

Figure 11.5 illustrates the default path and the cut-through connection finally established between two MPOA hosts. Remember, an ELAN is only a logical entity; physically all three ELANs can reside on a common ATM network.

11.3.3 MPOA control flows

As shown in Figure 11.6 four configuration flows can be differentiated. Flow (d) is used to obtain configuration information from the LECS.

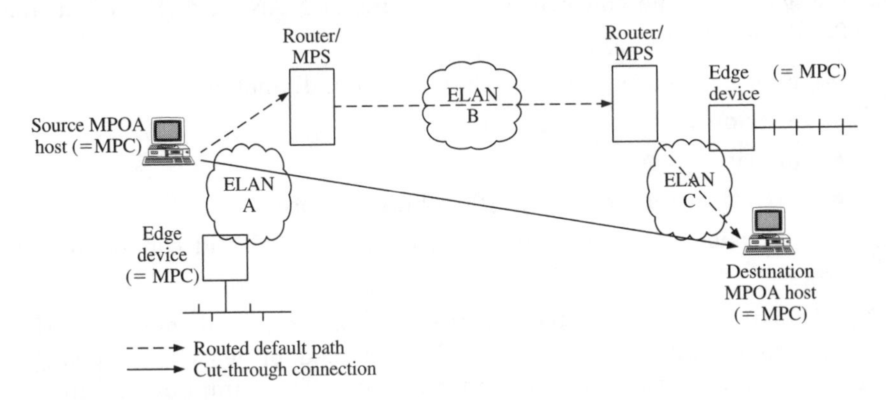

Figure 11.5 Routed default path and cut-through VCC in an MPOA environment.

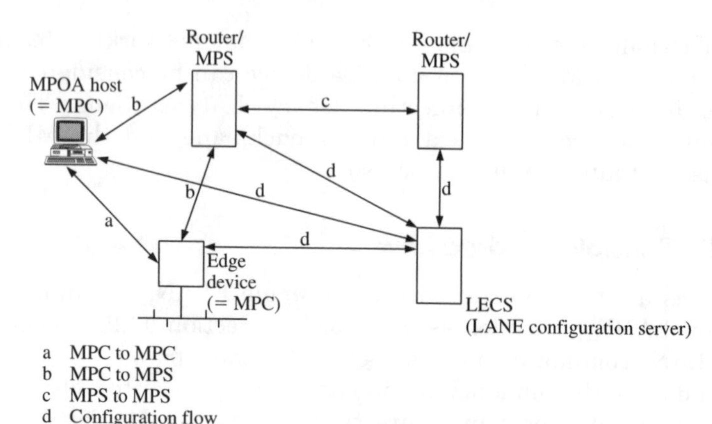

Figure 11.6 MPOA control flows.

MPOA components connect to a LECS as described for LANE. MPOA address resolution requests/replies are sent via flow (b). Among MPSs the standard routing protocols and the NHRP are operated via flow (c). If an MPC keeps invalid or wrong cache information, this might be detected by another MPC which can use flow (a) to purge this entry.

MPCs and MPSs automatically discover each other using a modified LANE ARP protocol. This reduces the administrative and operational overhead.

11.4 Metropolitan area networks

The increasing demand for data communication beyond the local area has led to the introduction of MANs. These can be considered as an evolution of LANs, and their main application is the interconnection of existing LANs. The characteristic features of MANs are (Biocca et al., 1990; Zitzen, 1990):

- coverage of areas of more than 50 km in diameter
- sharing a common medium
- distributed access control
- high transmission rates (100 Mbit/s or more)
- provision of isochronous, connection-oriented and connectionless services.

MAN installations can be divided into *private* and *public* MANs (Mollenauer, 1988). A private MAN is owned or leased by a single customer and carries that customer's traffic. This simplifies billing as well as privacy and security functions.

However, many MANs are shared by a number of customers. From the point of view of the network operator, accurate billing and additional management functions are essential. Security and privacy are also serious issues because customers would not like to see their confidential data passing through the building of a competitor.

11.4.1 Fiber-distributed data interface II

The **fiber-distributed data interface II** (FDDI-II) (ANSI, 1990) can be used for MAN implementation. This is an enhanced version of FDDI (ISO, 1989/90) which is a ring system using an optical fiber for transmission with a data rate of 100 Mbit/s. Access to the ring is controlled by a modified token-passing protocol which is specially designed for high-speed transmission systems.

For enhanced reliability, two rings with opposite transmission directions are used, making the system capable of surviving a cable break or station failure. In the event of such a failure, only one ring will be used. The total length of both rings is limited to 200 km. Up to 1000 stations with a maximum distance between stations of 2 km can be attached to the ring.

FDDI supports only packet-switched traffic. In addition, FDDI-II can handle isochronous traffic, including voice or video.

11.4.2 Distributed queue dual bus

The **distributed queue dual bus** (DQDB) (standardized for MAN/LAN applications) (IEEE, 1991) is the result of the continued development of the **queue packet and synchronous circuit exchange** (QPSX) (Newman et al., 1988). Isochronous, connection-oriented and connectionless services can be supported simultaneously.

The DQDB MAN consists of two unidirectional buses with opposite transmission directions to which multiple nodes are attached (see Figure 11.7). DQDB is independent of the underlying physical medium. This allows the use of existing PDH systems with transmission rates of, for example, 34, 45 and 140 Mbit/s (ITU-T, 1988a), as well as

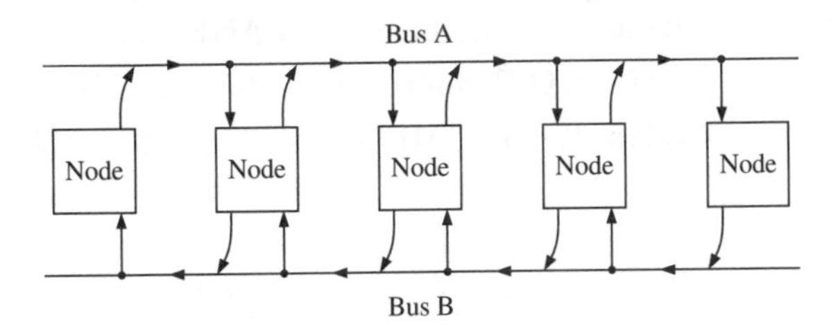

Figure 11.7 Distributed queue dual bus network.

SDH-based transmission systems according to ITU-T (1996c) and possibly future systems with transmission rates in the Gbit/s range.

The dual buses can be looped or open ended. In the looped bus case the head and tail of each bus are co-located but not interconnected. Looping allows reconfiguration of the bus system in the event of bus failure. Head and tail will then be interconnected, while new head and tail points will be generated close to the location of the failure. Each node can act as head or tail. This self-healing mechanism makes the looped bus configuration the preferred choice.

All information in the DQDB network is transported within **slots**. A slot consists of a header with five octets and an information field with 48 octets – the same as in an ATM cell. The commonality of DQDB PDUs and ATM PDUs simplifies the interconnection of these two networks.

A slot generator is located at the head of each bus. It creates empty slots and writes them on the bus. Isochronous services use **pre-arbitrated** slots. These slots are marked by the slot generator and have the appropriate VCI value inserted in the slot header.

All non-isochronous information is transported within **queue-arbitrated** slots. Pre-arbitrated and queue-arbitrated slots are distinguished by different values in the *slot type* field of the slot header. Queue-arbitrated slots are managed by the distributed queuing protocol (media access control). In contrast to the existing MAC procedures in a distributed queue system, each station buffers the actual number of slots waiting for access to the network. With this in mind, a station which has a slot ready to send determines its own position in the distributed queue. Its own slot will be transported after satisfying the queued slots.

Review questions

1. Which advantages can an ATM-based LAN provide in contrast to legacy LAN technologies?
2. Why is there a need for LANE?
3. Which layer of the OSI reference model is emulated by LANE?
4. What are the basic building blocks of LANE?
5. At which layer of the OSI reference model does MPOA operate?
6. What are the advantages of MPOA compared to LANE?

12

ATM switching

Two main tasks can be identified for an ATM switching or cross-connect node:

- VPI/VCI translation
- cell transport from its input to its dedicated output.

A switch fabric is necessary to establish a connection between an arbitrary pair of inputs and outputs within a switching node. In principle, a switch fabric can be implemented by a single switching element. Since such an element could not satisfy the requirements of a normal-size ATM switching node, larger switch fabrics are used, built up from a number of switching elements.

The throughput of a switching node will be in the Gbit/s range and the cross-node delay and cell loss should both be kept very low. Therefore, central control cannot be used to switch cells. Only switch fabrics with highly parallel architectures can meet these stringent requirements.

12.1 Switching elements

A **switching element** is the basic unit of a switch fabric. At the input port the routing information of an incoming cell is analyzed and the cell is then directed to the correct output port. In general, a switching element consists of an interconnection network, an input controller (IC) for each incoming line and an output controller (OC) for each outgoing line (see Figure 12.1). To prevent excessive cell loss in the case of internal collisions (two or more cells competing for the same output simultaneously), buffers have to be provided within the switching element.

Arriving cells will be synchronized to the internal clock by the IC. The OC transports cells which have been received from the interconnection network towards the destination. ICs and OCs are coupled by the interconnection network.

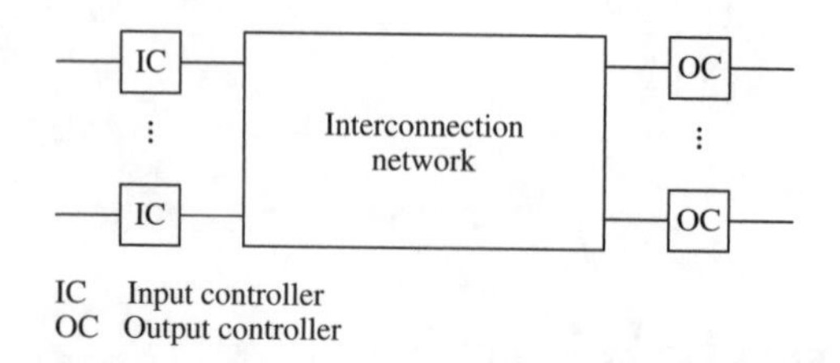

Figure 12.1 General model of a switching element.

12.1.1 Matrix-type switching elements

An internally non-blocking switching element can be constructed by using a rectangular matrix of crosspoints for the interconnection network (see Figure 12.2). It is always possible to connect any idle input/output pair. Whether or not a crosspoint connects an input to an output depends on the routing information of the cell as well as the occurrence of collisions.

Various buffer locations are possible within this switching element (Karol et al., 1987; Rathgeb et al., 1988, 1989):

- at the input controllers
- at the output controllers
- at the crosspoints.

Input buffers

The cell buffers are located at the input controller (see Figure 12.3). When using first-in first-out (FIFO) buffers, a collision occurs if two or more head-of-the-queue cells compete simultaneously for the same output. Then all but one of the cells are blocked. Cells following the

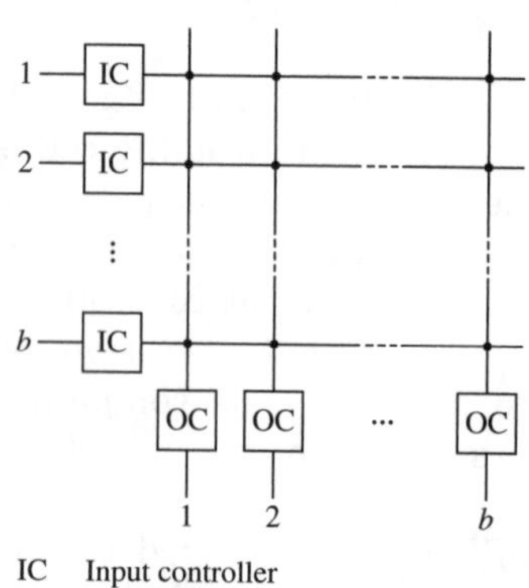

Figure 12.2 Matrix-type switching element.

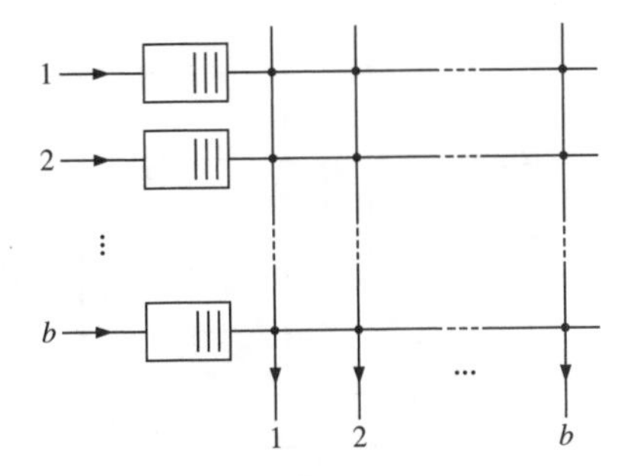

Figure 12.3 Switching matrix with input buffers.

blocked head-of-the-queue cell are also blocked, even if they are destined for another, available output.

To overcome this disadvantage, FIFO buffers can be replaced by random access memory (RAM) (Rothermel and Seeger, 1988). If the first cell in the buffer is blocked, the next cell destined for an idle output will be selected for transmission. However, this operation mode requires more complex buffer control to find a cell destined for an idle output, and to guarantee the correct sequence of cells destined for the same output. The total buffer capacity will logically be subdivided in a load-dependent manner into single FIFOs (one FIFO for each output).

Further enhancements can be achieved if more than one cell can be transferred simultaneously from one buffer to different outputs. This requires a buffer with multiple outputs (Lampe, 1988; Lutz, 1988) or a buffer with reduced access time.

Output buffers

Figure 12.4 shows a switching element consisting of a matrix with output buffers. Only if the matrix operates at the same speed as the incoming lines can collisions occur (several cells hunting simultaneously for the same output). This drawback can be compensated by reducing the buffer access time and by speeding up the switching matrix. These factors may lead to technological limitations on the size of the switching element.

A switching element with output buffers will be non-blocking only if the speed-up factor of the matrix is b (that is, b cells simultaneously hunting for the same output can be switched) for a $b \times b$ switching element. In all other cases, additional buffers are necessary at the input to avoid cell loss caused by internal blocking.

Crosspoint buffers

The buffers can also be located at the individual crosspoints of the matrix (see Figure 12.5). This is called a **butterfly** switching element

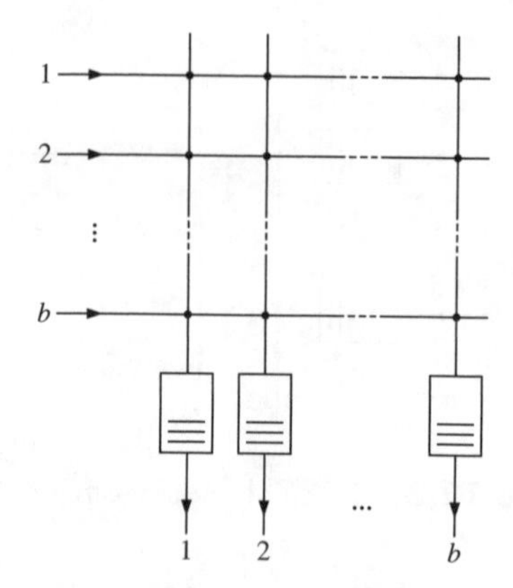

Figure 12.4 Switching matrix with output buffers.

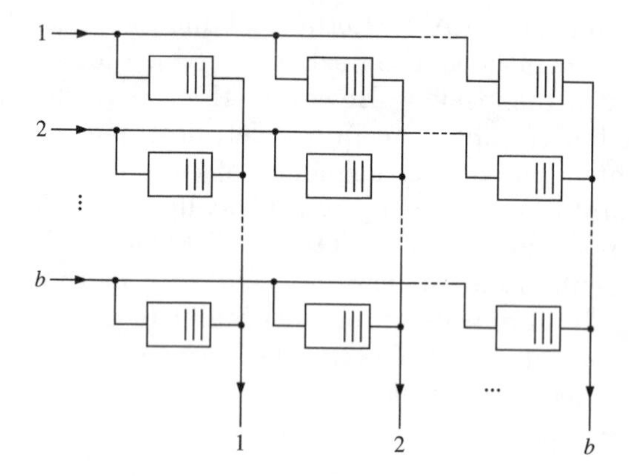

Figure 12.5 Switching matrix with crosspoint buffers.

(Brooks, 1987). It prevents cells that are hunting for different outputs affecting one another. If there are packets in more than one buffer belonging to the same output, control logic has to choose which buffer will be serviced first.

From the performance point of view, this buffer location strategy has the drawback that a small buffer is required at each crosspoint and no buffer sharing is possible. Therefore, it is not possible to achieve the same efficiency that a switching element with output buffers provides.

Arbitration strategies

If several cells compete simultaneously for the same output, only one cell can be transferred and all other cells will be delayed. An arbitration strategy is required to determine the 'winning' cell. Objectives for such a mechanism can be fairness or minimization of cell loss, or minimization of cell delay variations. The following strategies can be applied (Huber et al., 1988; Rathgeb et al., 1989):

- **Random:** The line which will be serviced first is chosen at random from all lines competing for the same output. This strategy requires only a small amount of implementation overhead.
- **Cyclic:** The buffers are serviced in a cyclic order. This approach also requires only a small overhead.
- **State dependent:** The first cell from the longest queue will be serviced first. For this algorithm, the lengths of the buffers hunting for the same output have to be compared.
- **Delay dependent:** This is a global FIFO strategy taking into account all the buffers that feed one output. However, it implies some overhead to store the relative order of arrival for competing cells.

A performance comparison of the different arbitration mechanisms is shown in Figure 12.6 (see also Rathgeb et al., 1989).

The random strategy has the highest delay variations. An insignificant improvement can be achieved by using the cyclic strategy. The optimum strategy with respect to cell delay variation is the delay-

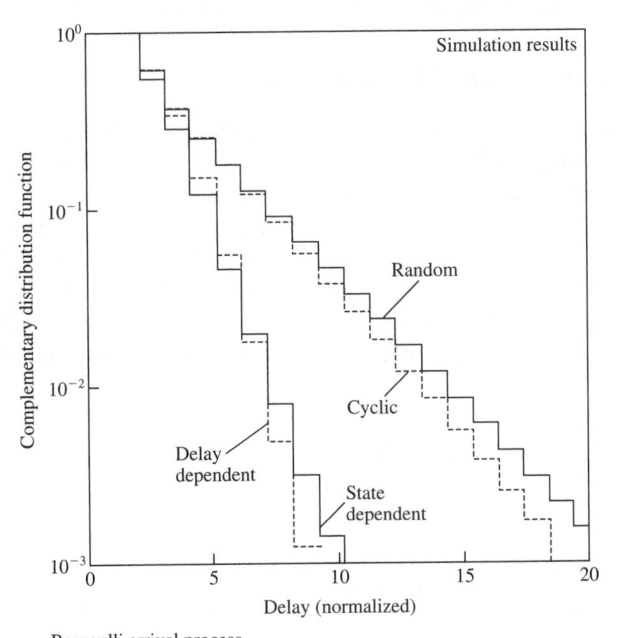

Figure 12.6 Influence of different arbitration strategies on the delay.

dependent strategy. Minimum cell loss can be achieved by implementing the state-dependent algorithm. The performance with respect to the delay requirements of this strategy is slightly worse, but still acceptable.

12.1.2 Central memory switching element

The principle of a central memory switching element is shown in Figure 12.7. All input and output controllers are directly attached to a common memory which can be written to by all input controllers and read by all output controllers.

The first example of such a switching element was used in the PRELUDE experiment (Coudreuse and Servel, 1987). The common memory can be organized to provide logical input as well as logical output buffers. Research and Development of Advanced Communication in Europe (RACE) project 1012 'Broadband Local Network Technology' used the *Sigma* switch, which was based on a common memory structure with logical output buffers. A possible realization of such a switching element is described in Section 12.2.3.

Since all the switching element buffers share one common memory, a significant reduction of the total memory requirements can be achieved in comparison with physically separated buffers. On the other hand, a high degree of internal parallelism is necessary to keep the frequency of memory access within a realizable range.

12.1.3 Bus-type switching element

The interconnection network can be realized by a high-speed **time-division multiplexing** (TDM) bus (see Figure 12.8). Conflict-free transmission can only be guaranteed if the total capacity of the bus is at least the sum of the capacities of all input links (De Prycker and De Somer, 1987). Bit-parallel data transmission (for example, 16 or 32 bit) on the bus system is required to achieve this high capacity.

Normally, a bus access algorithm is applied which allocates the bus to the individual input controllers at constant intervals. Each input

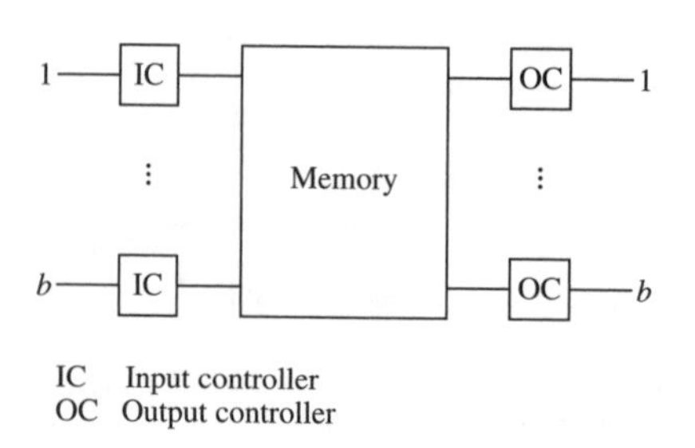

Figure 12.7 Central memory switching element.

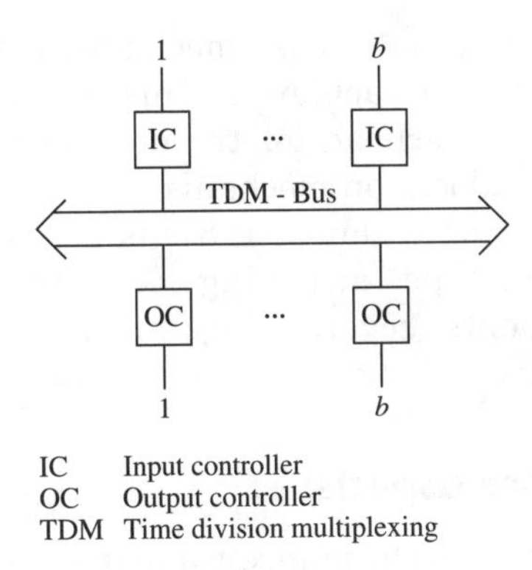

Figure 12.8 Bus-type switching element.

controller is able to transfer its cell towards the destination before the arrival of the next cell is completed. No buffers are required at the input controller. However, several cells may arrive at the same output controller, whereas only one cell can leave the controller. Therefore, buffers are required at the output controller. This switching element has the same performance as the matrix-type switching element with output buffers.

12.1.4 Ring-type switching element

The ring-type switching element is shown in Figure 12.9. All input and output controllers are interconnected via a ring network, which should be operated in a slotted fashion to minimize the overhead. In principle, a fixed time-slot allocation scheme can be used but this requires a ring capacity which is the sum of the capacities of all input links. If the ring capacity is less than the total input capacity, a flexible allocation scheme is necessary, which results in an additional overhead.

The ring structure has the advantage over the bus structure in that a time-slot can be used several times within one rotation. However, this requires that the output controller empties a received time-slot.

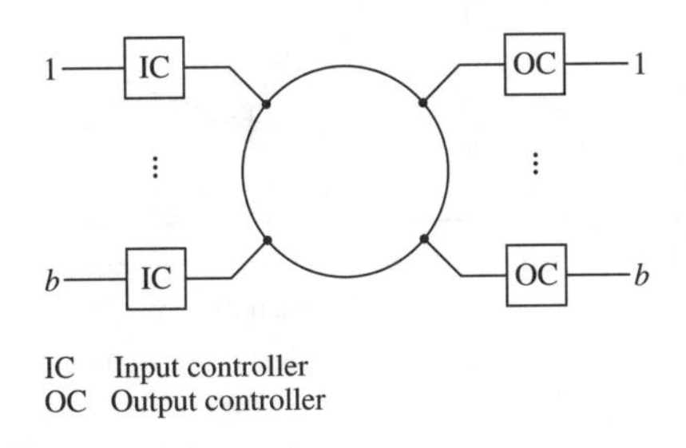

Figure 12.9 Ring-type switching element.

When using this destination release mechanism, an effective utilization of more than 100% can be achieved. This advantage has to be offset against the additional overhead for the destination release mechanism and flexible time-slot allocation mechanism.

The ORWELL ring (Adams, 1987) is one approach to the implementation of a ring-type switching element. To meet the high throughput requirements, several rings are used in parallel forming a so-called torus of rings.

12.1.5 Performance aspects

Many publications offer performance comparisons of different buffering strategies (Karol et al., 1987; Lutz, 1988; Oie et al., 1989; Rathgeb et al., 1988, 1989). Figure 12.10 (see also Rathgeb et al., 1989) shows the mean cell delay of a 16×16 switching element with different buffer locations and buffer operation modes.

The results for the switching element with a simple FIFO input buffer correspond to the curve with speed-up factor 1. A speed-up factor i means that the buffer access time is reduced by the factor i, or i cells from one buffer can be transferred simultaneously. The maximum throughput

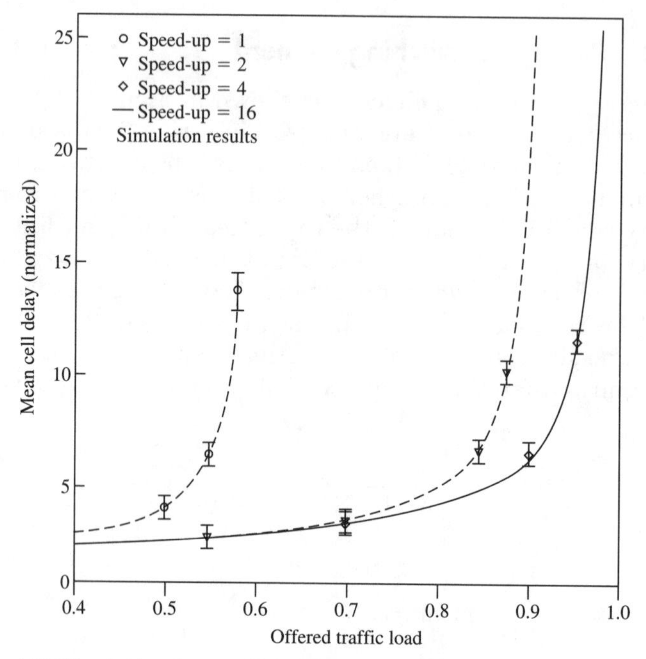

Figure 12.10 Performance comparison of buffering strategies.

of this element is limited to about 58% of the total capacity (Karol et al., 1987). A good performance improvement will be achieved using a speed-up factor of 2. The best results are obtained for switching elements with a speed-up factor of 16. In this case the behavior is identical to a switching element with output buffers. However, the ideal throughput can almost be obtained even with a speed-up factor of 4 (Rathgeb et al., 1989). This may significantly simplify the implementation.

12.1.6 Technological aspects

The performance results shown in Figure 12.10 are valid for infinite buffer sizes. However, only finite buffer sizes are possible. Table 12.1 presents the buffer sizes (in cells) for switching elements of different sizes and different buffer locations, assuming an average load of 85% at each input and a permissible cell loss probability of 10^{-9} (Lutz, 1988).

The central memory switching element requires the least memory capacity as a result of buffer sharing. The required memory capacity for a switching element with input buffers is low compared with the switching element with output buffers. This can be ascribed to the fact that in a switching element with input buffers only, one cell can be written into the buffer and several cells can be read from the buffer, whereas in the switching element with output buffers, several cells can simultaneously arrive but only one cell can leave.

In the following discussion it is assumed that a 16×16 or 32×32 switching element can be implemented in a single integrated circuit with complementary metal-oxide semiconductor (CMOS) or bipolar CMOS (BICMOS) technology. The chip area can be subdivided into the memory part and the random logic part (for example, serial-to-parallel converter). The memory area is smallest in the central switching element, whereas the required random logic area will be, without any doubt, larger than for the other two types.

Figure 12.11 shows the relationship between chip size and power dissipation for the three types of switching element (switching matrix with output buffer, switching matrix with input buffer and central memory switching element). It must be pointed out that specific implementation principles and a certain type of CMOS technology were used, which is why no scales are provided on the axes.

Table 12.1 Buffer size requirements.

	Size	
Type	*16 × 16*	*32 × 32*
Central memory	113	199
Input buffer	320	640
Output buffer	896	1824

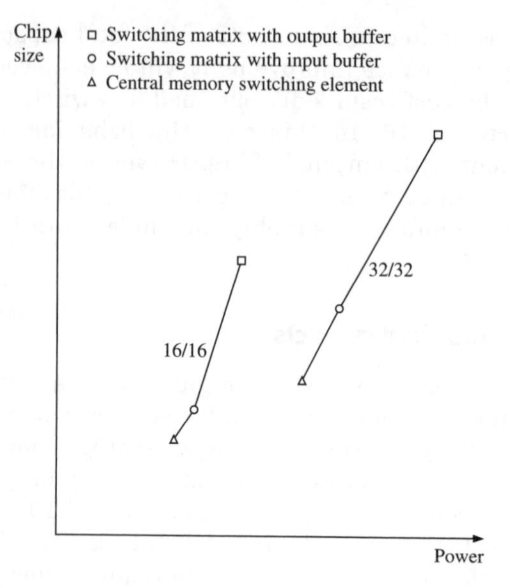

Figure 12.11 Relationship between chip size and power dissipation.

While the power dissipation of CMOS memories is very small, high-speed random logic consumes a relatively high amount of power. This results in a nonlinear relationship between chip size and power dissipation of the switching elements considered. From all these discussions it is evident that, as regards chip size and power dissipation, the central memory switching element has clear advantages, while the switching element with output buffers is the least favorable.

12.2 Switching networks

This section deals with the general classification of switching networks. Existing and proposed ATM switch architectures of various manufacturers and research institutes will not be presented.

Figure 12.12 gives an overview of the networks presented.

12.2.1 Single-stage networks

A single-stage network is characterized by a single stage of switching elements which are connected to the inputs and outputs of a switching network.

Extended switching matrix

Figure 12.13 shows an example of an extended switching matrix which is formed from $b \times b$ switching elements. Basically, any desired size of switching network can be implemented with this approach.

To realize an extended switching matrix the switching elements described in Section 12.1 have to be extended by adding b inputs and b outputs. Input signals are relayed to the next column of the matrix via the additional outputs. The additional inputs are connected to the normal outputs of the switching element in the same column but in the row above.

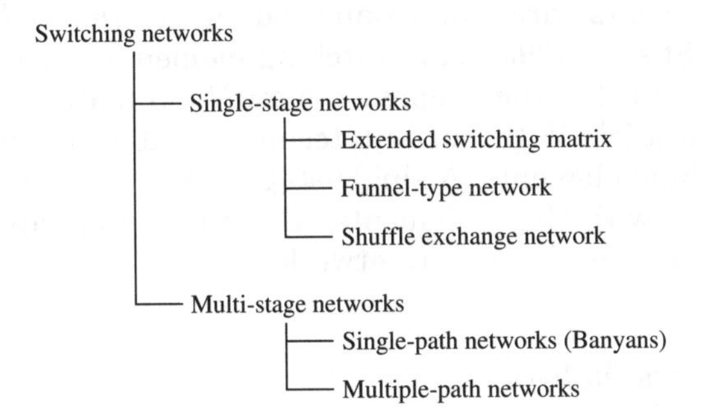

Figure 12.12 Classification of switching networks.

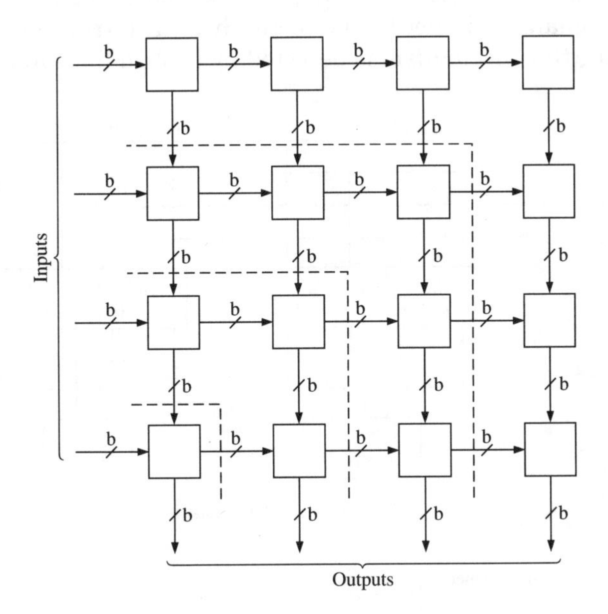

Figure 12.13 Extended switching matrix.

The advantage of the extended switching element is the small cross-delay because cells will only be buffered once when crossing the network. It should be noted that the cross-delay is dependent upon the location of the input. The fact that the number of switching elements increases with the number of required inputs limits the size of an extended switching matrix. It is certainly possible to form a 64×64 or 128×128 single-stage network, but multi-stage networks will be preferred for larger systems.

Funnel-type network

In the $N \times N$ non-blocking switching network shown in Figure 12.14, switching elements are interconnected in a *funnel-like* structure (Fischer and Stiefel, 1993). All switching elements consist of $2b$ inputs and b outputs. Each funnel represents an $N \times b$ matrix, of which there are N/b in parallel. With current technology it is possible to realize 32×16 switching elements. A single-stage 128×128 switching network can be realized with these elements. The multi-stage approach will be the preferred solution for larger networks.

Shuffle exchange network

The **shuffle exchange** network (Chen et al., 1981) belongs to the class of single-stage networks. It is based on a perfect shuffle permutation which is connected to a stage of switching elements (see Figure 12.15). A feedback mechanism is necessary to reach an arbitrary output from a given output (this mechanism is depicted by dashed lines in the figure.

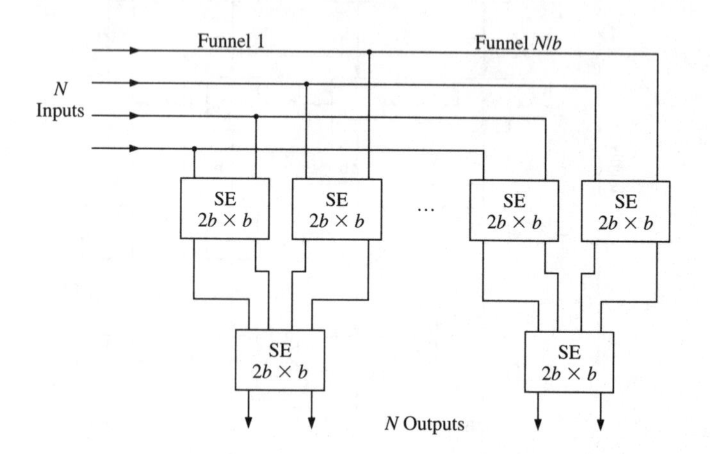

Figure 12.14 Funnel-type network.

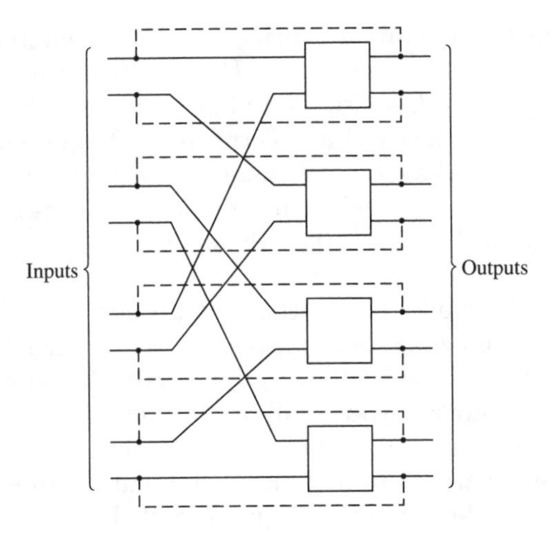

Figure 12.15 Example of a shuffle exchange network.

It is evident that a cell may pass through the network several times before reaching its proper destination. Therefore, this is also called a **recirculating** network. At the output of a switching element it has to be decided whether a cell can leave the network or has to be fed back to the input.

This type of network requires only a small number of switching elements but the performance is not very good. The cross-delay depends on the number of switching elements that have to be passed through.

12.2.2 Multi-stage networks

Multi-stage networks can be used to avoid the drawbacks of single-stage networks. They are built of several stages which are interconnected by a certain link pattern. According to the number of paths which are available for reaching a destination output from a given input, these networks can be subdivided into two groups called **single-path** and **multiple-path** networks.

Single-path networks

In single-path networks, there is only one path to the destination from a given input. These networks are also called **Banyan** networks (Goke and Lipovski, 1973). Because only one path exists to the proper output, routing is very simple. Banyan networks have the disadvantage that internal blocking can occur. This results from the property that an internal link can be used simultaneously by different inputs. According to Huber et al. (1988), Banyan networks can be classified into subgroups.

In **(L)-level** Banyan networks, only the switching elements of adjacent stages are interconnected. Each path through the network passes through exactly L stages. Furthermore, this class is subdivided into **regular** and **irregular Banyans**. Regular Banyans are constructed of identical switching elements, whereas irregular Banyans can use different types of switching element. The **generalized delta** network (Dias and Kumar, 1984) belongs to the class of irregular Banyans.

Regular Banyans have the advantage that they can be implemented economically because they are constructed of identical switching elements. In the following, only **SW-Banyans**, which are a subclass of the regular Banyans, will be considered.

SW-Banyans can be constructed recursively from the basic switching element with F input links and S output links. The simplest SW-Banyan is a single switching element (called a (1)-level Banyan). An (L)-level SW-Banyan will be obtained by connecting several (L – 1)-level SW-Banyans with an additional stage of ($F \times S$) switching elements. These extra switching elements are connected in a regular manner to the SW-Banyans.

Delta networks (Patel, 1981) are a special implementation of SW-Banyans. L-level networks which are constructed of ($F \times S$) switching elements have S^L outputs. Each output can be identified by a unique destination address, which is a number of base S with L digits. Each digit specifies the destination output of the switching element in a specific stage. This allows simple routing of cells through the delta network, which is called **self-routing**.

In **rectangular delta** networks the switching elements have the same number of inputs and outputs ($S = L$). Consequently, the number of network inputs is equal to the number of network outputs. These networks are also called delta-S networks. Figure 12.16 shows a delta-2 network with four stages which has the topology of a **baseline** network (Wu and Feng, 1980). The thick line indicates the path from input 5 to output 13 (binary destination address 1101).

Bidelta networks are a special class of delta networks. They remain delta networks even if the inputs are interpreted as outputs and vice versa. All bidelta networks are topologically equivalent and can be transformed into each other by renaming the switching elements and the links (Dias and Kumar, 1984).

Multiple-path networks

In multiple-path networks, a multiplicity of alternative paths exist to the destination output from a given input. This property has the advantage that internal blocking can be reduced or even avoided.

In most multiple-path networks the internal path will be determined during the connection set-up phase. All cells on the

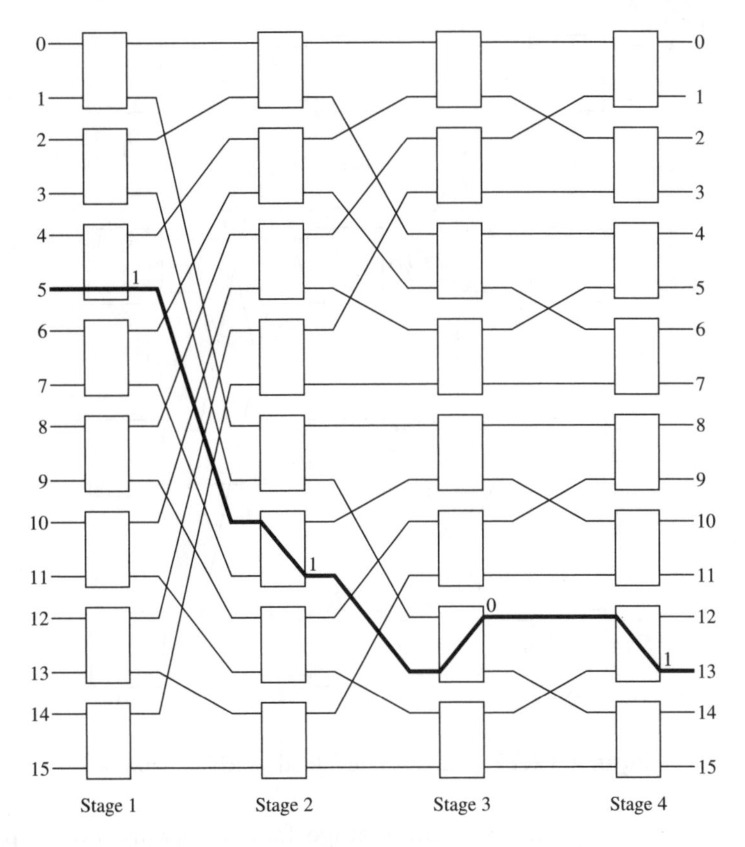

Figure 12.16 Delta-2 network with four stages.

connection will use the same internal path. If FIFOs are provided in the individual switching elements, cell sequence integrity can be guaranteed and no resequencing is necessary.

In this classification, multiple-path networks can be subdivided into **folded** and **unfolded** networks.

Figure 12.17 shows a three-stage folded network. In folded networks, all inputs and outputs are located at the same side of the switching network and the network's internal links are operated in a bidirectional manner (each link in Figure 12.17 represents the physical lines for both directions).

Folded networks have the advantage that short paths (Lutz, 1988; Theimer, 1991) can be used. For example, if the input line and the output line are connected to the same switching element, cells can be reflected at the switching element and need not be passed to the last stage. The number of switching elements that the cells of a connection have to pass through depends on the location of the input and output lines.

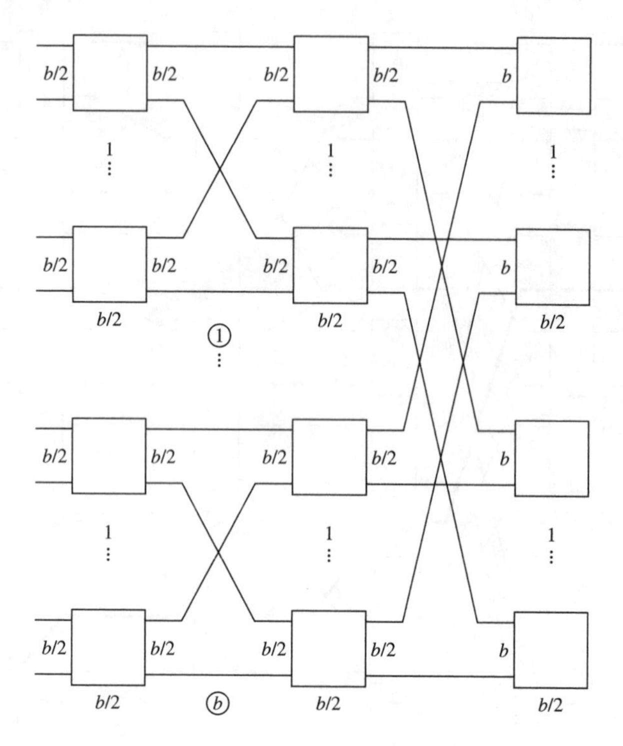

Figure 12.17 Three-stage folded switching network.

The port capacity of a three-stage folded network built up from $b \times b$ switching elements is $(b/2)(b/2)b$. With today's technology, switching elements of size 16×16 and 32×32 can be realized, leading to three-stage networks with 1024 and 8192 ports, respectively.

In unfolded networks, the inputs are located on one side and the outputs on the opposite side of the network. Internal links are unidirectional and all cells have to pass through the same number of switching elements.

Multiple-path unfolded network structures will be based on single-path network structures. Again, the basis for these networks is $b \times b$ switching elements. For simplicity, only 2×2 switching elements are presented in the figures.

Turner (1987) describes a switching network which consists of a buffered Banyan network and a preceding **distribution network** (see Figure 12.18). The distribution network has the purpose of distributing cells as evenly as possible over all inputs of the Banyan network. This approach can reduce internal blocking. However, the cell sequence integrity of a connection cannot be maintained and therefore an additional resequencing mechanism is required at the output.

Another realization of such networks is the combination of a **sorting network** (Batcher, 1968) and a **trap network** in front of a

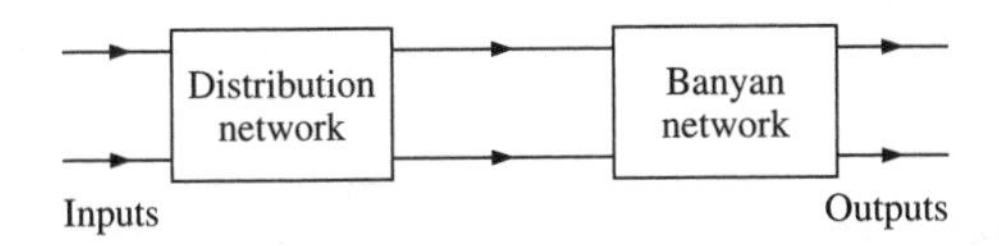

Figure 12.18 Basic structure of a distribution/Banyan network.

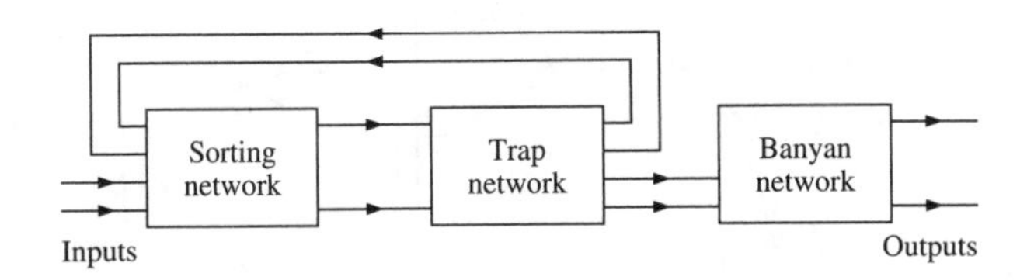

Figure 12.19 Basic structure of a sorting/trap/Banyan network.

Banyan network (Huang et al., 1984) (see Figure 12.19). The sorting network arranges arriving cells in a monotonous sequence depending on the network internal destination address. Cells with identical addresses are detected by the trap network and all but one of these cells are fed back to the input of the sorting network. Cells that have to pass through the sorting network again will be assigned a higher priority in order to maintain cell sequence integrity. Cells entering the Banyan network can be transported to their destination without any internal blocking.

Multiple-path networks can also be realized by using several planes of Banyan networks in parallel (see Figure 12.20). This is called **vertical stacking** (Lea, 1989).

All cells belonging to the same connection will pass through the same plane. This will be determined during the connection set-up phase. An incoming cell will be switched to its appropriate plane by the distribution unit which is located at each input line. At the switch output, a statistical multiplexer collects cells from all planes. In Theimer (1991) it is shown that even with two planes in parallel a virtually non-blocking switching structure can be achieved.

Adding a number of stages to a given Banyan network is called **horizontal stacking** (Lea, 1989). A **multi-path interconnection network** (MIN) (Amido and Seeto, 1988) is realized by adding a baseline network with reversed topology to an existing baseline network (see Figure 12.21). The baseline network has already been presented in Figure 12.16.

Assuming $b \times b$ switching elements the $N \times N$ network has $2 \log_b N$ stages. In an $N \times N$ network, N internal paths are available to a given output from an arbitrary input. From a particular input the internal path can be selected arbitrarily until the output of the baseline network is reached. Then the way through the reversed baseline network is fixed.

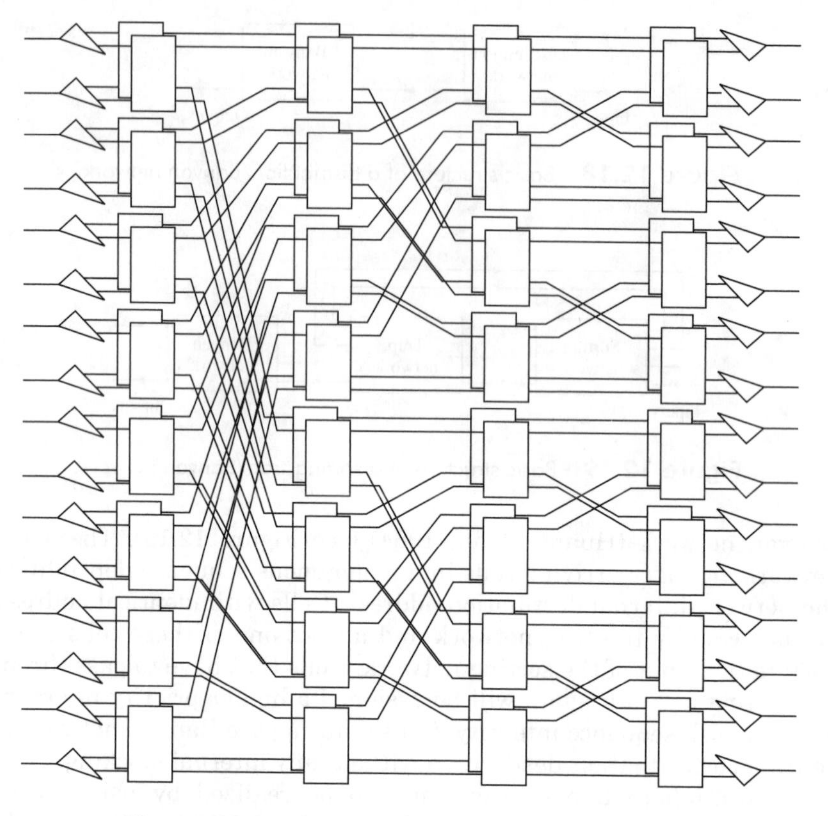

Figure 12.20 Example of a parallel Banyan network.

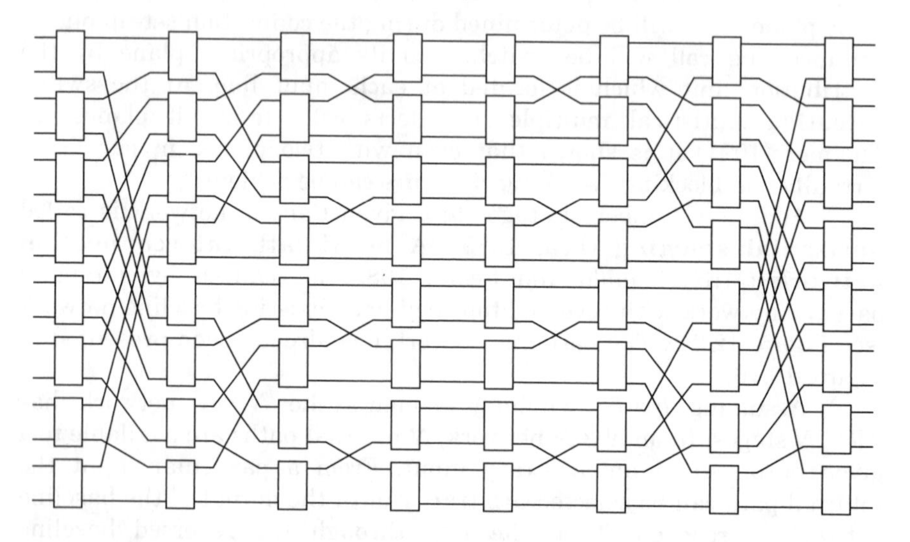

Figure 12.21 Example of a multi-path interconnection network.

A Beneš network (Beneš, 1965) is very similar to a MIN. The difference is that the last stage of the baseline network coincides with the first stage of the reversed baseline network. Therefore, the number of stages is reduced by one compared with the MIN. Figure 12.22 shows a seven-stage Beneš network.

Again, assuming $b \times b$ switching elements, only N/b alternative paths are available to any output from any input. Each path is uniquely determined by the switching element passed in the center stage.

12.2.3 Cell header processing in switch fabrics

The main tasks of ATM switching nodes are:

- VPI/VCI translation
- transport of cells from the input to the appropriate output.

In order to fulfill these tasks, two approaches can be applied (Schaffer, 1990):

- **self-routing principle**
- **table-controlled principle**.

Self-routing switching elements

When using self-routing switching elements, VPI/VCI translation only has to be performed at the input of the switching network. After translation the cell is extended by a switching network internal header.

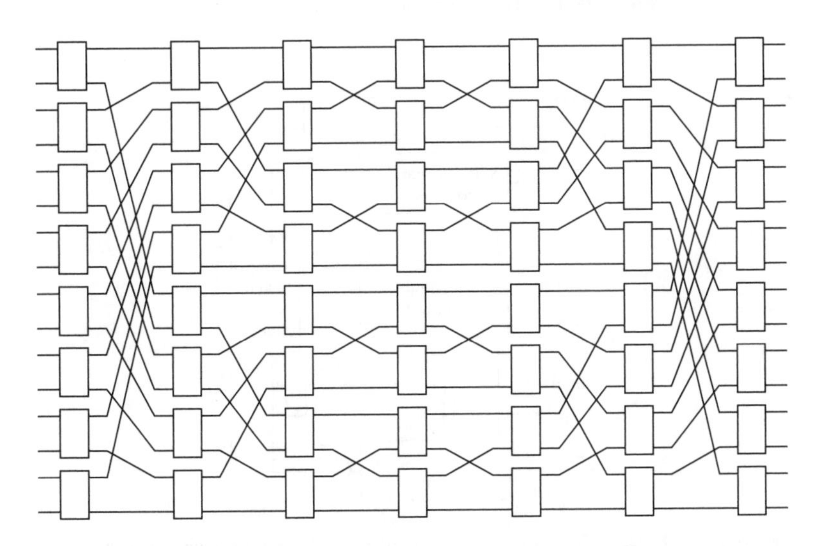

Figure 12.22 Example of a seven-stage Beneš network.

This header precedes the cell header. Cell header extension requires an increased internal network speed.

In a network with k stages the internal header is subdivided into k subfields. Subfield i contains the destination output number of the switching element in stage i. Figure 12.23 shows the cell header processing in a switching network built up of self-routing switching elements.

Figure 12.24 depicts a generic realization of the self-routing switching element (Fischer et al., 1991). It contains a central memory (see Section 12.1.2) with logical output queues and is controlled by the routing information included in the internal cell header. In order to keep the buffer access speed within the range given by technological constraints, a wide parallel memory interface is used requiring serial-to-parallel conversion at the inlet and parallel-to-serial conversion at the outlet. The central control with address queues assigned to each outlet takes care of correct delivery of the cells.

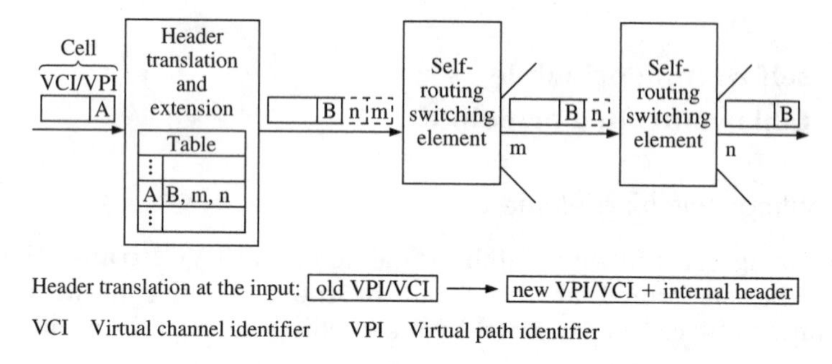

Figure 12.23 Self-routing switching elements.

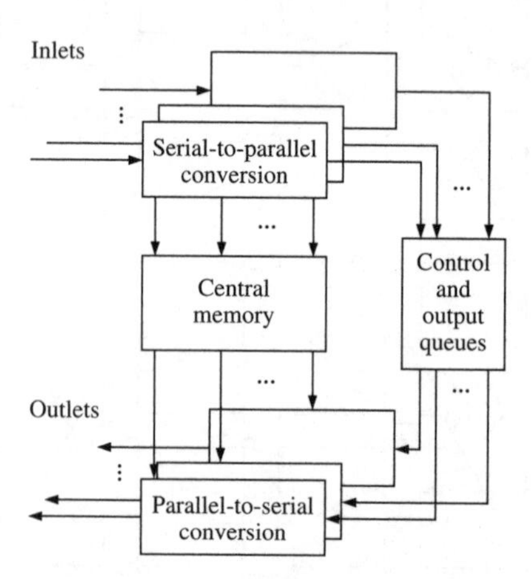

Figure 12.24 Generic realization of the self-routing switching element with central memory.

Table-controlled switching elements

When using the table-controlled principle, the VPI/VCI of the cell header will be translated into a new one in each switching element. Therefore, the cell length need not be altered. Figure 12.25 shows header processing in a switching network which consists of table-controlled switching elements.

The contents of the tables are updated during the connection set-up phase. Each table entry consists of the new VPI/VCI and the number of the appropriate output.

Extensive studies have been made to decide which principle is superior (Schaffer, 1990). For large multi-stage switching networks the self-routing principle will be preferred because it is superior in terms of control complexity and failure behavior. The need for a higher internal bit rate because of the cell extension is not critical.

12.2.4 Multicast functionality

Some data services and distributive services are characterized by point-to-multipoint communication. This capability can be supported by an ATM switch fabric. For this purpose an ATM switching element must be able to transmit copies of an incoming cell to different outlets. In the case of the shared buffer switching element which minimizes the number of buffers, a cell is only deleted from the memory when all the necessary copies have been transmitted through the outlets.

In Section 12.2.3 it was shown that the self-routing switching element offers a lot of advantages. However, the principle shown in Figure 12.23 results in a low performance for multicast communication. Header translation is performed at the input of the switching network. All cell copies made within the switch fabric possess the same VPI/VCI value. This is an unnecessary restriction which can be overcome by marking 'multicast' cells and performing header translation for that type of cell at the output of the switching network.

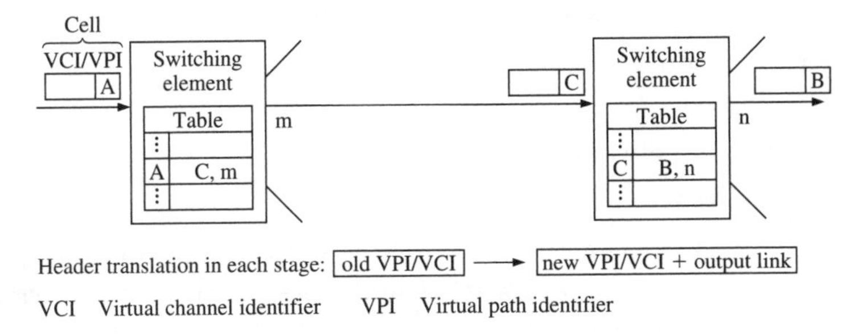

Figure 12.25 Table-controlled switching elements.

12.3 Switches and cross-connects

In the previous sections we have discussed the basic elements of ATM switches and cross-connects. However, more subsystems are required for the implementation of a switch/cross-connect. (The main difference between the switch and the cross-connect is related to the control function. A switch is under the control of signaling, whereas a cross-connect is controlled by network management.)

12.3.1 Generic system structure

The generic structure of an ATM system (switch or cross-connect) is shown in Figure 12.26 (Fischer and Stiefel, 1993). It has been designed according to the following principles:

- The system can be used either as a cross-connect or as a switch. The hardware can be identical, but the software will be different. This architecture also provides the possibility of performing switch and cross-connect functions in the same node.
- The switching network employs the self-routing principle because it is the most promising one (see Section 12.2.3).
- Connection-related information is stored in those peripheral units that are affected by the particular connection. This allows quick access to the per-connection information.
- The ATM switching network is non-blocking. It can easily be tested by sending node-internal test cells.
- A unique cell rate is used within the switching network/ multiplexer. Adaptation of different cell rates used on subscriber/trunk lines is performed in the interface modules.

12.3.2 System building blocks

The generic switch/cross-connect depicted in Figure 12.26 consists of the following modules:

- subscriber line module broadband (SLMB)
- trunk module broadband (TMB)
- multiplexer
- switching network
- system control.

A subscriber is connected to the switching network or multiplexer via an SLMB. Connection to other switches and cross-connects is performed via the TMB. It supports existing plesiochronous systems as well as SDH transmission with bit rates up to 2.4 Gbit/s.

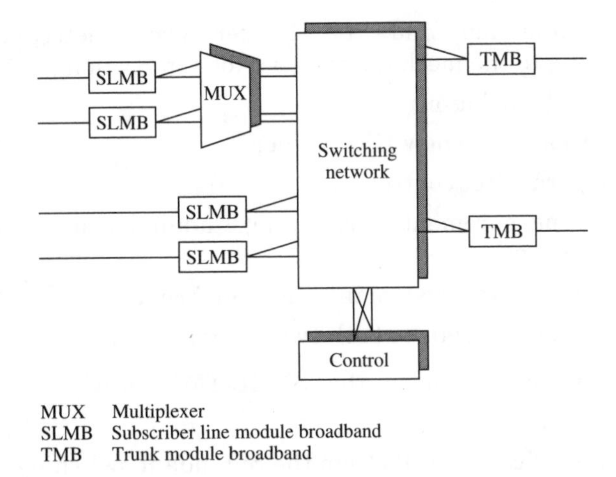

Figure 12.26 Generic switch/cross-connect architecture.

The multiplexer is used for local concentration of subscriber traffic within the switch.

The switching network connects the interface modules, the multiplexer and the control processor. It is also used for internal communication between the node subsystems.

The control processor is responsible for system control. It may also perform functions which are related to signaling or network management.

To achieve high system availability, the multiplexer, switching network and control processor are fully redundant. At the input an interface module (SLMB or TMB) sends copies of each cell to both multiplexer/switching network planes. At the output the interface modules decide from which plane a cell will be transmitted. More details on this node-internal redundancy concept are given in Fischer et al. (1991).

As mentioned in Section 12.3.1, a unique cell rate is used for the switching network and the multiplexer. Therefore, all peripheral units (SLMB, TMB and control processor) have the same type of interface to the multiplexer and switching network. A proprietary protocol with system internal cell formats is applied at these interfaces. The system internal cell format consists of the 53 octets of the standard cell (see Section 5.6.1), the internal routing information (see Section 12.2.3) and information for synchronization, housekeeping, control and test.

At the input of the node the SLMB/TMB perform the following functions:

- cell extraction, for instance from the SDH frame;
- rate adaptation to the node-internal speed;

- cell delineation and cell header error detection/correction according to the mechanisms described in Section 5.3.1;
- VPI/VCI translation;
- generation of the new HEC value;
- usage parameter control;
- traffic measurement for billing, administration and traffic engineering;
- generation of the system internal cell format;
- transmission of cells to both switching planes.

At the output of the node the SLMB/TMB provide the following functions:

- selection of correct cells from the redundant switching plane;
- conversion of the internal cell format to the standardized format;
- adaptation of the cell rate to the outgoing transmission speed by insertion of idle cells;
- filling of cells in the transmission system (for example, SDH frame).

Review questions

1. What are the two main tasks of an ATM switch/cross-connect?
2. Which classes of switching element can in principle be used for ATM switching?
3. How can switching networks be constructed from switching elements?
4. Describe *self-routing* and *table-controlled* cell switching.
5. What is multicast switching?

13

ATM transmission

Sections 4.1 and 4.2.1 addressed general network aspects of the ATM-based B-ISDN. A main feature of an ATM network is the establishment of VCCs and VPCs as described in Section 4.2.1. Switching of ATM cells was discussed in Chapter 12. The present chapter discusses the transmission aspects of ATM networks. After a brief overview of network elements, like ATM multiplexers and cross-connects, transmission systems for ATM cell transport and network synchronization will be discussed. Emphasis will be put on possible local loop implementations.

13.1 Overview

13.1.1 Cell transfer functions

The transfer of cells through an ATM network is supported by the following functions:

- generation of cells (packetizer, or more precisely, ATM-izer, for example in a B-ISDN terminal);
- transmission of cells;
- multiplexing/concentrating of cells;
- cross-connecting of cells;
- switching of cells.

Cell transmission will be addressed in Section 13.1.2, while the other functions will be discussed in this section.

Generation of cells

A B-ISDN terminal will send all its information mapped into cells (besides transmission overhead for instance in the SDH option, see Section 5.3.1), in which case no additional packetizing function is required in the ATM network.

Packetizers will be required, however, whenever interworking with non-ATM equipment has to be performed. An ATM packetizer either cuts STM channels into pieces that fit the ATM cell format or adapts non-ATM packets to ATM cells. This signal conversion is required at a B-NT2 (see Section 5.2.4), which provides the user with both ATM and non-ATM interfaces, or at any other place in the network where the connection of non-ATM traffic is foreseen (for example, in an ATM multiplexer or switch). The reverse procedure (depacketizing) has, of course, to be performed when ATM traffic is converted to non-ATM traffic.

The use of ATM cell packetizers in a customer network is shown in Figure 13.1. Figure 13.2 shows applications of ATM cell packetizers in the public network.

The need to convert non-ATM signals to the ATM cell format will arise especially during the introductory phase of ATM networks (see Section 2.2).

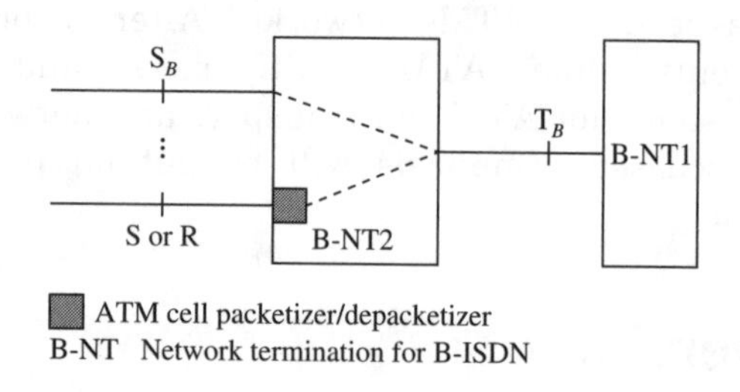

Figure 13.1 The use of cell packetizers/depacketizers in the customer network.

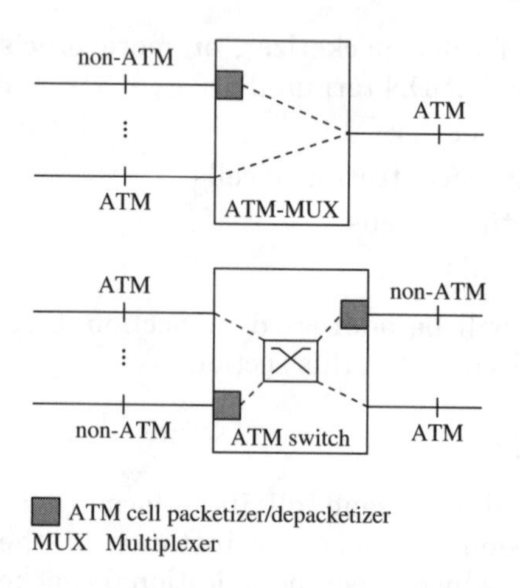

Figure 13.2 The use of cell packetizers/depacketizers in the public network.

The problems caused by packetization delay for ATM speech connections will be discussed in Section 14.1.

Multiplexing/concentration of cells

Figure 6.4 showed the use of an STM multiplexer which multiplexes several signals originating from different B-ISDN customers onto a single access line. In this STM-type multiplexer, no (cell) concentration takes place; that is, idle cells are not removed as the STM multiplexer does not process the signal payload carrying ATM cells. A simple example is shown in Figure 13.3.

All incoming idle cells will be sorted out in an ATM multiplexer, thus concentrating the ATM traffic. The achievable degree of concentration depends on the traffic characteristics and the requested quality of service. Figure 13.4 shows an example of how an ATM multiplexer is deployed to enable customers to share a common access line.

The ATM multiplexer unpacks arriving ATM cells from the transmission frame (the SDH STM-1 frame is shown in the figure), eliminates idle cells (and erroneous, uncorrectable ones) and multiplexes the valid cells into one STM-1 frame. In this example the ratio of the gross bit rates at both sides of the multiplexer is $1/m$. This can only work if m is a small number (up to eight) and the maximum sum bit rate used at any time by any tributary is small compared with 155.520 Mbit/s; that is, it does not exceed, say, 20–30 Mbit/s.

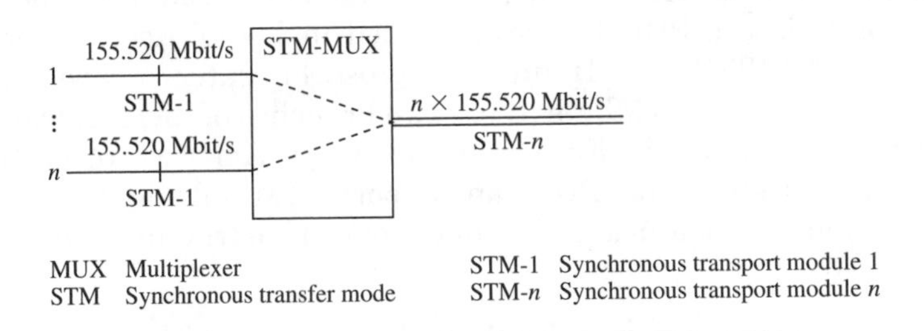

Figure 13.3 STM multiplexer.

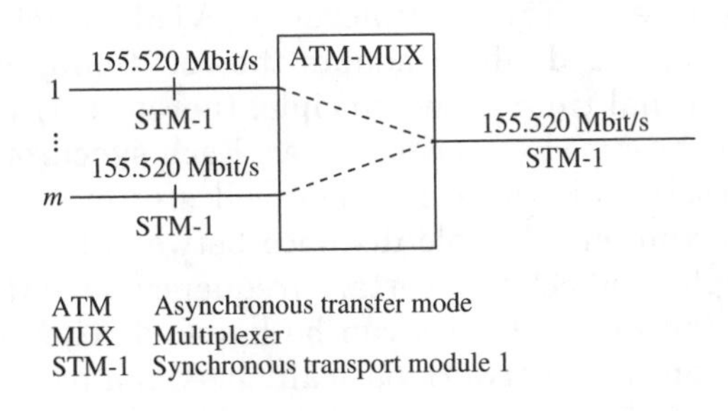

Figure 13.4 Example of an ATM multiplexer.

Cross-connecting of cells

An ATM cross-connect, like a VP/VC switch, can flexibly map incoming VPs/VCs onto outgoing VPs/VCs and thus enable VPCs/VCCs to be established through the ATM network. The cross-connect also concentrates ATM traffic as it eliminates idle cells. (Moreover, it performs necessary OAM functions at the physical layer and the ATM layer, as does the ATM multiplexer.)

A cross-connect can be used, for example, in the access network to separate customer traffic that is destined for the local switch from that to be transmitted on a fixed route through the network to a fixed endpoint (see Figure 13.5).

While a VP/VC switch establishes and releases connections according to a signaling protocol (see Chapter 8), a cross-connect is controlled through management operations.

13.1.2 Transmission systems

ATM cells can, in principle, be transported on many transmission systems. The only requirement is that bit sequence independence is guaranteed so that there are no restrictions on allowed cell information content.

ITU-T originally defined two options for the user–network interface, one based on SDH and the other on pure cell multiplexing (see Section 5.2.3). Other transmission systems may be used in the network. One example is the PDH as recommended in ITU-T Recommendation G.703 (ITU-T, 1991b). PDH provides gross bit rates of about 2, 34, 140 Mbit/s or 1.5 and 45 Mbit/s (these two branches of hierarchical levels are used, for example, in Europe and in North America, respectively). Transmission systems like PDH can support ATM cell transport if there is non-existent, or insufficient, SDH network infrastructure available.

13.1.3 Network synchronization

An ATM transport network needs bit timing and cell timing. At any entrance point to an ATM multiplexer or ATM switch, an individual synchronizer is provided which adapts the cell timing of the incoming signal to the internal timing. In principle, therefore, transmission links need not be synchronized with each other. Each synchronizer can adjust differing phases in units of a cell (idle cell stuffing/extraction). When limiting the maximum tolerable distance between two subsequent idle cells (for example, 256 cells), a certain frequency deviation between the link clock and the internal clock can be handled. The above considerations show that an ATM network basically need not be synchronous.

However, ATM networks must be able to integrate STM-based applications, including audio and video transmission, as long as they

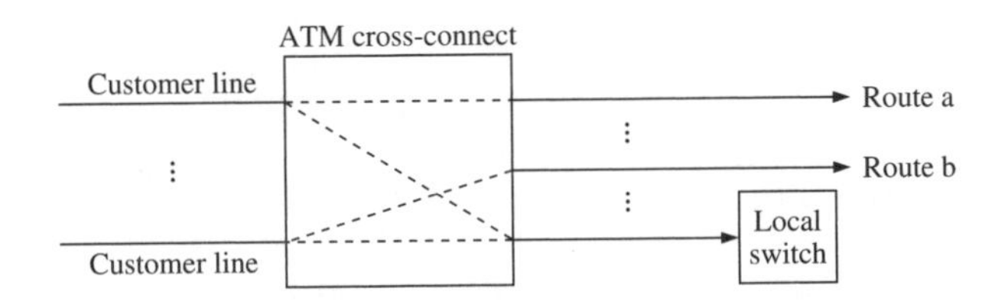

Figure 13.5 ATM cross-connect.

exist. To this end the sampling clock of the sender must be provided to the receiver in order to avoid slips. This implies requirements on the network in terms of support for synchronization of the access lines and tolerable slips in the case of synchronization failure.

One way of providing the necessary clock information in ATM-based B-ISDN is to make use of the existing clock distribution network for the 64 kbit/s ISDN. Performance requirements on the clock distribution network according to ITU-T Recommendations G.811 (ITU-T, 1988b) and G.821–824 (ITU-T, 1996d, 1988c, 1993c, 1993d) should then be considered as a basis for discussion on broadband networks and, if necessary, be redefined.

Derivation of the B-ISDN clock from the 64 kbit/s ISDN clock distribution network is shown in Figure 13.6.

The 64 kbit/s ISDN is structured into several hierarchical levels (four levels are indicated in the figure, as is the case in Germany). Each node receives its clock from the node of the higher level, the highest level node deriving its timing directly from the clock distribution network. In the event of failure, standby connections can be used to derive timing via alternative paths.

The simplest implementation for B-ISDN could be to derive the timing for each node from the corresponding 64 kbit/s ISDN node as shown in the figure (here, too, standby timing connections are provided to cover failure situations). Note that B-ISDN will most probably have fewer hierarchical levels than the 64 kbit/s ISDN. As clock accuracy decreases at the lower levels, it may become necessary to derive the clock for B-ISDN not from the corresponding 64 kbit/s ISDN level but from a higher one (or even directly from the clock distribution network).

Synchronization of terminals

For a constant bit rate service the source sampling frequency f_s and the filling rate R_c of cells pertaining to this service are tightly correlated:

$$R_c = f_s/48$$

(The ATM cell information field has 48 bytes.)

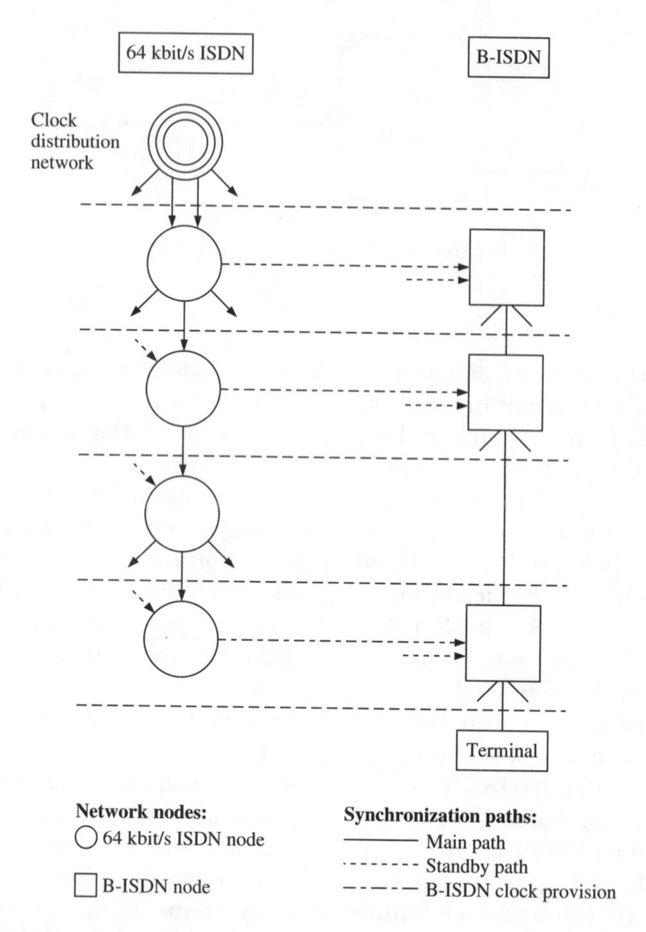

Figure 13.6 Example of network synchronization for B-ISDN.

Thus at the receiving side the original source sampling frequency can, in principle, be derived from the rate of those arriving cells belonging to the connection under consideration. However, because of the nature of the ATM network, arriving cells are subject to delay variations which are not so easy to compensate.

To overcome this problem, an STM-like mechanism can be adopted for ATM networks on an end-to-end basis provided that:

- timing of the sending side (terminal) is synchronized with network timing;
- at the receiving side the clock of the sending terminal is reproduced by virtue of the network clock.

To this end the terminals have to be provided with the network clock via the access line. In the case of SDH the clock will be derived from the signal across the user–network interface (see Section 5.3.2).

In the synchronous B-ISDN shown in Figure 13.6, no slips can occur as long as network synchronization works correctly. Should the synchronization fail, local exchanges provide the access lines with a clock of sufficient accuracy to sustain orderly operation of the accesses for a defined period of time. Terminals connected to different local exchanges must be able to remove any slips which might then occur.

In a synchronous network, source signals that are not correlated with the network clock (for example, video sources in free-running mode) can be transmitted by means of positive justification (this method is described in Bocker (1992)).

13.2 B-ISDN local network topology and technology

13.2.1 Local network structure

Conceptually the simplest realization of the B-ISDN local network is the star topology where there is one access line per customer. The remaining problem to be solved is to define the characteristics of the digital section extending between the T_B and V_B reference points (see Figure 13.7(a)).

However, other structures are also being investigated as candidates for implementation in the B-ISDN local network. These include multiple star, ring, bus and tree configurations. Some examples are shown in Figure 13.7. Case (b) shows the multiple star configuration in which several customers' signals are multiplexed onto one access line. Optionally, there might be an additional cross-connect to provide different paths through the network. Its functionality may comprise:

- connection of customers with high traffic volumes directly to the local exchange;
- separating traffic that is to be switched in the local exchange from traffic that is to be routed on a fixed path (permanent connection) through the network.

Configuration (c) is a ring structure in which many customers share a common transmission medium, such as a MAN. Customers with low traffic may share the access to the ring, as shown in the lower part of (c). The ring may be a single or double ring depending on:

- number and location of customers
- throughput requirements
- availability criteria
- cost aspects.

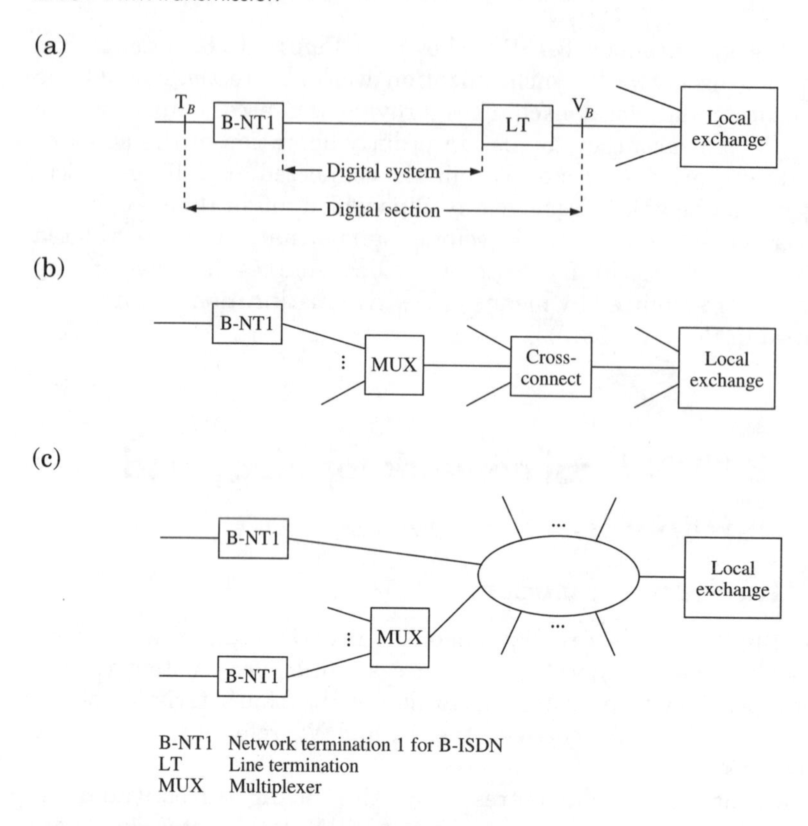

Figure 13.7 Examples of local network structures. (a) Star configuration; (b) multiple star configuration; (c) ring configuration.

When a customer requests a high performance and reliability level the network may be connected via multiple interfaces to different ring access nodes or even to different exchanges.

The ring structure saves transmission lines in terms of the length of cable to be laid in the ground but is restricted in terms of the bandwidth available to an individual user and in terms of upgradability (the same holds for bus configurations). In a star or multiple star configuration it is easier to connect additional customers to the B-ISDN.

Tree configurations will be discussed in the following section. For other (low-cost) ATM access systems see Section 14.3.

13.2.2 Fiber to the customer

The demand for broadband services will develop only gradually, and

> 'it therefore seems advisable if the introduction of fibre optic technology for the subscriber access is not coupled firmly to new interactive broadband services. Instead, solutions are initially

needed whereby optical fibres can be economically brought as close as possible to the subscriber, also for already existing services (television, telephony, data transfer)' (Siemens, 1989).

Thus, the magic spell is **fiber to the office** (FTTO) or **fiber to the home** (FTTH). A less ambitious interim goal is **fiber to the curb** (FTTC) (see Figure 13.10). In the case of business customers with large or rapidly expanding traffic volumes, it may already pay for network providers to install individual optical fiber lines. Because of their vast transmission capacities, they are deemed to be a future-proof investment. In the case of residential customers the hope for quick returns is not so justified. In this case, therefore, resource-sharing concepts have to be considered to allow cost-effective introduction of broadband services.

Passive optical network

One of these cost-effective concepts is the **passive optical network** (PON) (Faulkner and Ballance, 1988). A single fiber from the exchange feeds a number of customers via passive optical branching (see Figure 13.8). This technique allows the fiber and laser in the local exchange to be shared between several customers. A TDM signal is broadcast from the exchange to all terminals on a single optical wavelength. The signal is detected by an optical receiver, then each customer's equipment demultiplexes only the channels intended for that destination.

In the return direction, data from each customer is inserted at a predetermined time to arrive at the exchange within an allocated time-slot.

This simple PON architecture admittedly has some drawbacks, such as:

- limited bandwidth for interactive services per customer;
- multiplexing of upstream signals requires sophisticated measures;
- privacy and security problems may arise;
- restricted upgradability (how to overcome this problem will be discussed later on).

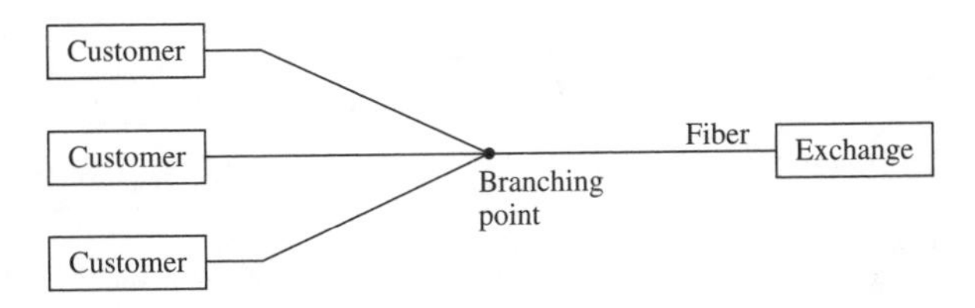

Figure 13.8 Passive optical network technology.

The PON concept can well support:

- unidirectional distribution services (TV and sound programs)
- telephony and other 64 kbit/s ISDN services.

Different optical wavelengths $\lambda_1, \dots, \lambda_n$ may be employed to separate services (and, possibly, transmission directions). For example, λ_1 for telephony, λ_2 for broadcast TV, λ_3 for video retrieval services, and so on, as illustrated in Figure 13.9. Note that different customers may wish to receive different service mixes.

In the long term it might be feasible, according to Faulkner and Ballance, to provide a separate wavelength to every customer with a specific mix of services multiplexed onto the wavelength as required by the customer. However, this concept can only work as long as the number of customers connected to one PON is not too large.

Several modifications of the above concept are conceivable. Instead of, or in addition to, WDM (use of several λ_i) more than one fiber could be installed, perhaps one for TV distribution and the other for interactive services. Implementation of both service categories could thus be decoupled to a great extent.

Figure 13.10 shows an access configuration which deploys two fibers, one for TV distribution and the other for telephony or 64 kbit/s ISDN services (fiber to the curb). After conversion of the optical signals into electrical ones close to the customer's premises (at the curb), a bus-structured coaxial distribution system is used to deliver TV programs to the customers (hybrid fiber-coax concept) while a star structure is used to provide a customer with copper-based twisted pair telephony access.

In Figure 13.10 the opto-electrical conversion that is necessary to serve present electrical terminal interfaces is done only once for several customers, thereby saving on the cost.

To facilitate later evolution from a PON architecture to a full B-ISDN, it may be advantageous to install additional 'dark' fibers from the outset. These are fibers which are not immediately used but can support later point-to-point connection of individual customers to the local exchange (see Möhrmann (1991)).

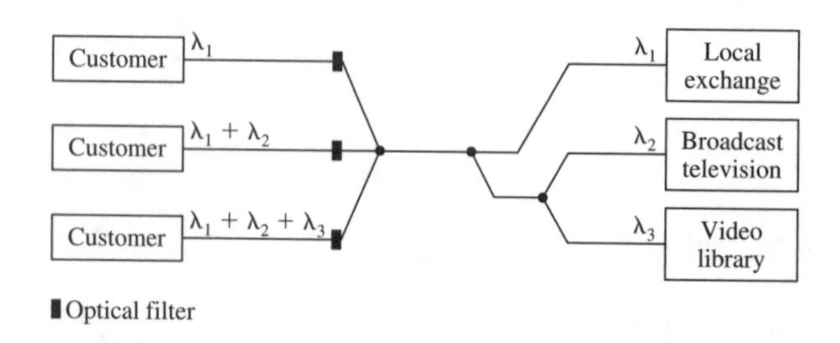

Figure 13.9 Upgraded passive optical network.

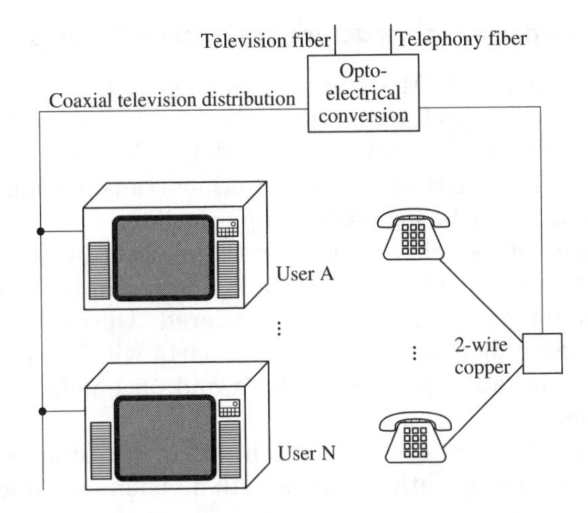

Figure 13.10 Alternative fiber to the home architecture.

ATM on a passive optical network

The previous section introduced the PON concept which will support narrowband services and distribution services like TV. This section deals with a new version of a PON transmitting ATM traffic. Several strategies for ATM-PONs have been proposed (De Prycker et al., 1992; du Chaffaut et al., 1992; Keller et al., 1993). An ATM-PON can be used either for upgrading existing PONs for broadband services or as a new and cost-efficient customer access network. An additional fiber or wavelength can be employed to broadcast cells to the customer and carry cells from individual customers to the node. The ATM-PON can be considered as a distributed flexible ATM multiplexer.

In an ATM-PON, several subscribers share a very large bandwidth (for example, 155.520 Mbit/s). The broadband communication requirements of a lot of subscribers can be simultaneously met by an ATM-PON because their required bit rates will predominantly be in the range of 10 to 30 Mbit/s (Hagenhaus and Überla, 1993). (Only a few subscribers need the full capacity of 155.520 Mbit/s.)

In the downstream direction (network to user), a TDM procedure is used to transport cells. In the opposite direction (upstream), a time division multiple access procedure is required. A suitable synchronization mechanism is necessary in the upstream direction to prevent cell collision and guarantee fair capacity sharing between subscribers according to their needs.

The hardware required at the user side is somewhat more than for a point-to-point connection, but when considering the overall network the PON has clear cost advantages because the network equipment is shared by several subscribers.

13.2.3 Transmission characteristics and technology

In the access network ATM signals carried on optical systems based on SDH (ITU-T Recommendation G.707 (ITU-T, 1996c)) will provide bit rates of approximately 155 Mbit/s (STM-1), 622 Mbit/s (STM-4) and 2.5 Gbit/s (STM-16). PDH systems and other transmission media, such as radio links, can also be used where appropriate.

The use of single-mode fibers in accordance with ITU-T Recommendation G.652 (ITU-T, 1993a) is favored in the access network as they allow longer distances to be covered. Optical signals can be generated by laser diodes. The electronic parts will be based on CMOS technology for bit rates up to 155 Mbit/s and on bipolar technology for higher bit rates.

Two optical fibers (one for each direction of transmission) may be employed. Alternatively optical wavelength division multiplexing (WDM) on a single fiber can be used to separate both transmission directions by using different wavelengths (for example 1530 nm and 1300 nm).

The bit error probability of the optical transmission link should be less than 10^{-9}. An ATM switch will guarantee a cell loss probability of about 10^{-9} (see Section 12.1.6). Cell loss in an ATM switch is primarily caused by buffer overflow in the case of traffic congestion or the detection of uncorrectable errors in the cell header. If the transmission bit error probability is less than 10^{-9}, the latter effect is negligibly small as it requires the occurrence of specific multiple-bit errors in one cell header.

Maintenance aspects of optical transmission

The following failures may occur in optical transmission systems:

- laser light emission failure
- laser sends continuous '1's
- receiver indicates continuous '0's or '1's.

In the event of such a failure the corresponding message will be delivered to a line maintenance entity (for example, located in the local exchange) which is responsible for setting the appropriate alarms. An urgent alarm will be generated if a unit is out of order or a warning issued if its performance is deteriorating.

13.3 Trunk network structure

A possible trunk network implementation is shown in Figure 13.11. (Realization of internodal signaling connections is not discussed in this chapter. Signaling may be established via the existing SS7 routes or, in the long term, over the ATM network itself.)

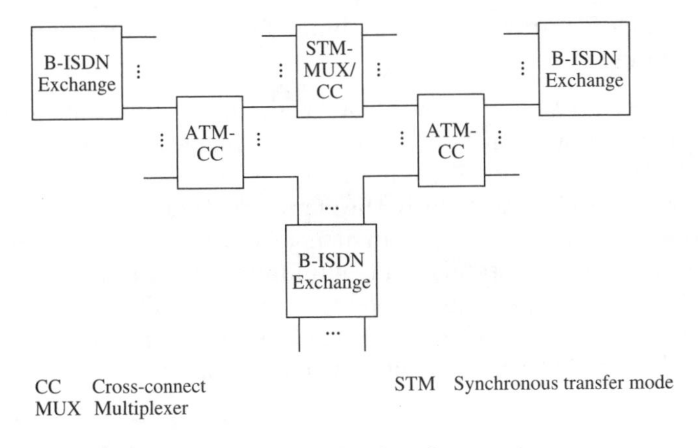

Figure 13.11 Example of trunk network structure.

Figure 13.11 shows the following network elements:

- B-ISDN exchange
- ATM cross-connect
- STM multiplexer/cross-connect.

ATM cross-connects act like VP switches and can flexibly provide VP connections through the network (see Section 4.2). STM cross-connects may also be deployed to facilitate rearrangement of the physical paths, for example in the event of transmission failures (protection switching). Finally, STM multiplexers merge, for example, 155.520 Mbit/s STM-1 signals into the higher bit rate signals STM-4 or STM-16 of about 622 Mbit/s or 2.5 Gbit/s, respectively.

ATM cross-connects process each arriving cell. In line with its VCI/VPI value, a cell is routed in the direction defined by the associated VCC/VPC. Therefore, VCCs/VPCs with arbitrary bit rates can be established and switched through an ATM cross-connect. Establishment and release of VCCs/VPCs can, in principle, be initiated by the user or the network provider by means of ATM layer management procedures. The time required to establish or release VCCs/VPCs will be much shorter than is needed to meet the request for a reserved, permanent channel in today's networks; that is, VCC/VPC establishment can be performed within a couple of seconds.

Multiplexers and cross-connects can be used to decouple the logical point-to-point configuration of the ATM switching network from the actual topology of the fiber-based transmission network and to achieve:

- flexibility in the realization of hierarchically structured networks or networks with any other structure (for example, meshed rings);
- easy provision of additional network capacity in the case of growing traffic;

- a means for providing redundancy for ATM connections between exchanges.

Segregation of ATM traffic with low/medium bit rates per connection from ATM traffic with high bit rates in the trunk network may help to manage ATM traffic more efficiently.

Individual channels with the same destination can be grouped and carried on the same VP. Transit nodes of the ATM network act as VP switches with pre-established (redundant) VPC capabilities between the B-ISDN exchanges and need not handle VC switching. This facilitates the operation of transit nodes, especially in the case of many simultaneous low bit rate connections on a trunk line.

13.4 ATM network implementation issues

ATM networks need a physical layer transmission capability. As long as ATM networks are rather thin overlay networks, it makes sense to use the existing PDH/SDH (SONET)-based transmission infrastructure; that is, to share the transmission network between STM and ATM traffic. (Cell-based transmission systems are not yet available; they only make sense when non-ATM traffic has decreased to a small quantity that can easily be adapted to ATM.) Figure 13.12 illustrates this principle.

With growing ATM traffic, dedicated ATM links (operating, for example, at STM-1, STM-4 or STM-16 bit rates of 155 Mbit/s, 622 Mbit/s and 2.5 Gbit/s, respectively) can be installed to connect ATM nodes directly. (This is indicated in Figure 13.12 by the dashed line.)

In the case of ATM on top of PDH/SDH, there is a strictly hierarchical relationship between the ATM network and the underlying transmission network: ATM is the client of the latter, the server network. If an ATM connection between two ATM nodes is to be established, the transmission network management will provide the necessary capacity, if possible (and monitor its performance). Should transmission fail, the physical layer is responsible for restoring a suitable link. However, if the ATM network cannot rely on the transmission network (perhaps because of a lack of supervision and restoration capabilities), it may provide redundant VPs so that in the event of any failure it can switch over from one VP to the other (VP protection switching).

When both the physical layer (PDH or SDH) and the ATM layer can provide protection switching, actions in both layers must be coordinated to avoid wasting resources or, even worse, the initiation of counteracting measures.

In an ATM network with dedicated ATM links, this problem seems to vanish as most, if not all, protection measures may be allocated

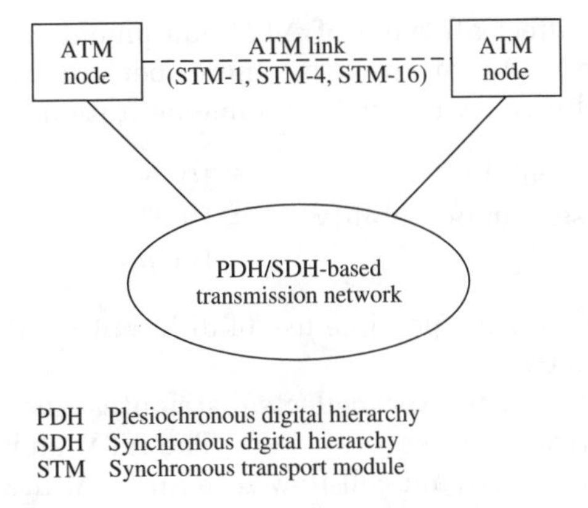

Figure 13.12 ATM overlay network on top of PDH/SDH.

to the ATM layer itself, that is, to the VP/VC level. More details on ATM network protection switching can be found in Händel (1992).

Another interesting question relating to ATM network implementation is the architectural concept for VP/VC handling. Although ITU-T allows cross-connection of both VPs and VCs (controlled by management) and signaling-controlled switching of VCs, for practical reasons (keep network management as simple as possible!) it may be better to restrict this functionality. It might be sensible to have a VP cross-connect network and, on top of it, a switched VC network. Leased line services would then be realized via the VP network and switched connections via the VC network.

13.5 ATM transmission network equipment

The ATM equipment required to perform the functions described in Section 12.3 has not yet been fully standardized by ITU-T in terms of functions, performance parameters and interfaces as is the case, for instance, for the standardized network equipment to be used in SDH transmission networks. Open issues relating to ATM equipment specification include:

- statistical multiplexing function
- multicast function
- OAM procedures: ATM-level protection switching
- ATM/physical layer relations
- performance requirements.

These problems affect all types of ATM equipment: multiplexer, switch and cross-connect. As to the equipment performance, the following values can be achieved with existing implementations:

cell loss probability	$< 10^{-10}$
cell misinsertion probability	$< 10^{-13}$
cell delay	$< 100\ \mu s$

Figure 13.13 outlines the possible use of different ATM equipment in an advanced ATM network.

The ATM multiplexer collects/distributes the ATM traffic (or ATM-ized STM traffic) for several users. The ATM add/drop multiplexer (ADM) inserts the upstream cell flow into an ATM access network ring and extracts the corresponding downstream cell flows from the ring. One, or for redundancy reasons, several, ADMs are connected with the trunk network cross-connects, which are intermeshed. (ATM cross-connects normally act at VP level, but VC cross-connection would also be possible.) The VC switch is operated as either a local exchange (terminating user-to-network signaling) or a transit exchange. The switching and cross-connecting functions may be integrated into a combined ATM cross-connect/switch, as indicated by the dashed box in Figure 13.13.

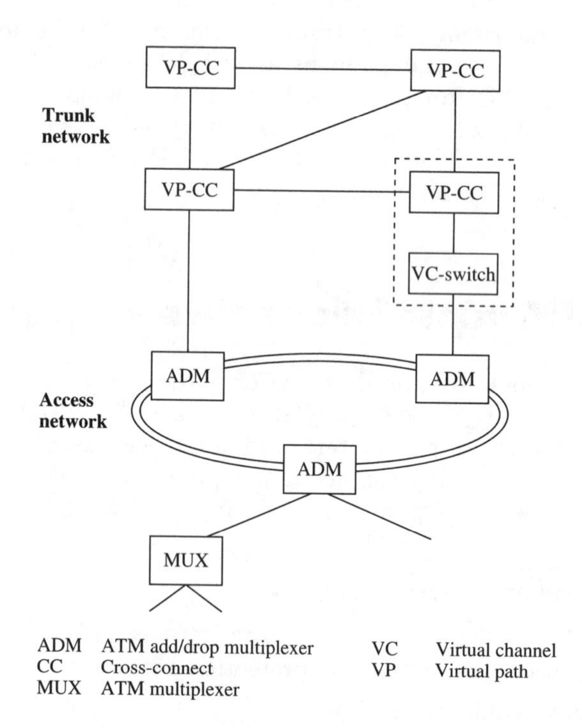

Figure 13.13 Example arrangement of ATM network equipment.

13.6 Optical networking and ATM

13.6.1 General aspects

Optical transmission and electronic switching are used for the B-ISDN as currently developed. For most of the services envisaged today, switched bit rates of several hundred Mbit/s seem to be more than sufficient. Only in the area of holographic imaging are Tbit/s (Tbit = 10^6 Mbit) rates being discussed.

Existing experimental switching systems use CMOS technology as well as emitter coupled logic techniques to realize systems that are capable of operating up to the Gbit/s range. In the future, speeds of up to a few Gbit/s will be achieved using gallium arsenide technology, but this seems to be the limit of electronic switching.

There are a number of reasons for considering optical switching networks (Nussbaum, 1988):

Historical perspective: Analog transmission was followed by electro-mechanical switching. Then digital transmission was introduced which was followed by digital switching. Now optical transmission is used, and if history repeats itself, optical switching networks will be the next step.

Speed limitation: As already mentioned, the speed of electronic switching networks will be limited to the lower Gbit/s range (approximately 10 Gbit/s). Higher speeds require optical switching networks.

Cost reduction: In systems with optical transmission and electronic switching, optical/electrical and electrical/optical interfaces are necessary. These expensive units can be avoided by staying entirely in the optical range.

As a first step, optical switching matrices using space or wavelength division multiplexing systems can be introduced in optical cross-connects. Information will not then need to be converted from optical to electrical and vice versa in the individual nodes. Control of the switching matrices is only necessary when a connection is set up or released, and this can be performed by electronics. The optical cross-connect network can be used for a **universal transport network**, as shown in Section 13.6.2. This universal network is available to other networks as well as to end users who need very high bit rate leased line services.

However, the B-ISDN uses asynchronous time division multiplexing and information is transported in cells. This requires the cell header to be processed in each switching element so that it can be forwarded to its proper destination output. Cell header processing requires extensive logic manipulations, and must at present be done electronically. Maybe in the future it will be possible to use optical computing and optical memories for cell header processing.

Other drawbacks of optical switching/cross-connecting, like optical losses, must be overcome. Furthermore, the regeneration and amplification inherent in digital electronic switching needs some counterpart in optical networks. Another very important point for the success of optical networking is the existence of an appropriate OAM concept.

Optical switching will only succeed in competing against electronic switching if it can be implemented economically or becomes absolutely necessary.

In conclusion, the large-scale implementation of optical switching will not take place in the near future and it may take several decades to replace today's electronic switching techniques. However, all-optical transport networks with space and wavelength division multiplexing may be introduced much more quickly (see the following section).

13.6.2 Optical networks

Currently, parallel to ATM networking there is a vision of all-optical networks within the private and public areas (Barnsley, et al., 1992; Brackett et al., 1993; Hinton, 1992; Janniello et al., 1993; Sato et al., 1993). Some of these networks are already running in field trials, while others are still in the planning phase.

Passive optical networks using WDM technology are one of the most favored solutions for the local/metropolitan area. Figure 13.14 shows two possible realizations of an optical star network based on the splitter/combiner approach (Green, 1993). All transmitted signals are broadcast to all receivers.

In both configurations the network consists of n nodes, which may be end users or electronic nodes (depending on the application of the network). In the left-hand configuration a node has a fixed receiving wavelength. Thus, each node needs only one filter with a fixed wavelength in the receiving direction. However, in the opposite direction each node needs a tunable sender which allows the connection of an arbitrary pair of nodes by selecting the appropriate wavelength. A suitable access protocol is necessary to prevent several senders attempting to transmit to the same receiver simultaneously.

The transmitting wavelengths in the right-hand configuration are fixed. To prevent information loss, each node has to receive $(n - 1)$ wavelengths simultaneously or a suitable access protocol is required. Multicast communication can be offered very easily in this way with different receivers receiving an optical signal simultaneously.

In the normal operation mode a node receives only one signal at a time. The fully meshed interconnection of all n nodes in such a passive optical network will require $n \times (n - 1)$ wavelengths. Technological

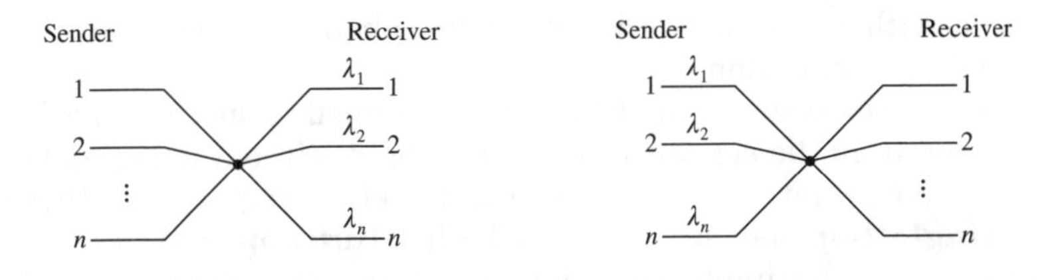

Figure 13.14 Passive optical networks with splitter/combiner.

limitations (the number of wavelengths on one fiber) restrict this network to a small number of nodes. This drawback can be avoided by using a passive optical star network, as shown in Figure 13.15 (McMahon, 1992).

This network is not based on the splitter/combiner concept. The star coupler consists of passive wavelength multiplexers with the given specific interconnection. Only $(n-1)$ wavelengths are required for the full meshing of all n nodes. However, it should be noted that the sending and receiving directions are decoupled by using two fibers.

A bus topology can be used instead of the star topology. In terms of the number of required wavelengths and the necessary access protocols to the shared fiber it behaves like a star network. However, from the optical power budget point of view it has real drawbacks

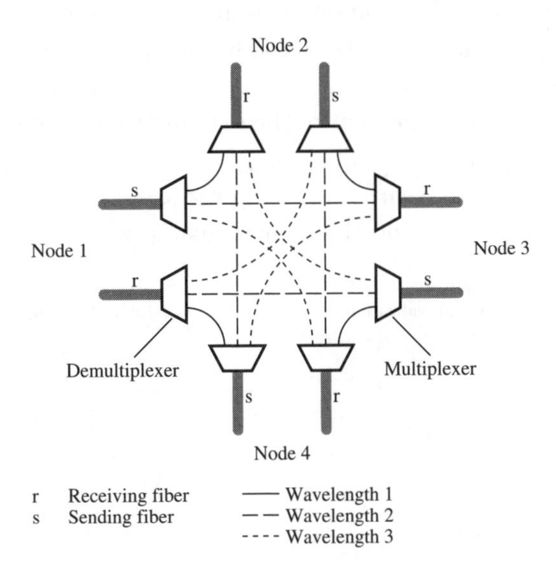

Figure 13.15 Passive optical networks with multiplexer/demultiplexer.

compared with a star network with splitter/combiner because it has a much higher attenuation.

In these systems it has been assumed that information is transmitted from the sender to receiver as light without being converted back to electrical form. The implementation effort required for these so-called **single hop** networks is very high. **Multihop** networks can be used to reduce the implementation complexity (Brackett et al., 1993; Green, 1993; Henry, 1989; Ramaswami, 1993). In this case the destination node is not reached directly by the optical signal transmitted by the source. The signal emitted by the source first passes through an intermediate node which converts the optical signal back to an electrical one and analyzes it. This node decides whether the signal is addressed to it or if it has to be transmitted to the next node. Figure 13.16 shows a passive optical star network and its associated logical topology for multihop communication. It should be noted that the physical network structure of a multihop network is not limited to the star topology.

In the configuration shown in the figure each node uses two wavelengths with fixed values in the receiving and transmitting directions. This reduces the implementation complexity for the individual nodes. In this example, node 1 has no direct optical link to node 2. Thus to send information to node 2, node 1 sends the information to node 3 using wavelength λ_1. Node 3 converts the signal to an electrical one and analyzes it, then converts the signal back to an optical one using wavelength λ_6. This optical signal is then received by node 2.

Multihop networks have the following drawbacks:

- optical–electrical conversion is required in intermediate nodes;
- additional delays are introduced by the intermediate nodes;
- enlargement of a system (for example, addition of a new node) requires a rearrangement of the existing connections.

In all the networks described above and later on, the electrical information may be SDH-frames, ATM-cells or have any other structure. The bit rate of the signal using wavelength λ_i may be

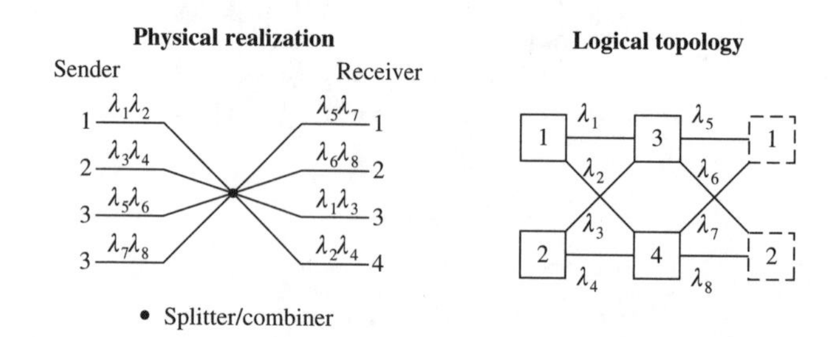

Figure 13.16 Multihop network.

different from the one which uses wavelength λ_j.

The star network should only be used in the access network because of its limitations in terms of transmission distance and number of nodes. Cross-connects (CCs) and ATM networks may be the preferred solutions for the core area of a WAN (Barnsley et al., 1992; Chidgey et al., 1992; Johansson et al., 1992). Figure 13.17 depicts a possible scenario for an all-optical wide area transport network and its association with the electronic parts. It consists of a meshed cross-connect network and WDM rings (Elrefaie, 1993; Huber and Osborne, 1994).

Such a CC/ADM network offers a lot of advantages to the network operators and users:

- nearly unlimited bit rates;
- universal transport network carrying signals with different formats and bit rates;
- flexibility based on the enhanced CC/ATM functionality and improved reliability as a result of using small nodes with an increased number of passive optical components and suitable protection mechanisms for the network and its systems;
- future-proof system because new applications can be integrated very simply by using an unoccupied wavelength;
- cost-effectiveness by avoiding unnecessary optical–electrical conversion and by reducing the efforts for electronic processing.

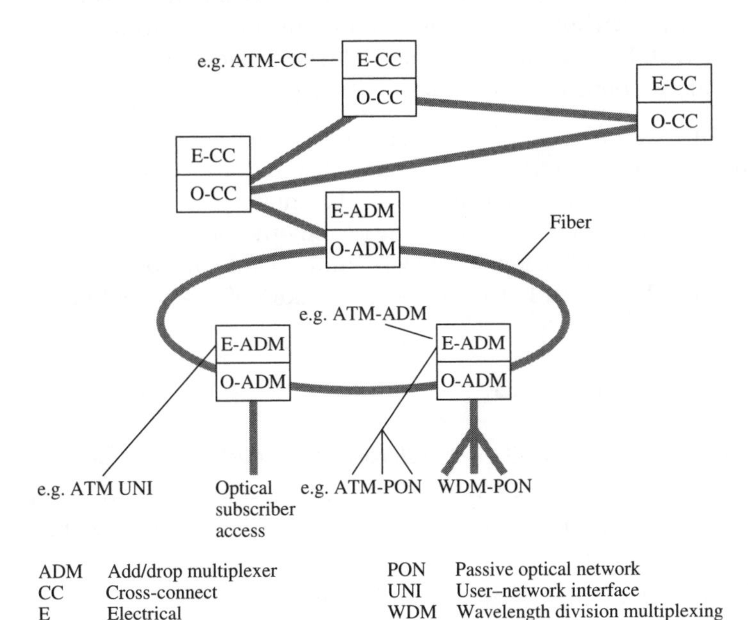

Figure 13.17 All-optical transport network.

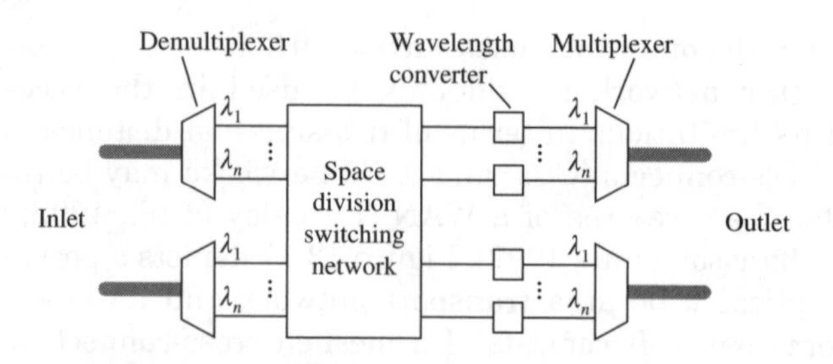

Figure 13.18 WDM cross-connect.

The generic architecture of a WDM-CC (or a WDM-ADM) is shown in Figure 13.18 (without the required node control). It consists of the following subsystems:

- **Multiplexer/demultiplexer**: The wavelengths used on the fiber are separated at the inlet of the CC so that they can be individually cross-connected. At the outlet, the multiplexer combines the different wavelengths onto the outgoing signal.
- **Switching network**: The space division multiplexing switching network performs the space switching of the individual wavelengths.
- **Wavelength converter**: This subsystem converts the wavelength λ_i at its inlet to the new wavelength λ_j for the outlet. Wavelength converter is a key function in WDM networks because without it the performance of the node would be very low. (A WDM-CC without wavelength conversion would be like a TDM node without time-slot interchange.)

WDM networks are still in an early phase. Although a lot of research into optical components has already been done, much work is still needed to improve the characteristic parameters of the components. In addition to the technological improvement of optical components, enhanced communication protocols and OAM protocols have to be developed to enable full advantage to be taken of all-optical networks.

Review questions

1. What do ATM multiplexers/cross-connects/switches perform?
2. What ATM access network topologies are possible and what criteria can be used to determine which one to use in a given environment?
3. What is a PON? What is an ATM-PON?
4. Which transmission systems can be used for ATM?

14

Miscellaneous

14.1 Telephony over ATM

In an ATM telephone connection, speech samples are collected until they fill the information field of a cell. As the information field consists of 48 octets, a delay of 48/8 kHz = 6 ms is encountered for 64 kbit/s telephony as a result of the packetization of speech. (If AAL type 1 is used for telephony, this delay can be slightly reduced to 47/8 kHz = 5.875 ms; as this is only a small effect it will be neglected here.) Depacketization of ATM speech also causes some delay. As individual ATM cells are subject to different transfer delays on their way through the network, this delay variation has to be smoothed out at the receiving side in order to generate the constant input required by the speech decoder. This procedure leads to an additional delay of approximately 1 ms.

The total delay may exceed the tolerable limit, especially when there are several transitions from STM to ATM, and vice versa, within one connection (for example, in the case of traversing three ATM islands the additional transfer delay resulting from packetization and depacketization would be $3 \times [6\,\text{ms} + 1\,\text{ms}] = 21\,\text{ms}$).

The main problem is not the one-way delay itself but the delayed speaker echo received at the sending side. This arises primarily at the far end of the transmission line, specifically at the hybrid junction between four-wire and two-wire analog circuits. The maximum one-way delay permissible for telephone connections is 400 ms according to ITU-T Recommendation G.114 (ITU-T, 1996a). This value has been chosen to take into account connections involving satellite links. A considerably lower one-way delay can be achieved for other connections, as is the case for ATM links.

However, echo received at the sending side will be perceived as disturbing by the speaker at a rather low value for the round-trip delay. As a rule of thumb, echo cancellation should be employed whenever a one-way delay of about 25 ms is exceeded.

Figure 14.1 (based on ITU-T (1996a)) gives more information on the echo problem related to ATM speech connections. It is assumed that one ATM-based subnetwork is inserted into an existing analog or digital telephone network. This insertion causes additional delay and aggravates the echo problem. The use of ATM reduces the transmission length that can be covered without echo control measures. The curves drawn in Figure 14.1 directly show the effect of (non-processed) echo delay on the perception of human beings in terms of 'probability of encountering objectionable echo' (for example, 1% means that 1% of the speakers complain about the quality of their telephone connection). According to ITU-T Recommendation G.131 (ITU-T, 1996b), no more than 1% objections to speech quality should be made in a network, although values up to 10% may be tolerated in exceptional cases.

The 1% limit of objections would confine the transmission length to about 300 km where one ATM area is inserted into an analog environment, and to about 650 km where one ATM area is inserted into a digital environment, unless echo control is provided.

In Figure 14.1 the loss of the echo path was assumed to be 28 dB. The two isolated spots in the figure have been calculated for local connections (transmission length up to about 50 km) and inter-local connections (transmission length of about 100 km) with specified losses of 21 dB and 24 dB, respectively, according to specific US standards (see Kammerl (1988)). As the low loss requirements in these cases may be reduced in future, the potential echo problems are expected to decrease. This is indicated in the figure by down-arrows.

To sum up:

- ATM introduces additional delay and thus increases echo problems on speech connections.
- The number of ATM/STM conversions should be kept to a minimum.
- Compared with today's telephone networks, more echo cancellation devices will have to be deployed with 'straightforward' ATM. This can be avoided, however, by using either partly filled cells or composite cells where several voice connections are multiplexed into one cell. For example, by filling only 24 octets of the 48-octet cell information field the packetization delay would be halved but half the bandwidth would be wasted which renders this method inefficient.

The use of composite cells has been standardized meanwhile (see AAL type 2, Section 5.7.3), especially to support voice over ATM using less than 64 kbit/s encoding which is indispensable in the case of voice over wireless/mobile ATM.

Voice trunking in ATM networks – supported by circuit emulation – was described in Section 10.2.

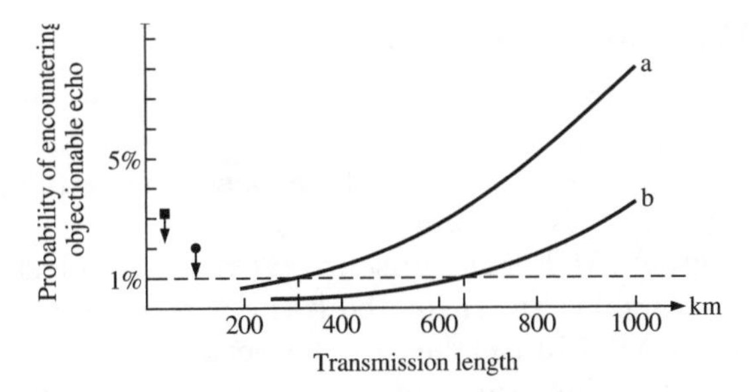

Figure 14.1 Echo tolerance curves of telephone connections.

14.2 Wireless ATM and mobile ATM

14.2.1 Rationale

There is increasing need for wireless network access for several reasons. In the LAN area, wireless access is easier to implement than conventional cabling, and easier to adapt to organizational changes. As for public network providers, newcomers often face a lack of access facilities to their switching nodes due to missing rights of way or simply because they cannot afford a full-blown subscriber access network infrastructure right from the beginning.

This holds true for ATM networks as well; however, as ATM is a high-speed technology, wireless ATM access is a special challenge. *Wireless* comprises different categories, for instance *fixed wireless*, *microwave*, *satellite* and *mobile*. Whereas standard ATM interfaces can be applied for the first two (wireless in these cases just means another physical layer), special requirements have to be met for satellite and mobile accesses. The problem with satellite is the increased delay: this can be taken care of by larger buffer sizes. Mobile networks require extra measures, for example in terms of authentication, handover and interworking with other (fixed) networks. In mobile networks, due to their specific bit error profile (caused, for example, by dust or moisture in the air), forward error correction does not work economically, so some selective retransmission at the cell level needs to be implemented. As a consequence, mobile ATM requires new medium access control and data link control functionality.

14.2.2 ATM Forum specifications

Architecture and protocols for wireless ATM (WATM) are still under discussion. Open issues are, for instance – as discussed in the ATM Forum – whether mobile ATM should be based on a telephony (SS7) approach or on an IP (P-NNI) approach.

The current ATM Forum Baseline Text for wireless ATM (ATM Forum, 1997i) intends 'to cover both near-term requirements for mobility control in ATM infrastructure networks, as well as longer-term requirements for seamless radio extensions of ATM to mobile devices'. The document contains a stage 1 description (from a user's perspective) of the *ATM mobility extension service* (AMES), a reference architecture for WATM and a section on system-level requirements, and it addresses both *mobile ATM* dealing with higher-layer control/signaling functions needed for generic mobility support and the *radio access layer* dealing with radio link protocols for wireless ATM access.

The system-level requirements part covers security issues (general aspects of ATM network security, see Section 14.6) and handover requirements.

Inclusion of the following security measures into WATM systems is suggested:

- terminal identity confidentiality (protection against tracing the location of a terminal)
- terminal identity authentication
- user data confidentiality on physical connections
- signaling information confidentiality
- authentication of security-related network entities (such as servers) and of the (fixed) network.

Handover requirements comprise:

- low latency (the duration of the handover procedure should be as short as possible);
- scalability (the handover algorithm should work well in any geographical location and apply both to public and private networks);
- quality of service guarantee;
- low signaling traffic (especially important where handovers are done frequently, for example in microcell environments);
- minimal buffering (buffering causes delay and may negatively influence handover);
- data integrity (which in the case of WATM means that ATM cells must not get out of sequence, be lost or duplicated); this is important in data connections;
- group handover (multiple connections from a single terminal must be handed over at the same time).

The mobile ATM issues part is to include summary requirements and technical solutions for handover, location management, routing, traffic management/QoS and network management.

The radio access layer part will specify the physical layer, medium access control (MAC), data link control (DLC) and radio resource control (RRC).

14.3 Residential broadband solutions

Concerning broadband for residential customers, cost is the crucial factor for acceptance, even more so than in the case of business usage. So cheap solutions in terms of broadband access, terminals and service provision should be sought. For residential users access to telephony and facsimile, to data services (for example, via the Internet) gradually evolving to true multimedia services, and to entertainment programs (such as TV/audio distribution, video on demand or near video on demand) at a cost comparable with today's cost for similar services is required. Therefore residential broadband will most often offer rather modest bit rates far below 155 Mbit/s per home or per small office/home office ('SOHO').

Appropriate fiber-based and hybrid fiber/coax access systems providing economic solutions were described in Section 13.2. As an alternative, ADSL (*asymmetric digital subscriber line*) or VDSL (*very high-speed digital subscriber line*) technology can be employed. Both utilize the legacy copper wire telephony access. ADSL (ANSI, 1997a) can provide 2–6 Mbit/s downstream (that is, from the network to the customer) and at least 64 kbit/s upstream (from the customer to the network) over a distance of about 4 km. VDSL (ANSI, 1997b) can offer up to about 50 Mbit/s downstream and 1.5 Mbit/s upstream; however, the distance to be covered is limited to some hundred meters.

Note that the prevalent encoding standard for digital video is MPEG-2 as defined in ITU Recommendation H.263 (ITU-T, 1996e). It allows video encoding with bit rates of 1–2 Mbit/s per channel for VHS quality, 4–5 Mbit/s for PAL quality, about 10 Mbit/s for enhanced definition TV, and 20–30 Mbit/s for high definition TV. In the following a few considerations on how to provide customers with TV programs are given.

TV programs could be offered to the B-ISDN customer as switched or non-switched services (see Figure 14.2).

Figure 14.2(a) shows the full integration of TV distribution into B-ISDN. TV programs are fed into the local exchange, which has sole responsibility for operation, administration and maintenance of the customer's access link. Program selection is done via the usual B-ISDN signaling channels and procedures.

While this solution fully complies with the idea of an integrated broadband network, it may have market drawbacks. Switched TV programs via optical fibers cannot compete in terms of cost with today's TV distribution to the customer via satellite or cable-based transmission systems. This fact may inhibit such a solution in many countries. Moreover, as the network provider knows what TV programs are watched by customers, some critics are concerned about possible infringements of privacy by the unauthorized transfer of such information to people who might be interested in making use of it. From a purely technical viewpoint the provision of switched TV programs has the highest flexibility as there is, in principle, no limit on the number of programs that can be offered.

Figure 14.2(b) shows an architecture in which a fixed block of TV programs is fed from the TV program provider directly into the access link. The customer can choose a specific program either in his or her TV set or, optionally, in a customer-owned TV switch (to be located in B-NT2).

This method limits the number of different TV programs that can be received as a result of the limited bandwidth of the access line. The bit rate required to transmit digital TV signals (including one or more sound channels and additional data channels for control purposes) is some Mbit/s for conventional TV quality and considerably higher for high definition pictures. In the case of HDTV, which requires around 20 to 30 Mbit/s per TV channel, such a rigid distribution scheme could fail unless more powerful methods, like coherent transmission, can be employed.

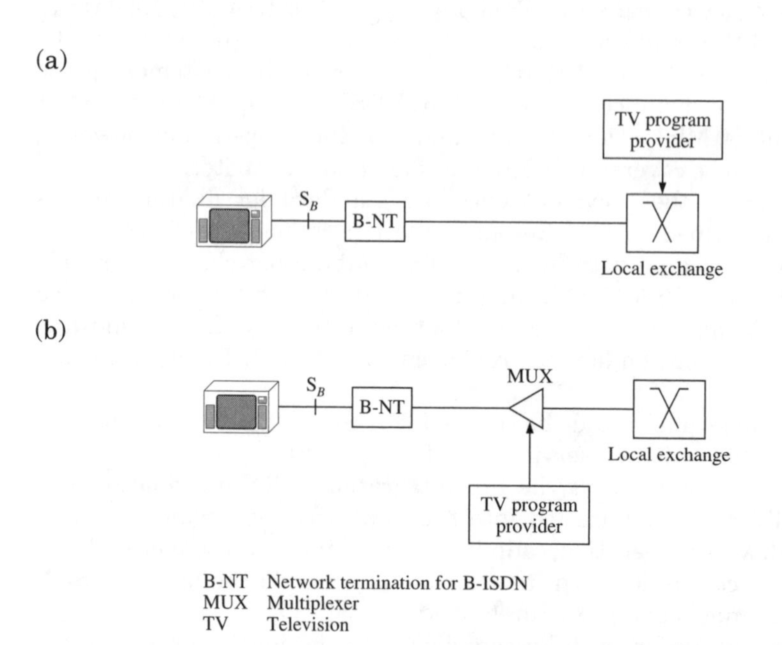

Figure 14.2 Provision of TV programs to the B-ISDN customer.

A mixture of switched and non-switched provision of TV program channels might also be realized: in addition to a block of fixed TV channels the user could be given the option to select from a TV program pool.

TV program selection (unless merely performed in the customer network) makes some demands on the signaling procedures. The network must be able to handle numerous simultaneous signaling messages rapidly. Changing the TV channel must not take longer than the time that people are used to nowadays. Perhaps this is easier to achieve by (in-band) end-to-end-signaling between user and TV program provider after the connection has been established. In this case, however, coordination between the customer equipment (TV sets and/or B-NT2), the local exchange and the TV program providing unit is necessary to avoid conflicts and to make economic use of the available access line bandwidth. (When, for example, a customer switches on a TV set and selects a TV channel, a new connection normally has to be established unless the TV channel required is already provided to another TV set. If a new connection is to be established, the network must be checked (for example, in the local exchange) to see whether this can be done without interfering with existing connections on the customer's access line.)

A comprehensive approach to residential broadband (RBB) is given in the ATM Forum's baseline text (ATM Forum, 1997j) which aims at a 'complete end-to-end ATM system both to and from the home and within the home, to a variety of devices, e.g. set-top box, personal computer and other home devices'. This document defines a generic reference architecture comprising the core ATM network, ATM access network, access network termination, home ATM network and terminal equipment.

As for home UNIs, a 25 Mbit/s plastic optical fiber interface is proposed.

Several options are specified for the access network:

- hybrid fiber coax (HFC)
- fiber to the curb (FTTC)
- fiber to the home (FTTH) via a passive optical network (PON)
- asymmetric digital subscriber loop (ADSL).

14.4 Intelligent network aspects of B-ISDN

The **intelligent network** (IN) concept has been developed for ISDNs but may also be incorporated in B-ISDNs. INs will provide existing and new service components that can be flexibly combined according to the user's wishes. INs require powerful signaling procedures, effective service control and management of service-related data. ATM-based networks offer excellent support for IN features.

14.4.1 Architectural model

The term **intelligent network** is used (according to ITU-T (1990))

> 'to describe an architectural concept for all telecommunication networks. IN aims to ease the introduction of supplementary services (universal personal telecommunication, free-phone, etc.) based on more flexibility and new capabilities'.

The IN concept for the creation and provision of services is characterized by:

- extensive use of information processing techniques;
- efficient use of network resources;
- modularity of network functions;
- integrated service creation and implementation by means of reusable standard network functions;
- flexible allocation of network functions to physical entities;
- portability of network functions among physical entities;
- standardized communication between network functions via service-independent interfaces;
- service provider access to the process developing services by combining of network functions;
- service subscriber control of subscriber-specific service attributes;
- standardized management of service logic.

The functions required for IN and their division into functional entities are depicted in Figure 14.3.

The main principle is the distribution of service control between call control/service switching and service control. To illustrate this concept, Figure 14.4 gives an example of how these functions are

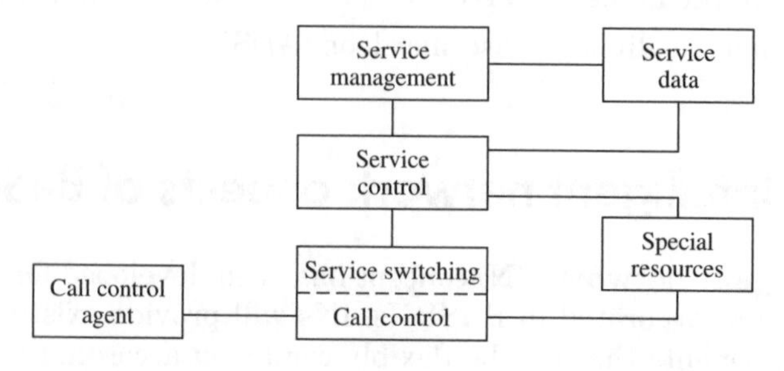

Figure 14.3 Intelligent network functional entities.

mapped onto the physical entities of a telecommunication network. Call control/service switching is located in the exchanges which use trigger tables to determine whether they can complete a call themselves or if it has to be handled by the service control point. Exchanges communicate with the service control point via the signaling network (for example SS7, as shown in the figure). The special resource functions (located in this example in an intelligent peripheral) may provide protocol conversion, speech recognition, synthesized speech provision, and so on.

The service control point contains the IN service logic and handles service-related processing. It uses the service data function which provides access to all necessary service-related data and network data and also carries out consistency checks on data.

Finally, the service management function involves service management control, service provision control and service deployment control. Example service management control functions are: collection of

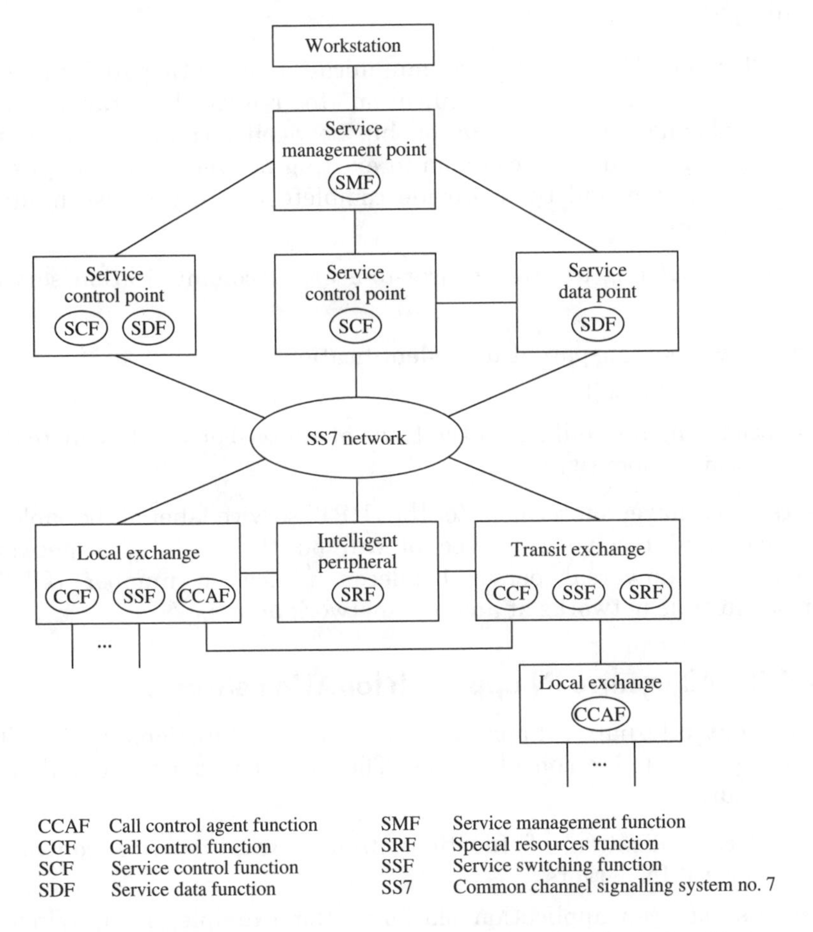

CCAF Call control agent function
CCF Call control function
SCF Service control function
SDF Service data function
SMF Service management function
SRF Special resources function
SSF Service switching function
SS7 Common channel signalling system no. 7

Figure 14.4 Example of functions mapping onto physical entities.

service statistics, reporting of usage of non-existent freephone numbers and reporting of unauthorized access in a virtual private network. Service provision control handles operation and administration for service provisioning (creating new subscribers, modifying subscription records, and so on). Service deployment control is invoked when a new service is introduced into the network. It deals with service logic allocation, signaling and routing definition, service data introduction, allocation of trigger capabilities, special service feature allocation, and so on.

14.4.2 Overview of IN services

IN services can be grouped into three categories according to the network capabilities that are required for their implementation (Frantzen et al., 1992). Examples of IN services are described in Table 14.1.

Another field of application for IN capabilities is **universal personal telecommunication** (UPT). It is described as follows (ITU-T, 1990):

> 'Universal personal telecommunication is anticipated to bring network personal identification to reality by transparently replacing the static relationship between terminal identity and subscriber identity common in existing networks with a dynamic association and thus provide complete mobility across multiple networks.'

The aim of UPT is to provide user-to-user telecommunication services with:

- network-transparent user identification
- personal mobility
- charging and billing on the basis of subscriber identity instead of terminal identity.

Users who have subscribed to the UPT service should be able to establish and receive any type of call on the basis of a network-transparent universal personal telecommunication number (UPTN) across multiple networks at any user-network access.

14.4.3 Alternative IN approach for ATM networks

At present alternative approaches towards IN functionality in ATM networks are under consideration. They are based on the following architecture:

- decentralization of functions from central service control to several IN servers
- use of open application platforms (for example, Java, Windows NT)

Table 14.1 Intelligent network services.

Category	*Service*	*Description*
B (called party) number services	Basic green number service (free-phone)	Toll-free service paid for by called party. A special access code, e.g. 800, serves as trigger.
A + B (calling and called party) number services	Alternate billing service	Allows the user to bill the call to a number other than his or her own number.
	Enhanced emergency response service	Dialing of a special countrywide emergency number serves as a trigger. The service control point determines the appropriate emergency response control number for call completion.
	Virtual private network	Provides the functionality of a private or dedicated network using the shared facilities of the public network.
	Area wide centrex	Offers dynamic resource allocation and uniform numbering plan over geographically dispersed locations. The centrex service utilizes a public network local exchange to provide PBX-like features to a group of business customers. Area wide centrex interconnects multiple customer locations as if they were connected to a single switch.
Interactive services	Interactive green number service	Allows the user to select one of a set of offered alternatives associated with a single green number.
	Voice messaging	Requires a user dialog for entering control commands, for example.
	Call completion	Comprises various features to assist the calling party in completing the call to the called party.

- any protocol simpler than SS7 between nodes (such as the IP protocol).

It is to be expected that specification of such IN solutions will start soon. Whether they will replace the 'classical' IN architecture also in the case of public ATM networks or will mainly be adopted for corporate ATM networks is not yet clear.

14.5 Tariffing in B-ISDN

Tariffing of services in future ATM networks is a crucial factor that will strongly influence the acceptance of B-ISDN by customers. The following questions may be raised:

- How can tariffs for ATM networks be made cost-beneficial to operators' expenses (investments and operational costs)? What tariff components should be defined, and what parameters have to be measured for charging?
- How can a smooth transition from tariffing of services in existing networks to future ATM tariffs be achieved?

A brief overview of tariffing in existing networks is given before moving on to the topic of how to charge for ATM services.

14.5.1 Tariffing in existing networks

Two different charging categories are used at present:

- **Basic rate** (per month, for example) which is charged for network access irrespective of a customer's actual traffic load.
- **Traffic charges** which are determined by one or more of the following components:
  - charges for establishing/releasing connections
  - charging per connection time
  - charging per volume
  - flat rate.

For instance, in today's telephone networks calls are usually charged according to the connection time. Tariffs may vary during the day (higher charges at busy hours) and may depend on the distances to be covered. Volume-dependent tariffing is often chosen for data packet networks; that is, a charge is made for each packet delivered. In addition, small charges may be levied for establishing and maintaining a connection.

These two examples show that different networks may require different tariffing schemes to comply with the general requirements for cost-adequate charging and user acceptance (the latter sometimes reflects the historical and social situation in a country and may outweigh the former).

14.5.2 Tariffing in public ATM networks

An ATM-based public B-ISDN may supply both stream and packet-type applications. Charging in such a network is therefore a complex issue to resolve.

Although all information in an ATM network is transmitted in cells, simple (linear) volume-dependent tariffing, as used in conventional packet networks, may not be accepted. If a per-cell charge was introduced which led to phone call charges comparable to those of today, a broadband stream communication with a bit rate of, say, 10 Mbit/s would be so expensive that it would hardly be used.

Based on these considerations, ATM charging must meet the following requirements (Hagenhaus, 1990):

- no change in charges for existing mass services like telephony, facsimile and data transmission;
- service-independent charging;
- flexibility with respect to the introduction of new services;
- transparency of tariffs;
- due consideration of peak bit rates (as these strongly affect network dimensioning and costs);
- fair charging for variable bit rate services;
- kept administrative efforts as low as possible (for example, avoid cell counting for telephone connections).

A proposal was made in Hagenhaus (1990) that tries to take into account all these requirements on ATM charging. It is based on:

- a basic rate
- traffic-dependent charges comprising two components:
  - charging for call establishment
  - charging according to the duration and bit rate of a call.

The basic rate reflects the cost of providing the customer access. When a customer wishes permanently to restrict access to a bit rate below the maximum possible (only 30 Mbit/s, say, of a 155 Mbit/s interface), and this actually reduces network access costs, the basic rate may be adjusted accordingly.

As the processing effort for setting up a connection is almost independent of the requested bit rate, call establishment may be charged at the same rate in all cases. However, the setting up of virtual connections implies the reservation of network resources. By raising different charges for call establishment in the case of prioritized, high quality of service or high bit rate connections, users could be dissuaded from reserving network resources that would not be needed.

The second component of the traffic-dependent charges – charging according to call duration and bit rate – takes into account the cost of using transmission systems in the network. This charge would normally increase with the bit rate in a nonlinear way. The time unit for charging should be sufficiently low to ensure adequate charging for variable bit rate services. The peak bit rate during this short time unit could be used to determine the charge for such a connection.

A charging mechanism of this type would reflect the network efforts to support ATM connections better than a plain volume-dependent tariff (charging per cell).

Discussion within ITU-T on charging in ATM networks has yielded some general charging principles as outlined above, but no detailed charging procedures are yet available. The following ATM-specific items may have to be taken into account:

- existence of two cell loss priorities
- usage parameter control capabilities
- actual resource allocation in the network.

14.6 Security in ATM networks

Security is quite a general network issue in which both operators and customers have a fundamental interest. Additional security issues arise in the context of ATM networks. In the following we first discuss general network security requirements and then address a special ATM network security problem.

The main *security objectives* of networks are:

- confidentiality (exclusively authorized access to all stored and transferred information);
- data integrity (protection of information against malicious or inadvertent corruption);
- accountability (whoever initiates any action should be held responsible for it);
- availability (of network facilities).

To achieve these security goals, a set of security measures must be implemented to prevent:

- 'spoofing' (an entity pretends to be someone else);
- eavesdropping (unauthorized monitoring of a communication);
- unauthorized access to data or services;
- loss or corruption of information;
- repudiation or forgery (an entity involved in a communication afterwards refuses responsibility or claims someone else was the author);
- denial of service.

It is obvious that in open network environments much more stringent security measures have to be taken than in closed groups. The following actions may have to be applied (partly or in total):

- user authentication
- access control and authorization
- activity logging (to be able to trace offenses)
- audit (analyzing and exploiting logged data on security-relevant events)

- alarming in the case of attacks on security
- recovery from attempted breaches on security.

Different technical realizations of the above measures are possible: for instance, to enforce authentication, either symmetric algorithms can be applied, where the two parties concerned must first share a secret key, or public algorithms where each node only has to know the other's public key. (For further details see, for example, the ATM Forum specification on security ('Security 1.0', work in progress).)

An ATM-specific security issue is described in (Ginsburg, 1996):

> 'A VCC terminating on an end-system internal to a corporate network will bypass the organization's firewall. Firewall functionality on each and every end-system is of course not practical. One alternative method may be to include higher-layer protocol information via information elements in the connection signaling message. Intermediate systems (for instance, an ATM switch acting as a corporate firewall) could then filter on this information.'

14.7 ATM application programming interface (API)

As Figure 14.5 shows, the API is a standard interface over which applications have access to the ATM layer to map their QoS requirements onto appropriate ATM parameters. The ATM Forum has

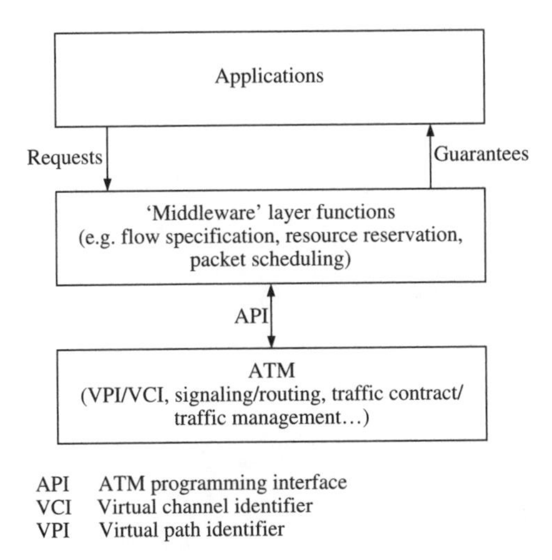

Figure 14.5 ATM application programming interface (API).

only defined requirements on APIs, so no single open API for ATM will exist. Examples of APIs are: *WinSock 2* for Microsoft and *X/Open transport interface* for UNIX and Apple.

WinSock 2, a network programming and service interface for the Windows environment, is independent of a specific layer 3 protocol or a specific layer 2 network. WinSock 2 supports QoS, the usage of both point-to-point and point-to-multipoint VCCs, and ATM addressing.

Review questions

1. What is the impact of ATM on voice connections and how can one get rid of this?
2. What are the requirements of residentials/SOHOs on ATM access and which access technologies exist to meet them?
3. Which are the functional entities of the 'classical' IN model and which are the physical entities they can be mapped onto?
4. Which security objectives should be met by ATM networks? Which security measures may be used?
5. What is an API?

15

ATM implementations

15.1 Early broadband/ATM trials

The obvious interest in new services, applications and systems is always accompanied by elements of uncertainty (acceptance, demand, technology, economy and organization). Laboratory implementations, field trials with pilot applications and test networks make it possible to overcome these uncertainties at an early stage; they can assist in (Ambrüster and Schneider, 1990):

- the design of realistic communication scenarios for supporting important applications;
- requirements specification of commercial services and terminals;
- influencing international standardization;
- information and motivation of potential users, service providers, network operators and manufacturers.

All around the world, field trials with pilot applications and test networks have been run (Armbrüster and Schneider, 1990; *Communications Weekly International*, 1990; Coudreuse and Servel, 1987; David et al., 1990; Göldner, 1990; Schaffer, 1990). A very early broadband trial was the BERKOM (*Ber*liner *Kom*munikationssystem) project started in Berlin in 1986 with the objective to design, develop and demonstrate applications for B-ISDN. The following items were included in the BERKOM project (Popescu-Zeletin et al., 1988):

- text and document processing in an office environment;
- high-quality printing and publishing;
- high-speed transfer of medical images (X-ray images, for example);
- applications in the field of computer-aided design (CAD) and computer-integrated manufacturing (CIM);

- distribution of HDTV programs;
- access to information bases containing text, pictures and video;
- use of high-quality audio and video information in the residential area.

In autumn 1989, Siemens installed a pre-standard ATM switch as part of the BERKOM trial (Göldner, 1990). The switch fabric transferred cells (2-octet header and 30-octet information field) from 16 inlets to 16 outlets, all operating at 140 Mbit/s. The participants were located in Berlin and were attached to the ATM switching node via single-mode fibers.

Typical applications using this ATM switch were (see also Figure 15.1):

- interconnection of LANs;
- video communications at 64 kbit/s, 2 Mbit/s and 34 Mbit/s;
- joint editing over broadband links;
- interconnection of HICOM PBXs via ATM (private networking);
- access to 64 kbit/s ISDN via an EWSD exchange.

15.2 ATM services of public network operators

At present a strong push towards ATM can be observed in the field of corporate networking. The reason is simply that ATM for the first time offers a unique, standard networking technique which both supports existing and upcoming data transmission protocols (X.25, frame relay, connectionless data services, and so on), and promises to accommodate real-time service such as voice and/or video transmission or arbitrary combinations thereof (multimedia applications). In ATM the transport link capacity and the service bit rate are decoupled, so performance-limited shared-medium configurations (rings, bus systems) can be replaced with a star-configured topology with ATM switches/hubs where each user is entitled to establish a (switched) connection at the full interface bit rate to anyone on the customer network.

Public network operators also have good reasons for implementing ATM networks:

- create a common backbone network for data services (with an option later on to use it for all types of services including telephony);
- be competitive against private networking (for example, offer ATM-based virtual private networks);

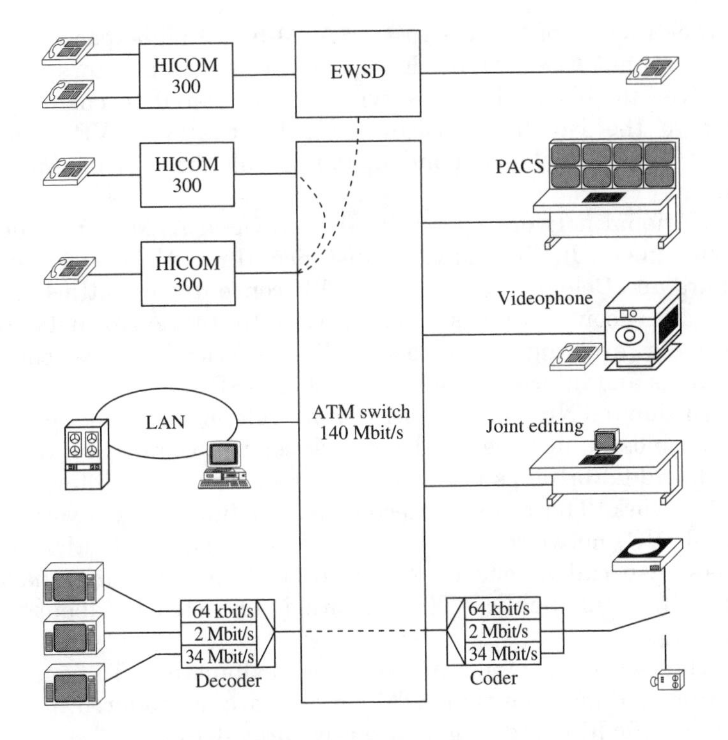

Figure 15.1 ATM applications of the BERKOM trial.

- test the acceptance of new services, especially multimedia services;
- use leading-edge technology.

Examples of some public ATM network implementations are highlighted here. In North America all the major telecommunications network operators have been running ATM trials with the intention of making commercial use of this installed basis as quickly as possible. The services mainly offered by these ATM networks are:

- ATM bearer service (first VP-based, then additionally switched VCs)
- frame relay
- SMDS
- LAN emulation
- circuit emulation (1.5/45 Mbit/s, for example)
- video applications.

In Europe the era of border-crossing ATM-based high-speed networking started in mid-1994. Many European network operators committed themselves to this goal. The services to be offered to customers were similar to the North American ones. International VP nodes were foreseen, for example, in London, Paris, Cologne, Berlin, Madrid and Milan.

National network operators also provided ATM infrastructure to their customers. In Germany, for instance, three ATM nodes in Berlin, Hamburg and Cologne, with remote ATM concentrators attached to each node, gave about 200 customers access to the ATM network. The network is evolving from leased line services to switched ATM connections and interworking with 64 kbit/s ISDN.

In Japan, Nippon Telegraph and Telephone Corporation (NTT) drafted a most ambitious ATM network concept comprising ATM-based 2.5 Gbit/s add/drop rings in the local access area and an intermeshed VP trunk network. The cross-connected (leased line) VP network and the switched VC network were clearly separated. Sophisticated OAM concepts (especially concerning VP protection switching) have been introduced. A full public ATM network is expected by about the year 2015.

The above examples are far from exhaustive. To conclude this topic we would mention that ATM services are also offered in Australia and the Pacific Rim and by global service providers.

Appendix A

ATM standardization

A.1 ITU-T recommendations

A variety of standards have to be agreed worldwide in order that any two customers can indeed communicate with each other via B-ISDN. These standards are set by recognized standards bodies, amongst which the ITU-T (previously CCITT) plays an outstanding role in the context of B-ISDN as it produces standards (called **recommendations**) which are acknowledged worldwide.

Table A.1 summarizes the current B-ISDN-related ITU-T I-series recommendations. The scope of application of these recommendations is illustrated by Figure A.1.

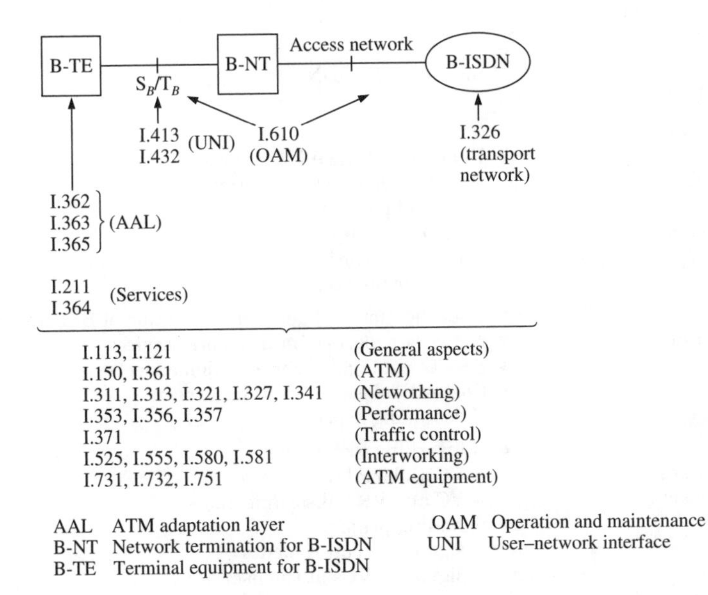

Figure A.1 Application scope of B-ISDN I-series recommendations.

The following recommendations are in preparation:

- I.313: B-ISDN network requirements
- I.341: B-ISDN connection types
- I.351: ISDN performance
- I.371.1: Conformance definitions for ABT and ABR
- I.375: Network capabilities to support multimedia services
- I.581: General arrangements of B-ISDN interworking.

Other ITU-T recommendations (cf. http://www.itu.ch/publications) relevant to B-ISDN/ATM are part of the

- F series (broadband service descriptions)
- G series (transport networks)
- Q series (signaling).

Also E.164 (numbering plan for ISDNs) is of importance for B-ISDNs.

Table A.1 ITU-T I-series recommendations for B-ISDN.

Number: Title	*Contents*
I.113: Vocabulary of terms for broadband aspects of ISDN	Definition of terms used for B-ISDN
I.121: Broadband aspects of ISDN	Principles of B-ISDN
I.150: B-ISDN ATM functional characteristics	• Functions of the ATM layer (cell header): – virtual channel connection – virtual path connection – payload type – cell loss priority – generic flow control
I.211: B-ISDN service aspects	• Classification and general description of B-ISDN services • Network aspects of multimedia services • Service timing, synchronization aspects • Connectionless data service aspects • Videocoding aspects
I.311: B-ISDN general network aspects	• Networking techniques: – VC and VP level – VC and VP links/connections • Signaling principles: – signaling requirements – signaling VCs at the user access – meta-signaling channel

Table A.1 *(continued)* ITU-T I-series recommendations for B-ISDN.

Number: Title	*Contents*
I.321: B-ISDN protocol reference model and its application	• Extension of I.320 • Overview of functions in layers and sublayers • Distinction between physical layer and ATM layer functions
I.326: Functional architecture of transport networks based on ATM	• Client–server associations • Topology • Connection supervision
I.327: B-ISDN functional architecture	• Enhancement of I.324 • Basic architecture model, network capabilities • Reference connection, connection elements • Connectionless service function, access via VCCs or VPCs
I.353: Reference events for defining ISDN and B-ISDN performance parameters	• Definition of relevant measurement points and associated performance-significant reference events
I.356: B-ISDN ATM cell layer transfer performance	• Performance model • Performance parameters • Performance measurement methods
I.357: B-ISDN semi-permanent connection availability	• Availability model and parameters • Availability performance objectives
I.361: B-ISDN ATM layer specification	• ATM cell structure: – 5-octet header + 48-octet information field • Coding of cell header
I.362: B-ISDN ATM adaptation layer functional description	• Basic principles of the AAL: – service dependent – may be empty for some applications • Sublayering of the AAL: – SAR and CS • 4 AAL service classes based on: – timing relation source/sink – constant/variable bit rate – connection-oriented/connectionless mode
I.363.1/.2/.3/.5: B-ISDN ATM adaptation layer specification	• AAL protocol types defined with respect to: – services – SAR/CS functions – SAR-PDU structure and coding
I.364: Support of the broadband connectionless data bearer service by the B-ISDN	• Framework for service provision • Protocols (UNI/NNI)
I.365.1–4: B-ISDN ATM adaptation layer sublayers	• Frame relaying service-specific convergence sublayer • Service-specific coordination function to provide the connection-oriented network service • Service-specific coordination function to provide the connection-oriented transport service • Service-specific coordination function for HDLC applications

Table A.1 *(continued)* ITU-T I-series recommendations for B-ISDN.

Number: Title	*Contents*
I.371: Traffic control and congestion control in B-ISDN	• Traffic descriptors and parameters • Functions and procedures for traffic control and congestion control
I.374: Framework recommendation on 'network capabilities to support multimedia services'	• Functional architecture of multimedia services • Network capabilities • Multimedia resource management
I.413: B-ISDN user–network interface	• Reference configurations • Examples of physical configurations • Interfaces at T_B and S_B
1.432.1–5: B-ISDN user–network interface – physical layer specification	• Electrical/optical parameters • Interface structure (e.g. SDH frame) • Header error control generation/verification • Cell delineation
I.525: Interworking between networks operating at bit rates less than 64 kbit/s with 64 kbit/s-based ISDN and B-ISDN	• Interworking configurations and examples
I.555: Frame relaying bearer service interworking	• Interworking between frame relaying bearer service and B-ISDN
I.580: General arrangements for interworking between B-ISDN and 64 kbit/s based ISDN	• Interworking configurations • Interworking functional requirements
I.610: B-ISDN operation and maintenance principles and functions	• OAM principles: – hierarchical structure of OAM functions and information flows • OAM functions for B-ISDN UNI and access network • Implementation of OAM functions of the physical layer and ATM layer
I.731: Types and general characteristics of ATM equipment	• Classification of ATM equipment types • Generic performance requirements, quality of service aspects, and timing synchronization requirements applicable to ATM equipment
I.732: Functional characteristics of ATM equipment	• Transfer functions and layer management functions • Configuration/fault/performance/accounting/security management • Adaptation functions
I.751: ATM management of the network element view	• ATM network management architecture • ATM network element plane management requirements • Management information model

A.2 ATM Forum

Table A.2 is a (non-exhaustive) list of important ATM Forum specifications either already issued or in preparation (cf. http://www.atmforum.com/atmforum/specs).

Table A.2 ATM Forum specifications.

Document name: Title or subject		*Contents*
AF-UNI: ATM UNI Specification (V.2.0/3.0/3.1/4.0)		UNI description (V.2.0/3.0/3.1 include signaling and OAM)
AF-PHY:	Physical layer specifications	UNIs at 1.5/2/6/25.6/34/45/50/155/622 Mbit/s
	ATM inverse multiplexing	describes transmission of ATM via n times 1.5/2 Mbit/s
	UTOPIA	Defines an interface between the physical layer and the ATM layer
	WIRE	Defines an interface between the PMD and TC sublayers of the physical layer
AF-DXI: ATM data exchange interface		Defines a frame-based ATM access
AF-PNNI: PNNI V.1.0		Private network-to-network interface
AF-BICI: B-ICI		Service interface between ATM networks
AF-SAA:	Frame UNI (FUNI)	Frame-based UNI supporting signaling and QoS
	Circuit emulation service interoperability specification	Defines a CES via ATM to support interconnection of PBXs and video codecs
AF-LANE: LAN emulation over ATM		Defines LAN emulation service across ATM
AF-MPOA: MPOA V.1.0		Multi-protocol over ATM
AF-VTOA: Voice and telephony over ATM		Defines trunking for voice and telephony over ATM
AF-TM: ATM traffic management specifications		Includes ABR support
AF-NM: Public ATM network management specifications		Describe requirements and protocols for the M3, M4, M5 interfaces
AF-ILMI: ILMI 4.0		Integrated local management interface
AF-TEST: ATM Forum test specifications		Define conformance, performance and interoperability testing
AF-SEC: Security 1.0		Defines a security framework for ATM
IPNNI: Integrated PNNI		PNNI which supports IP routing
AINI: Public/private ATM interworking		Defines an interface for the interconnection of ATM networks which may internally use either PNNI or B-ISUP or even other protocols
RBB: Residential broadband		Defines ATM access for residential broadband users
WATM: Wireless and mobile ATM		Defines requirements and architecture of wireless/mobile ATM

Appendix B

Glossary of basic ATM-related terms

Asynchronous Transfer Mode (ATM) Multiplexing scheme and switching technique tailored for high-speed networks. ATM basically supports all kinds of applications: constant bit rate and burst-type traffic, real-time services and data services with less stringent quality of service requirements. ATM is asynchronous in the sense that transmission capacity (*ATM cell*) is provided for an individual connection only if actually demanded.

ATM adaptation layer (AAL) A limited number of AAL types (AAL 1 – 5) has been defined to link the ATM layer to higher layers. AAL functions are segmentation and reassembly of information and other, service-class-specific functions.

ATM address resolution protocol (ATM-ARP) This protocol is used for address resolution between IP addresses and ATM addresses.

ATM cell Information unit of ATM: a cell is a fixed-size slot comprising 53 octets five of which are allocated to the cell header and 48 to the information field.

ATM cell header The first five octets of an ATM cell containing address information and additional control data.

ATM connection-oriented service Prior to the exchange of data between two entities an end-to-end connection is established (either via network management procedure or signaling) onto which all data are sent.

ATM connectionless service All data are sent via a pre-established ATM virtual connection to a connectionless server which then establishes connections to the destinations. (Other implementations are currently being investigated.)

ATM Forum An organization of end-users, carriers and equipment vendors who define a common set of 'implementation agreements' for ATM based networking.

ATM network-node/network-to-network interfaces Interfaces between ATM nodes/networks. Several types exist: NNI between ATM nodes of a public network, PNNI between private network nodes, BICI between public networks, AINI to connect any two ATM networks which may employ different internal network protocols.

ATM protocol reference model (ATM PRM) It was derived from the ISDN PRM which itself was derived from the OSI layered model. The ATM PRM puts the ATM layer (which performs cell handling) on top of the physical layer (which is responsible for the proper bit transport). Above the ATM layer are the AAL and the higher OSI layers. Note that it proved impossible to map the ATM layer onto a single OSI layer due to the ATM layer functionality which encompasses OSI layer 1 to 3 aspects.

ATM transfer capability See QoS.

ATM user-network interface (UNI) Interface providing access to an ATM network. Besides physical-layer-specific information, all information is carried in ATM cells across an ATM UNI. UNIs in the range of 1.5 Mbit/s up to 622 Mbit/s have been standardized. Different physical layers may be used (for example electrical/optical; SDH/PDH framing).

ATM virtual connection (VC) ATM is a connection-oriented networking technique. Virtual connections are established link by link either by signaling protocols or by subscription. ATM has a two-tier network architecture: the basic connection is a virtual channel connection (VCC), and several virtual channels can be grouped into a virtual path. A virtual path connection (VPC) is handled as a single entity inside the network.

Available bit rate (ABR) An ATM transfer capability where the network controls the sender's cell emission rate through the use of resource management cells. As long as the cell sender complies with the information obtained from the network a very low cell loss ratio can be guaranteed. However, a maximum cell transfer delay is not specified.

Broadband-ISDN (B-ISDN) ATM-based high-speed network based on certain ISDN principles such as common access to many services via standardized user–network interfaces, out-of-band signaling, and ISDN numbering scheme.

Broadcast and unknown server (BUS) A server used for LAN emulation which transmits all broadcast/multicast and unknown traffic over the emulated LAN.

Cell delay variation (CDV) A quality of service parameter describing the variations of the cell transfer delays of a single connection. The cell delay variation tolerance (CDVT) is the maximum acceptable CDV for a connection.

Cell loss priority (CLP) A bit in the ATM cell header marking those cells which should be discarded prior to other cells in case of congestion.

Cell loss ratio (CLR) Ratio of lost cells to sent cells.

Cell transfer delay (CTD) The time needed by a cell from its source to its destination.

Connection admission control (CAC) An algorithm used in ATM switches which checks if enough resources within the switch and/or the network are available for carrying the requested connection without affecting other, already established, connections.

Constant bit rate (CBR) An ATM transfer capability with guaranteed cell loss ratio and cell transfer delay. Normally, CBR connections have the highest priority compared to the other ATM transfer capabilities.

Circuit emulation service (CES) Transport of a constant bit rate signal, e.g. 15/2 Mbits meeting the specified requirements on delay bit error rate etc., for such signals.

E.164 An addressing scheme specified by ITU-T.

Early packet discard (EPD) In case of congestion ATM switches may discard entire data frames instead of single ATM cells. This mechanism can only be used on AAL type 5 connections which use the PTI field for indication of the end of a higher layer data frame.

Explicit forward congestion indication (EFCI) A bit in the ATM cell header which might be set by network elements in case of congestion. End-terminals are expected to lower their cell emission rate if this bit is set.

Header error control (HEC) A cyclic redundancy check code which allows the correction of single bit errors in the ATM cell header. In case of multiple bit errors the cell is discarded to prevent it from being misrouted.

Integrated services digital network (ISDN) Circuit-switched network based on 64 kbit/s channels (derived from digitized telephony requirements) with a common user–network interface for access to a variety of services. ISDNs have out-of-band signaling capabilities which allow for provision of new and improved services relative to 'plain old telephony'.

Internet engineering task force (IETF) International organization specifying the Internet protocol suite. Specifications are published in 'request for comments' (RFC).

Internet protocol (IP) The network layer of the Internet communications protocol providing a connectionless data delivery service. Its basic information unit is the Internet datagram.

ITU-T Formerly CCITT, an international organization delivering recommendations for (mostly) public communication relevant aspects.

LAN emulation (LANE) In LANE ATM emulates the medium access protocol. It is based on a set of ATM Forum specifications.

Management information base (MIB) An abstract description of a specific network element or connection and transmission parameters which can be manipulated and/or evaluated for network management purposes.

Meta-signaling: ATM uses out-of-band signaling via specific signaling (virtual) channels. Unless pre-established, signaling channels are allocated by a meta-signaling protocol via a predefined meta-signaling (virtual) channel.

Multiprotocol over ATM (MPOA) Based on a set of ATM Forum specifications for running layer 3 protocols over ATM exploiting ATM's quality of service offerings.

Next hop routing protocol (NHRP) A protocol for routing over ATM (and other) networks for optimized end-to-end connectivity.

Operation and maintenance (OAM) Specific measures to keep the network running and delivering a specified level of service. As an example dedicated ATM OAM cells are defined for this purpose.

Payload type identifier (PTI) A three bit field in the ATM cell header identifying user data cells, OAM cells, EFCI indications and data frame boundaries of AAL type 5 connections.

Permanent virtual connection (PVC) A connection which is established (and released) by network management commands and not through signaling.

Policing The set of actions taken by the network to delimit the actual cell rate of a connection entering that network to the negotiated cell rate.

Private network-to-network interface (PNNI) Based on a set of ATM Forum specifications for addressing, signaling and routing within a private network. Nowadays, also used in some public networks.

Quality of service (QoS) A description/specification of the 'quality' of an ATM connection using a set of parameters like CLR, CDVT, and CTD bit rate. Different end-user applications require ATM connections with different levels of QoS. Several ATM transfer capabilities have been specified each offering a different QoS.

Resource management All actions taken for sharing network 'resources' like transmission capacity, buffer space etc. among connections.

Signaling A means of establishing connections on demand by end-user equipment or network equipment. Dialing a number for establishing a telephone call is one form of signaling.

Switched virtual connection (SVC) A connection established via signaling procedures.

Synchronous digital hierarchy (SDH) ITU-defined transmission standard – derived from the North American SONET – which allows the hierarchical mapping of quite different information streams (for example, fixed bit rate channels, ATM cells) into a common transmission frame.

TCP (transmission control protocol) The transport layer of the Internet communications protocol providing a connection-oriented full duplex service. TCP normally uses the IP protocol to transmit data.

Unspecified bit rate service (UBR) 'Best effort' type of ATM service with no explicit commitment of the network to support the connection.

Usage/network parameter control (UPC/NPC) See policing.

Variable bit rate (VBR) An ATM transfer capability which exists in two variations, VBR-rt and VBR-nrt for real-time and non-real-time services, respectively.

Virtual channel connection (VCC)/Virtual path connection (VPC) See ATM virtual connection.

Appendix C

Acronyms

AAL	ATM adaptation layer
AAL-CU	AAL composite user
ABR	Available bit rate
ABT	ATM block transfer
ACR	Allowed cell rate
ADM	Add/drop multiplexer
ADSL	Asymmetric digital subscriber line
AINI	ATM inter-network interface
AIS	Alarm indication signal
AL	Alignment
ANP	AAL type-2 negotiation procedure
ANS	ATM name server
ANSI	American National Standards Institute
API	Application programming interface
APS	Automatic protection switching
ARP	Address resolution protocol
ASN.1	Abstract syntax notation no. 1
ATM	Asynchronous transfer mode
AU	Administrative unit
AUU	ATM layer user to ATM layer user
BASize	Buffer size
BEDC	Block error detection code
B-ICI	B-ISDN intercarrier interface
B-ISDN	Broadband integrated services digital network
B-ISUP	Broadband ISDN user part
BLER	Block error result
B-LLI	Broadband lower-layer information
B-NT	Network termination for B-ISDN
B-TA	Terminal adaptor for B-ISDN
B-TE	Terminal equipment for B-ISDN
Bc	Committed burst size
BCI	Bearer control identifier

Be	Excess burst size
BECN	Backward explicit congestion notification
BERKOM	Berliner Kommunikationssystem
BGP	Boundary Gateway Protocol
BICMOS	Bipolar complementary metal-oxide semiconductor
BIP	Bit interleaved parity
BOM	Beginning of message
BSVC	Broadcast signaling virtual channel
BSVCI	Broadcast signaling virtual channel identifier
BT	Burst tolerance
Btag	Beginning tag
BUS	Broadcast and unknown server
C-*i*	Container *i*
CAC	Connection admission control
CAD	Computer-aided design
CATV	Cable television
CAU	Cause
CBDS	Connectionless broadband data services
CBR	Constant bit rate
CC	Cross-connect
CCITT	Comité Consultatif International Télégraphique et Téléphonique
CDV	Cell delay variation
CDVT	Cell delay variation tolerance
CEI	Connection element identifier
CES	Circuit emulation service
CI	Congestion indication
CIC	Circuit identification code
CID	Channel identifier
CIM	Computer-integrated manufacturing
CIR	Committed information rate
CL	Connectionless
CLNAP	Connectionless network access protocol
CLNIP	Connectionless network interface protocol
CLP	Cell loss priority
CLSF	Connectionless service function
CMI	Coded mark inversion
CMI	Cluster member identifier
CMIP	Common management information protocol
CMIS	Common management information service
CMOS	Complementary metal-oxide semiconductor
CN	Customer network
CN	Congestion notification
CO	Connection oriented
COM	Continuation of message
CP	Common part

CPS	Common part sublayer
CPCS	Common part convergence sublayer
CPI	Common part indicator
CPN	Customer premises network
CPS-PH	CPS packet header
CPS-PP	CPS packet payload
CRC	Cyclic redundancy check
CRMA	Cyclic reservation multiple access
CS	Convergence sublayer
CS	Capability set (for signaling)
CSI	Convergence sublayer indication
CSI	Carrier scale internetworking
CSMA/CD	Carrier sense multiple access with collision detection
CSPDN	Circuit-switched public data network
DAN	Desk area network
DBR	Deterministic bit rate
DC	Direct current
DCE	Data circuit-terminating equipment
DE	Discard eligibility
DFA	DXI frame address
DLC	Data link control
DLCI	Data link connection identifier
DLPI	Data link provider interface
DMDD	Distributed multiplexing distributed demultiplexing
DQDB	Distributed queue dual bus
DTE	Data terminal equipment
DXI	Data exchange interface
ELAN	Emulated LAN
EOM	End of message
EPD	Early packet discard
ER	Explicit rate
ET	Exchange termination
Etag	Ending tag
ETSI	European Telecommunication Standards Institute
EWSD	Elektronisches Wählvermittlungssystem Digital
FA	Frame address
FCS	Frame check sequence
FDDI	Fiber-distributed data interface
FEC	Forward error correction
FECN	Forward explicit congestion notification
FIFO	First-in first-out
FR	Frame relay
FR-SSCS	Frame relaying service-specific convergence sublayer
FRSF	Frame relay service function
FTP	File transfer protocol
FTTC	Fiber to the curb

FTTH	Fiber to the home
FTTO	Fiber to the office
FUNI	Frame-based user-to-network interface
GCAC	Generic call admission control
GCRA	Generic cell rate algorithm
GFC	Generic flow control
GSMP	General switch management protocol
GSVCI	Global signaling virtual channel identifier
HDSL	High-speed digital subscriber line
HDTV	High-definition television
HEC	Header error control
HFC	Hybrid fiber coax
HICOM	High technology communication
HIPPI	High-performance parallel interface
HLPI	Higher-layer protocol identifier
HSLAN	High-speed local area network
IC	Input controller
ICI	Intercarrier interface
ICMP	Internet control message protocol
ICP	IMA control protocol
ICR	Initial cell rate
ID	Identifier
IE	Information element
IEEE	Institute of Electrical and Electronic Engineers
IGMP	Internet group management protocol
ILMI	Integrated local management interface
IMA	Inverse multiplexing ATM
IME	ATM interface management entities
IN	Intelligent network
InvARP	Inverse ARP
IP	Internet protocol
IPNNI	Integrated private network-to-network interface
ISDN	Integrated services digital network
ISO	International Standards Organization
ISUP	ISDN user part
ITU	International Telecommunication Union
IWF	Interworking function
IWU	Interworking unit
LAN	Local area network
LANE	LAN emulation
LEC	LAN emulation client
LECS	LAN emulation configuration server
LES	LAN emulation server
LI	Length indicator
LIS	Logical IP sub-network
LLC	Logical link control

LLS	LAN-like switching
LMI	Local management interface
LNNI	LANE–NNI
LSB	Least significant bit
LUNI	LANE–UNI
LT	Line termination
MAC	Media access control
MAN	Metropolitan area network
MARS	Multicast address resolution server
MBS	Maximum burst size
MCR	Minimum cell rate
MCS	Multicast server
MCSN	Monitoring cell sequence number
MIB	Management information base
MID	Multiplexing identifier
MIN	Multi-path interconnection network
MIR	Maximum information rate
MPC	MPOA client
MPEG	Motion picture experts group
MPLS	Multi protocol label switching
MPOA	Multi-protocol over ATM
MPS	MPOA server
MSB	Most significant bit
MSVC	Meta-signaling virtual channel
MT	Message type
MTP	Message transfer part
MTU	Maximum transmission unit
NDIS	Network driver interface specification
NHRP	Next hop resolution protocol
NHS	Next hop server
NI	No increase
NIC	Network interface card
NNI	Network–node interface
NPC	Network parameter control
NSAP	Network service access point
NRZ	Non-return-to-zero code
NTT	Nippon Telegraph and Telephone Corporation
NTSC	North American Television Standards Committee
OAM	Operation and maintenance
OC	Output controller
ODI	Open data link interface
OSF	Offset field
OSFP	Offset field parity
OSI	Open system interconnection
OSPF	Open shortest path first
P	Parity

PAD	Padding
PAL	Phase alternating line
PBX	Private branch exchange
PC	Personal computer
PCI	Protocol control information
PCI	Peripheral component interconnect
PCR	Point-to-point SVC cell rate; peak cell rate
PD	Protocol discriminator
PDH	Plesiochronous digital hierarchy
PDU	Protocol data unit
PL	Physical layer
PLCP	Physical layer convergence protocol
PM	Physical medium
PNNI	Private network-to-network interface
POH	Path overhead
PON	Passive optical network
PPD	Partial packet discard
PPT	Packet payload type
PPTU	PDUs per time unit
PRM	Protocol reference model
PSPDN	Packet-switched public data network
PSVC	Point-to-point signaling virtual channel
PSVCI	Point-to-point signaling virtual channel identifier
PT	Payload type
PTI	Payload type identifier
PV	Protocol version
PVC	Permanent virtual connection
QoS	Quality of service
QPSX	Queued packet and synchronous circuit exchange
RACE	Research and Development of Advanced Communication in Europe
RAM	Random access memory
RDI	Remote defect indication
REI	Remote error indication
RI	Reference identifier
RIP	Routing information protocol
RM	Resource management
RRC	Radio resource control
RSVP	Resource reservation protocol
RTS	Residual time stamp
S-AAL	ATM adaptation layer for signaling
SAP	Service access point
SAR	Segmentation and reassembly
SBR	Statistical bit rate
SCR	Sustainable cell rate
SCSP	Server cache synchronization protocol

SDH	Synchronous digital hierarchy
SDU	Service data unit
SECAM	Système en Couleur avec Mémoire
SIR	Sustained information rate
SLMB	Subscriber line module broadband
SMDS	Switched multi-megabit data service
SMS	Special multicast server
SMTP	Simple mail transfer protocol
SN	Sequence number
SNAP	Subnetwork attachment point
SNMP	Simple network management protocol
SNP	Sequence number protection
SOH	Section overhead
SONET	Synchronous optical network
SPID	Service profile identifier
SPN	Subscriber premises network
SRTS	Synchronous residual time stamp
SS7	Common channel signaling system no. 7
SSCF	Service-specific coordination function
SSCOP	Service-specific connection-oriented protocol
SSCS	Service-specific convergence sublayer
SSM	Single-segment message
SSP	Service-specific part
ST	Segment type
STF	Start-field
STM	Synchronous transfer mode
STM-*i*	Synchronous transport module *i*
STP	Shielded twisted pair
SVC	Switched virtual connection
SVC	Signaling virtual channel
SVCI	Signaling virtual channel identifier
Tc	Committed rate measurement interval
TC	Transmission convergence sublayer
TCAP	Transaction capabilities application part
TCP	Transmission control protocol
TDM	Time division multiplexing
TDMA	Time-division multiple access
TDP	Tag distribution protocol
TE	Terminal equipment
TELNET	Remote login protocol
TMB	Trunk module broadband
TMN	Telecommunications management network
TRCC	Total received cell count
TSTP	Time stamp
TUC	Total user cell number
TULIP	TCP and UDP over lightweight IP

TUNIC	TCP and UDP over a nonexistent IP connection
UBR	Unspecified bit rate
UDF	User defined field
UDP	User datagram protocol
UME	UNI management entity
UNI	User–network interface
UPC	Usage parameter control
UPT	Universal personal telecommunications
UPTN	Universal personal telecommunications number
UTOPIA	Universal test and operations physical interface for ATM
UU	CPCS user-to-user indication
UUI	User-to-user indication
VBR	Variable bit rate
VBR-nrt	Variable bit rate – non real time
VBR-rt	Variable bit rate – real time
VC	Virtual channel
VC-*i*	Virtual container *i*
VCC	Virtual channel connection
VCI	Virtual channel identifier
VD	Virtual destination
VDSL	Very high-speed digital subscriber line
VLAN	Virtual LAN
VP	Virtual path
VPC	Virtual path connection
VPCI	Virtual path connection identifier
VPI	Virtual path identifier
VPN	Virtual private network
VS	Virtual source
WAN	Wide area network
WATM	Wireless ATM
WDM	Wavelength division multiplexing
WIRE	Workable interface requirements example
WWW	World wide web

References

Acampora, A.S. and Karol, M.J. (1989). An overview of lightwave packet networks. *IEEE Network*, **3**(1), January, 29–40

Adams, J.L. (1987). The Orwell Torus Communication Switch. In *Proc. GLSB Seminar on Broadband Switching*, Albufeira, 215–24

Anido, G.J. and Seeto, A.W. (1988). Multipath interconnection: a technique for reducing congestion within fast packet switching fabrics. *IEEE Journal on Selected Areas in Communications*, **6**(9), December, 1480–8

ANSI (1988). ANSI Standard T1.105-1988 'SONET Optical Interface Rates and Formats'

ANSI (1990). ANSI Standard 'Hybrid Ring Control'. Revision 6, May

ANSI (1997a). ANSI Draft Standard T1E1.413 'ADSL'

ANSI (1997b). ANSI Standard T1E1.4.497–131.R2 'VDSL'

Armbrüster, H. and Schneider H. (1990). Phasing-in the universal broadband ISDN: initial trials for examining ATM applications and ATM systems. In *Proc. Int. Conf. on Integrated Broadband Services and Networks*, London, 200–5

van As, H.R., Lemppenau, W.W., Zafiropulo, P. and Zurfluh, E. (1991). CRMA-II: A Gbit/s MAC protocol for ring and bus networks with immediate access capability. In *Proc. European Fibre Optic and Local Area Networks Conference*, London, 162–9

ATM Forum (1993b). ATM Data Exchange Interface (DXI) Specification, af-dxi-0014.000 Version 1.0.

ATM Forum (1993c). ATM User-Network Interface Specification V3.0, af-uni-0010.001

ATM Forum (1993d). B-ICI 1.0, af-bici-0013.000

ATM Forum (1994a). *Newsletter ('53 Bytes')*, **2**(1), January, 2–3

ATM Forum (1994b). UTOPIA, An ATM-PHY Interface Specification Level 1, Version 2, af-phy-0017.000

ATM Forum (1994c). ATM User-Network Interface Specification V3.1, af-uni-0010.002

ATM Forum (1994d). Interim Inter-Switch Signaling Protocol, af-pnni-0026.000

ATM Forum (1994e). Mid-range Physical Layer Specification for Category 3 UTP, af-phy-0018.000

ATM Forum (1994f). B-ICI 1.1, af-bici-0013.001

ATM Forum (1995a). UTOPIA, An ATM-PHY Interface Specification Level 2, Version 1.0, af-phy-0039.000.

ATM Forum (1995b). LAN Emulation over ATM 1.0. af-lane-0021.000

ATM Forum (1995c). 6,312 kbps UNI Specification, af-phy-0029.000

ATM Forum (1995d). Physical Interface Specification for 25.6 Mb/s over Twisted Pair, af-phy-0040.000

ATM Forum (1995e). B-ICI 2.0, af-bici-0013.003

ATM Forum (1996a). Frame-based User-to-Network Interface Specification v2, str-saa-funi-01.01.

ATM Forum (1996b). WIRE, Workable Interface Requirements Example, af-phy-0063.000

ATM Forum (1996c). Traffic Management 4.0, af-tm-0056.000

ATM Forum (1996d). UNI Signaling 4.0, af-sig-0061.000

ATM Forum (1996e). P-NNI V1.0, af-pnni-0055.000

ATM Forum (1996f). P-NNI 1.0 Addendum (soft PVC MIB), af-pnni-0066.000

ATM Forum (1996g). B-ICI 2.0 Addendum, af-bici-0068.000

ATM Forum (1996h). Integrated Local Management Interface (ILMI) Specification 4.0, af-ilmi-0065.000

ATM Forum (1997a). LAN Emulation over ATM Version 2 – LUNI Specification Draft 6. February

ATM Forum (1997c). LAN Emulation over ATM Version 2 – LNNI Specification – Draft 8. February

ATM Forum (1997d). Multi-Protocol Over ATM Version 1.0. Straw Ballot. February

ATM Forum (1997e). Inverse ATM Mux, af-phy-0086.000

ATM Forum (1997f). P-NNI ABR Addendum, af-pnni-0075.000

ATM Forum (1997g). Proposed text for the AINI Specification, 1996–1523 R1

ATM Forum (1997h). Circuit Emulation Service 2.0, af-vtoa-0078.000

ATM Forum (1997i). Baseline Text for Wireless ATM Specifications, BTD-WATM-01.01

ATM Forum (1997j). Baseline Text for Residential Broadband, BTD-RBB-001.01

ATM Forum (1997k). Multi-Protocol Over ATM Specification V1.0, af-mpoa-0087.000, July

Baireuther, O. (1991). Überlegungen zum Breitband-ISDN. *Der Fernmelde = Ingenieur*, **45**(2/3), February/March, 1–62

Barnsley, P.E., Wickes, H.J. and Wickens, G.E. (1992). Switching in future ultra-high capacity all-optical networks. In *Proc. XIV Int. Switching Sym.*, Yokohama, paper B10.1

Batcher, K.E. (1968). Sorting networks and their applications. In *AFIPS Proc. Spring Joint Computer Conf.* vol. 32, 307–14

Bellcore (1989). Technical Advisory TA-TSY-000772 'Generic System Requirements in Support of Switched Multi-Megabit Data Service'. Issue 3, October

Beneš, V. (1965) *Mathematical Theory of Connection Networks.* New York: Academic Press

Besier, H.A. (1988). SPN (Subscriber Premises Network), an essential part of the broadband communication network. In *Proc. GLOBECOM'88*, Hollywood, 102–6

Biersack, E. (1989). Principles of network interconnection. In *Proc. 7th European Fibre Optic Communications & Local Area Networks Exposition (EFOC/LAN)*, Amsterdam, 37–43

Biocca, A., Freschi, G., Forcina, A. and Melen, R. (1990). Architectural issues in the interoperability between MANs and the ATM network. In *Proc. XIII Int. Switching Sym.*, Stockholm, vol. II, 23–8

Bocker, P. (1992). *ISDN – The Integrated Services Digital Network – Concepts, Methods, Systems.* Berlin-Springer: Verlag

Brackett, C.A., Acampora, A.S., Sweitzer, J., Tnaonan, G., Smith, M.T., Lennon, W., Wang, K.C. and Hobbs, R.H. (1993). A scalable multiwavelength multihop optical network: a proposal for research on all-optical networks. *IEEE Journal of Lightwave Technology*, **11**(5/6), May/June, 736–53

Breuer, H.-J. and Hellström, B. (1990). Synchronous transmission networks. *Ericsson Review*, no. 2, 60–71

Brill, A., Huber, M.N. and Petri, B. (1993). Designing signalling for broadband. *Telecommunications*, July, 29–34

Brooks, E.D. (1987). A butterfly-memory interconnection for a vector processing environment. *Parallel Computing*, no. 4

Byrne, W.R., Kafka, H.J., Luderer, G.W.R., Nelson, B.L. and Clapp, G.H. (1990). Evolution of metropolitan area networks to broadband ISDN. In *Proc. XIII Switching Sym.*, Stockholm, vol. II, 15–22

Chen, P.Y., Lawrie, D.H., Yew, P.C. and Padua, D.A. (1981). Interconnection networks using shuffles. *IEEE Computer*, **14**(12), December, 55–63

Chidgey, P.J., Hill, G.R. and Saxtoft, C. (1992). Wavelength and space switched optical networks and nodes. In *Proc. XIV Int. Switching Sym.*, Yokohama, paper B9.3

Communications Weekly International, (1990). More broadband for BellSouth. p. 12

Coudreuse, J.P. and Servel, M. (1987). PRELUDE: an asynchronous time-division switched network. In *Proc. Int. Conf. on Communications*, Seattle, paper 22.2

David, R., Fastrez, M., Bauwens, J., De Vleeschouwer, A., Christiaens, M. and van Vyve, J. (1990). A Belgian broadband ATM experiment. In *Proc. XIII Int. Switching Sym.*, Stockholm, vol. III, pp. 1–6

Denzel, W.E., Le Boudec, J.-H., Port, E. and Troung, H.L. (1993). FALCON: a switched-based ATM LAN. In *Proc. European Fibre Optic Communications and Networks*, The Hague, 72–7

De Prycker, M. and De Somer, M. (1987). Performance of a service independent switching network with distributed control. *IEEE Journal on Selected Areas in Communications*, **5**(8), October, 1293–302

De Prycker, M., Verbiest, W. and Mestagh D. (1992). ATM passive optical networks: preparing the access network for BISDN. In *Proc. XIV Int. Switching Sym.*, Yokohama, paper B4.1

De Smedt, A., De Vleeschouwer, A. and Theeuws R. (1990). Subscribers' premises networks for the Belgian broadband experiment. In *Proc. XIII Int. Switching Sym.*, Stockholm, vol. VI, 105–9

Dias, D.M. and Kumar, M. (1984). Packet switching in n log n multistage networks. In *Proc. GLOBECOM'84*, Atlanta, 114–20

Dobrowski, G.H., Estes, G.H., Spears, D.R. and Walters, S.M. (1990). Implications of BISDN services on network architecture and switching. In *Proc. XIII Int. Switching Sym.*, Stockholm, vol. I, 91–8

du Chaffaut, G., Borgne, M., Coatanea, P., Ballance, J.W., Lee, R.F. and Faulkner, D.W. (1992). ATM PON based networks and related service requirements. In *Proc. IEEE Workshop on Local Optical Networks*, Versailles, September

Elrefaie, A.F. (1993). Multiwavelength survivable ring network architecture. In *Proc. Int. Conf. on Communications*, Geneva, 1245–51

ETSI (1990). ETSI Sub Technical Committee NA5, Report of the Rome Meeting. March

ETSI (1993). ETR DE/NA-53203: 'Network Aspects (NA); CBDS over ATM', May

Falconer, R.M. and Adams, J.L. (1985). Orwell: A protocol for an integrated service local network. *British Telecom Technology Journal*, **3**(4), October, 27–35

Faulkner, D.W. and Ballance, J.W. (1988). Passive optical networks for local telephony and cable TV provision. *International Journal of Digital and Analog Cabled Systems*, **1**(3), July, 159–63

Fischer, W., Fundneider, O., Goeldner, E.-H. and Lutz, K.A. (1991). A scalable ATM switching system structure. *IEEE Journal on Selected Areas in Communications*, **9**(8), October, 1299–307

Fischer, W. and Stiefel, R. (1993) A flexible ATM system architecture for world-wide field trials. In *Proc. Int. Conf. on Communications*, Geneva, 86–90

Frame Relay Forum (1994). FRF.5 Frame Relay/ATM Network Interworking. December

Frame Relay Forum (1995). FRF.8 Frame Relay/ATM PVC Sertvice Interworking. April

Frantzen, V., Maher A. and Eske Christensen, B. (1989). Towards the intelligent ISDN. In *Proc. Int. Conf. on Intelligent Networks*, Bordeaux, 152–6

Frantzen, V., Huber, M.N. and Maegerl, G. (1992). Evolutionary steps from ISDN signalling towards B-ISDN signalling. In *Proc. GLOBECOM'92*, Orlando, 1161–5

Ginsburg. D. (1996). *ATM: Solutions for Enterprise Internetworking.* Harlow: Addison-Wesley

Göldner, E.H. (1990). The network evolution towards B-ISDN: applications, network aspects, trials (e.g. BERKOM). In *Proc. Int. Conf. on Communications*, Atlanta, paper 212.2

Goke, L.R. and Lipovski, G.J. (1973). Banyan networks for partitioning multiprocessor systems. In *First Ann. Sym. on Computer Architecture*, 21–8

Green Jr., P.E. (1993). *Fibre Optic Networks.* Englewood Cliffs, NJ: Prentice-Hall

Hagenhaus, L. (1990) Gebührenerfassung bei ATM-Netzen. *Internal Siemens Report*, September

Hagenhaus, L. and Überla, A. (1993). Users' bandwidth needs versus 155.520 Mbit/s broadband interface. In *Proc. X Int. Sym. on Subscriber Loops & Services*, Vancouver, 44–9

Händel, R. (1992). Operation and maintenance issues of ATM networks. In *Proc. 1992 Int. Conf. on Communication Technology*, Beijing, vol. 1, 12.07.1–4

Hauber, C. and Wallmeier, E. (1991) Blocking probabilities in ATM pipes controlled by a connection acceptance algorithm based on mean and peak bit rates. In *Proc. XIII Int. Teletraffic Congress, ITC Workshops 'Queueing, Performance and Control in ATM'*, Copenhagen, 137–42

Henry, P.S. (1989). High-capacity lightwave local area networks. *IEEE Communications Magazine*, **27**(10), October, 20–6

Hinton, H.S. (1992). Photonics in switching. *IEEE LTS The Magazine of Light-wave Telecommunication Systems*, **3**(3), August, 26–35

Huang, A. and Knauer, S. (1984). STARLITE: a wideband digital switch. In *Proc. GLOBECOM'84*, Atlanta, 121–5

Huber, M.N., Rathgeb, E.P. and Theimer, T.H. (1988). Self routing Banyan networks in an ATM-environment. In *Proc. Int. Conf. on Computer Communication*, Tel Aviv, 167–74

Huber, M.N., Frantzen, V. and Maegerl, G. (1992). Proposed evolutionary paths for B-ISDN signalling. In *Proc. XIV Int. Switching Sym.*, Yokohama, paper C3.3

Huber, M.N. and Tegtmeyer, V.H. (1993). Bandwidth management of virtual paths – performance versus control aspects. *Tagungsband Messung, Modellierung und Bewertung von Rechen- und Kommunikationssystemen*, Aachen, 226–38

Huber, M.N. and Osborne, R. (1994). Architecture of an all-optical ring network. *Proc. European Fibre Optic Communications and Networks*, Heidelberg, 102–5

Hui, J.Y. (1988). Resource allocation for broadband networks. *IEEE Journal on Selected Areas in Communications*, **6**(9), December, 1598–608

IEC Publication 825 (1993). Radiation Safety of laser products, equipment classification, requirements and user's guide

IEEE (1988). IEEE Std 802.1 D – 1988: 'MAC Bridges'

IEEE (1989). IEEE Std 802.2 – 1989: 'Information Processing Systems – Local Area Networks – Part 2: Logical Link Control'

IEEE (1991). IEEE Std 802.6 – 1991: 'Distributed Queue Dual Bus (DQDB) Subnetwork of a Metropolitan Area Network (MAN)'

Internet Engineering Task Force (IETF) (1981). Internet Protocol and Transmission Control Protocol *RFCs 791 and 793*, September

Internet Engineering Task Force (IETF) (1982). An Ethernet Address Resolution Protocol *RFC 826*, November

Internet Engineering Task Force (IETF) (1989). Host Extensions for Multicasting *RFC 1112*, August

Internet Engineering Task Force (IETF) (1990a). Path MTU Discovery *RFC 1191*, November

Internet Engineering Task Force (IETF) (1990). Simple Network Management Protocol, *RFC 1157* (J.D. Case, M.S. Fedore, M.L. Schoffstall and J.R. Davin), May, DDN Network Information Center

Internet Engineering Task Force (IETF) (1991). Management Information Base for Network Management of TCP/IP-based internets;

MIB-II, *RFC 1213* (K. McCloughrie and M.T. Rose (eds)), March, DDN Network Information Center

Internet Engineering Task Force (IETF) (1992). TCP Extensions for High Performance *RFC 1323*, May

Internet Engineering Task Force (IETF) (1993a). IESG Advice from Experience with Path MTU Discovery *RFC 1435*, March

Internet Engineering Task Force (IETF) (1993). Multiprotocol Encapsulation over ATM Adaptation Layer 5 *RFC 1483*, July

Internet Engineering Task Force (IETF) (1994a). Classical IP and ARP over ATM *RFC 1577*, January

Internet Engineering Task Force (IETF) (1994b). Default IP MTU for use over ATM AAL'5 *RFC 1626*, May

Internet Engineering Task Force (IETF) (1995). ATM Signaling Support for IP over ATM *RFC 1755*, February

Internet Engineering Task Force (IETF). (1996a) Support for Multicast over UNI 3.0/3.1 based ATM networks *RFC 2022*, November

Internet Engineering Task Force (IETF) (1996b). IP Broadcast over ATM networks. Internet Draft. November

Internet Engineering Task Force (IETF) (1996c). Classical IP and ARP over ATM. Internet Draft. November

Internet Engineering Task Force (IETF) (1996d). 'Local/Remote' Forwarding Decisions in Switched Data Link Subnetworks *RFC 1937*. Informational. May

Internet Engineering Task Force (IETF) (1996e). ATM Signaling Support for IP over ATM – UNI 4.0 Update. Internet Draft, November

Internet Engineering Task Force (IETF) (1996f). Ipv6 over NBMA Network. Internet Draft. November

Internet Engineering Task Force (IETF) (1996g). Internet Protocol, Version 6 Specification *RFC 1883*. January

Internet Engineering Task Force (IETF) (1996h). Transient Neighbors for Ipv6 over ATM. Internet Draft. November

Internet Engineering Task Force (IETF) (1996i). NHRP for destinations off the NBMA subnetwork. Y. Rekhter, work in progress

Internet Engineering Task Force (IETF) (1996j). Server Cache Synchronization Protocol – SCSP. Internet Draft

Internet Engineering Task Force (IETF) (1996k). Next Hop Resolution Protocol (NHRP)

Internet Engineering Task Force (IETF) (1996l). Ipsilon's General Switch Management Protocol Specification, Version 1.1, *RFC 1987*, August

Internet Engineering Task Force (IETF) (1996m). Ipsilon Flow Management Protocol Specification for IPv4, Version 1.0, *RFC 1953*, May

Internet Engineering Task Force (IETF) (1996n). Transmission of Flow Labelled IPv4 on ATM Data Links, Version 1.0, *RFC 1954*, May

Internet Engineering Task Force (IETF) (1997). Multicast Server Architectures for MARS-based ATM Multicasting. Internet Draft. January

Internet Engineering Task Force (IETF) (1997a). A Framework for Multiprotocol Label Switching. Internet Draft. May

ISO (1984). ISO 7498 – 1984: 'Information Processing Systems – Open System Interconnection – Basic Reference Model'. New York: American National Standards Association

ISO (1987). ISO 8824 – 1987: 'Information Processing Systems – Open System Interconnection – Specification of Abstract Syntax Notation One (ASN.1)'.

ISO (1989/90). ISO 9314-1,-2,-3: 'Fibre Distributed Data Interface (FDDI)'. New York: American National Standards Association

ISO (1990a). ISO 8802-3 – 1990: 'Carrier Sense Multiple Access with Collision Detection (CSMA/CD) Access Method and Physical Layer Specifications'. New York: American National Standards Association

ISO (1990b). ISO 8802-4 – 1990: 'Token-passing Bus Access Method and Physical Layer Specifications'. New York: American National Standards Association

ISO (1991). ISO 8802-5 – 1991: 'Token Ring Access Method'. New York: American National Standards Association

ITU-T (1984). COM XVIII-228-E, Geneva, March

ITU-T (1987). COM XVIII-D1109, Hamburg, July

ITU-T (1988a). Recommendation G.702. 'Digital Hierarchy Bit Rates'. *Blue Book*, Fascicle III.4. Geneva

ITU-T (1988b). Recommendation G.811. 'Timing Requirements at the Outputs of Primary Reference Clocks suitable for Plesiochronous Operation of International Digital Links'. *Blue Book*, Fascicle III.5. Geneva

ITU-T (1988c). Recommendation G.822. 'Controlled Slip Rate Objectives on an International Digital Connection'. *Blue Book*, Fascicle III.5. Geneva

ITU-T (1988d). Recommendation I.412. 'ISDN User–Network Interfaces – Interface Structures and Access Capabilities'. *Blue Book*, Fascicle III.8. Geneva

ITU-T (1988e). Recommendation I.601. 'General Maintenance Principles of ISDN Subscriber Access and Subscriber Installation'. *Blue Book*, Fascicle III.9. Geneva

ITU-T (1988f). Recommendation Q.940. 'ISDN User–Network Interface Protocol for Management – General Aspects'. *Blue Book*, Fascicle VI.11. Geneva

ITU-T (1988g). Recommendation X.208. 'Specification of Abstract Syntax Notation One (ASN.1)'. *Blue Book*, Fascicle VIII.4. Geneva

ITU-T (1988h). Recommendation X.21 bis. 'Use on public data networks of Data Terminal Equipment (DTE) which is designed for interfacing to synchronous V-Series modems'. Blue Book Fascicle VIII.2. Geneva

ITU-T (1988i). Recommendation V.35. 'Data Transmission at 48 kbit/s Using 60-108 kHz Group Band Circuits'. Blue Book, Fascicle VIII.1. Geneva

ITU-T (1990). COM XVIII-R41-E, Geneva, June

ITU-T (1991a). Recommendation E.164: 'Numbering Plan for the ISDN Era'. Geneva

ITU-T (1991b). Recommendation G.703. 'Physical/Electrical Characteristics of Hierarchical Digital Interfaces'. Geneva

ITU-T (1991c). Recommendation I.121. 'Broadband Aspects of ISDN'. Geneva

ITU-T (1991d). Recommendation I.233. 'Frame Mode Bearer Services'. Geneva

ITU-T (1991e). Recommendation I.321. 'B-ISDN Protocol Reference Model and its Application'. Geneva

ITU-T (1991f). Recommendation I.370. 'Congestion Management for the ISDN Frame Relaying Bearer Service'. Geneva

ITU-T (1992a). Recommendation M.20. 'Maintenance Philosophy for Telecommunications Networks'. Geneva

ITU-T (1992b). Recommendation M.3600. 'Principles for the Management of ISDNs'. Geneva

ITU-T (1992c). Recommendation Q.922. 'ISDN Data Link Layer Specification for Frame Mode Bearer Services'. Geneva

ITU-T (1992d). Recommendation X.21. 'Interface between Data Terminal Equipment and Data Circuit-terminating Equiment for synchronous operation on public data networks'. Geneva

ITU-T (1993a). Recommendation G.652. 'Characteristics of a Single-Mode Optical Fibre Cable'. Geneva

ITU-T (1993b). Recommendation G.804. 'ATM Cell Mapping into Plesiochronous Digital Hierarchy (PDH)'. Geneva

ITU-T (1993c). Recommendation G.823. 'The Control of Jitter and Wander within Digital Networks which are based on the 2048 kbit/s Hierarchy'. Geneva

ITU-T (1993d). Recommendation G.824. 'The Control of Jitter and Wander within Digital Networks which are based on the 1544 kbit/s Hierarchy'. Geneva

ITU-T (1993e). Recommendation I.113. 'Vocabulary of Terms for Broadband Aspects of ISDN'. Geneva

ITU-T (1993f). Recommendation I.120. 'Integrated Services Digital Networks (ISDNs)'. Geneva

ITU-T (1993g). Recommendation I.211. 'B-ISDN Service Aspects'. Geneva

ITU-T (1993h). Recommendation I.320. 'ISDN Protocol Reference Model'. Geneva

ITU-T (1993i). Recommendation I.327. 'B-ISDN Functional Architecture'. Geneva

ITU-T (1993j). Recommendation I.362. 'B-ISDN ATM Adaptation Layer (AAL) Functional Description'. Geneva

ITU-T (1993l). Recommendation I.411. 'ISDN User–Network Interfaces – Reference Configurations'. Geneva

ITU-T (1993m). Recommendation I.413. 'B-ISDN User–Network Interface'. Geneva

ITU-T (1993n). Recommendation I.365. 'B-ISDN ATM Adaptation Layer Sublayers'. Geneva

ITU-T (1993o). Recommendation I.431. 'Primary Rate User–Network Interface – Layer 1 Specification'. Geneva

ITU-T (1993p). Recommendation I.555. 'Frame Relay Bearer Service Interworking'. Geneva

ITU-T (1993q). Recommendation M.60. 'Maintenance Terminology and Definitions'. Geneva

ITU-T (1993r). Recommendation Q.761. 'Functional Description of the ISDN User Part of Signalling System No. 7'. Geneva

ITU-T (1993s). Recommendation Q.762. 'General Functions of Messages and Signals'. Geneva

ITU-T (1993t). Recommendation Q.763. 'Formats and Codes'. Geneva

ITU-T (1993u). Recommendation Q.764. 'Signalling Procedures'. Geneva

ITU-T (1993v). Recommendation Q.931. 'ISDN User–Network Interface Layer 3 Specification for Basic Call Control'. Geneva

ITU-T (1993w). Recommendation Q.1400. 'Architecture Framework for the Development of Signalling and Organisation, Administration and Maintenance Protocols Using OSI Concepts'. Geneva

ITU-T (1993x). Recommendation X.28. 'DTE/DCE Interface for a Start–Stop Mode Data Terminal Equipment accessing the Packet Assembly/Disassembly Facility (PAD) in a Public Data Network situated in the same Country'. Geneva

ITU-T (1994a). Recommendation Q.2100. 'B-ISDN Signalling ATM Adaptation Layer (SAAL) Overview Description'. Geneva

ITU-T (1994b). Recommendation Q.2110. 'B-ISDN Signalling ATM Adaptation Layer (SAAL) – Service Specific Connection Oriented Protocol (SSCOP)'. Geneva

ITU-T (1994c). Recommendation Q.2130. 'B-ISDN Signalling ATM Adaptation Layer (SAAL) – Service Specific Coordination Function for support of Signalling at the User-to-Network Interface (SSCF at UNI)'. Geneva

ITU-T (1994d). Recommendation X.200. 'Information Technology – Open Systems Interconnection – Basic Reference Model'. Geneva

ITU-T (1995a). Recommendation G.704. 'Synchronous Frame Structures Used at 1544, 6322, 2048, 8488 and 44 736 kbit/s Hierarchical Levels'. Geneva

ITU-T (1995b). Recommendation G.832. 'Transport of SDH Elements on PDH Networks: Frame and Multiplexing Structures'. Geneva

ITU-T (1995c). Recommendation G.957. 'Optical Interfaces for Equipments and Systems Relating to the Synchronous Digital Hierarchy'. Geneva

ITU-T (1995d). Recommendation I.150. 'B-ISDN ATM Functional Characteristics'. Geneva

ITU-T (1995e). Recommendation I.361. 'B-ISDN ATM Layer Specification'. Geneva

ITU-T (1995f). Recommendation I.430. 'Basic User–Network Interface – Layer 1 Specification'. Geneva

ITU-T (1995g). Recommendation I.610. 'B-ISDN Operation and Maintenance Principles and Functions'. Geneva

ITU-T (1995h). Recommendation Q.2010. 'B-ISDN Overview – Signalling Capability Set 1, Release 1'. Geneva

ITU-T (1995i). Recommendation Q.2120. 'B-ISDN Meta-signalling Protocol'. Geneva

ITU-T (1995j). Recommendation Q.2140. 'B-ISDN Signalling ATM Adaptation Layer – Service Specific Coordination Function for Signalling at the Network–Node Interface (SSCF at NNI)'. Geneva

ITU-T (1995k). Recommendation Q.2761. 'Functional Description of the B-ISDN User Part of Signalling System No. 7'. Geneva

ITU-T (1995l). Recommendation Q.2762. 'General Functions of Messages and Signals'. Geneva

ITU-T (1995m). Recommendation Q.2763. 'Formats and Codes'. Geneva

ITU-T (1995n). Recommendation Q.2764. 'Basic Call Procedures'. Geneva

ITU-T (1995o). Recommendation Q.2931. 'Digital Subscriber Signalling System No. 2 – UNI layer 3 Specification for Basic Call/Connection Control'. Geneva

ITU-T (1995p). Recommendation X.213. 'Information technology - Open Systems Interconnection - Network service definition'. Geneva

ITU-T (1996a). Recommendation G.114. 'One-way Transmission Time'. Geneva

ITU-T (1996b). Recommendation G.131. 'Control of Talker Echo'. Geneva

ITU-T (1996c). Recommendation G.707. 'Network Node Interface for the Synchronous Digital Hierarchy'. (G.707 (1996) replaces G.707–709 (1993)). Geneva

ITU-T (1996d). Recommendation G.821. 'Error Performance of an International Digital Connection operating at a bit rate below the primary rate and forming Part of an Integrated Services Digital Network'. Geneva

ITU-T (1996e). Recommendation H.263. 'Video coding for low bit rate communication'. Geneva

ITU-T (1996f). Recommendation I.311. 'B-ISDN General Network Aspects'. Geneva

ITU-T (1996g). Recommendation I.356. 'B-ISDN ATM Layer Cell Transfer Performance'. Geneva

ITU-T (1996h). Recommendation I.363. 'B-ISDN ATM Adaptation Layer Specification'. Geneva

ITU-T (1996i). Recommendation I.371. 'Traffic Control and Congestion Control in B-ISDN'. Geneva

ITU-T (1996j). Recommendation I.432. 'B-ISDN User–Network Interface – Physical Layer Specification'. Geneva

ITU-T (1996k). Recommendation M.3010. 'Principles for a Telecommunications Management Network'. Geneva

ITU-T (1996l). Recommendation Q.704. 'Signalling Network Functions and Messages'. Geneva

ITU-T (1996m). Recommendation X.1. 'International User Classes of Service in, and Categories of Access to, Public Data Networks and Integrated Services Digital Networks'. Geneva

ITU-T (1996n). Recommendation X.25. 'Interface between Data Terminal Equipment (DTE) and Data Circuit-terminating Equipment (DCE) for Terminals Operating in the Packet Mode and Connected to Public Data Networks by Dedicated Circuit'. Geneva

ITU-T (1996o). Recommendation X.32. 'Interface between Data Terminal Equipment (DTE) and Data Circuit-terminating Equipment (DCE) for terminals operating in the packet mode and accessing a Packet-Switched Public Data Network through a public switched telephone network or an Integrated Services Digital Network or a Circuit-Switched Public Data Network'. Geneva

ITU-T (1996p). Recommendation X.38. 'G3 facsimile equipment/DCE interface for G3 facsimile equipment accessing the Facsimile Packet Assembly/Diassembly facility (FPAD) in a public data network situated in the same country'. Geneva

ITU-T (1996q). Recommendation I.353. 'Reference events for defining ISDN and B-ISDN performance parameters'. Geneva

ITU-T (1996r). Recommendation Q.2119. 'B-ISDN ATM adaptation layer – Convergence function for SSCOP above the frame relay core service'. Geneva

ITU-T (1996s). Recommendation Q.2721.1. 'B-ISDN user part – Overview of the B-ISDN network node interface signaling capability set 2, step 1'. Geneva

ITU-T (1996t). Recommendation M.3610. 'Principles for Applying the TMN Concept to the Management of B-ISDN'. Geneva

Janniello, F.J., Ramaswami, R. and Steinberg, D.G. (1993). A prototype circuit-switched multi-wavelength optical metropolitan-area network. *IEEE Journal of Lightwave Technology*, **11**(5/6), May/June, 777–82

Johansson, S., Lindblom, M., Buhrgard, M., Granestrand, P., Lagerström, B., Thylén, L. and Wosinska, L. (1992). Photonic switching in high capacity transport networks. In *Proc. XIV Int. Switching Sym.*, Yokohama, paper B9.1

JTC (1990). JTC1/SC6 WG6 N71: 'Specification of the Asynchronous Transfer Mode Ring (ATMR) Protocol'. Japanese National Body, Japan

Kammerl, A. (1988). Contribution on echo control. Submitted to CEPT/NA5 Meeting in Madeira, October

Karol, M.J., Hluchyi, M.G. and Morgan, S.P. (1987). Input versus output queueing on a space-division packet switch. In *IEEE Trans. on Communications*, **35**(12), December, 1347–56

Keller, H., Glade, B., Hartl, B. and Horbach, C. (1993) Optical broadband access using ATM on a passive optical network. *International Journal of Digital and Analog Communication Systems*, **6**, 143–9

Koljonen, J. (1992). HDSL boosts the value of the copper-based network. *Discovery*, **27**, 34–7

Kröner, H. (1990). Comparative performance study of space priority mechanisms for ATM networks. In *Proc. INFOCOM'90*, San Francisco, 1136–43

Kuehn, P.J. (1989). From ISDN to IBCN (Integrated Broadband Communication Network). In *Proc. World Computer Congress IFIP'89*, San Francisco, 479–86

Lampe, D. (1988). Transfer delay deviation of packets in ATD switching matrices and its effect on dimensioning a depacketizer buffer. In *Proc. Int. Conf. on Computer Communication*, Tel Aviv, 55–60

Lea, C.T. (1989). Multi-$\log_2 N$ self-routing networks and their applications in high speed electronic and photonic switching systems. In *Proc. INFOCOM'89*, Ottawa, 877–86

Lemppenau, W.W., Goetzer, M. and Seibolt, W. (1993). Access protocols for high-speed LANs and MANs. In *Proc. European Fibre Optic Communications and Networks*, The Hague, 253–9

Lutz, K.A. (1988). Considerations on ATM switching techniques. *International Journal of Digital and Analog Cabled Systems*, **1**(4), October, 237–43

Lyles, B. and Swinehart, D. (1992). The emerging gigabit environment and the role of local ATM. *IEEE Communications Magazine*, **30**(4), April, 52–8

McMahon, D.H. (1992). Doing wavelength-division multiplexing with today's technology. *IEEE LTS The Magazine of Lightwave Telecommunication Systems*, **3**(1), February, 40–50

Möhrmann, K.H. (1991). Kupfer oder Glasfaser zum Teilnehmer – Wettbewerb oder Ergänzung? *Tagungsband zur VDI/VDE-Tagung Verbindungstechnik'91*, Karlsruhe

Mollenauer, J.F. (1988). Standards of metropolitan area networks. *IEEE Communications Magazine*, **26**(4), April, 15–19

Müller, H.R., Nassehi, M.M., Wong, J.W., Zurfluh, E., Bux, W. and Zafiropulo, P. (1990). DQMA and CRMA: new access schemes for Gbit/s LANs and MANs. In *Proc. INFOCOM'90*, San Francisco, 185–91

Newman, P. (1992). ATM technology for corporate networks. *IEEE Communications Magazine*, **30**(4), April, 90–101

Newman, R.M., Budrikis, Z.L. and Hullett, J.L. (1988). The QPSX MAN. *IEEE Communications Magazine*, **26**(4), April, 20–8

Nussbaum, E. (1988). Communication network needs and technologies – a place for photonic switching? *IEEE Journal on Selected Areas in Communications*, **6**(7), August, 1036–43

Oie, Y., Murata, M., Kubota, K. and Miyahara, H. (1989). Effect of speedup in nonblocking packet switches. In *Proc. Int. Conf. on Communications*, Boston, 410–14

Partridge, C. (1994). *Gigabit Networking*. Wokingham: Addison-Wesley .

Patel, J.H. (1981). Performance of processor-memory interconnection for multiprocessors. *IEEE Transactions on Computers*, **30**(10), October, 771–80

Popescu-Zeletin, R., Egloff, P. and Butscher, B. (1988). BERKOM – a broadband ISDN project. In *Proc. Int. Zurich Seminar*, Zürich, paper B5

RACE (1989). RACE Project R1044/2.10 Broadband User/Network Interface BUNI, 4th Deliverable 'Interface Specification – Draft A and Rationale'. *British Telecom Research Laboratories*, Martlesham Heath, November

Ramaswami, R. (1993). Multiwavelength lightwave networks for computer communication. *IEEE Communications Magazine*, **31**(2), February, 78–88

Rathgeb, E.P. (1990). Policing mechanisms for ATM networks – modelling and performance comparison. In *Proc. 7th Int. Teletraffic Congress Seminar on 'Broadband Technologies: Architectures, Applications, Control and Performance'*, Morristown, paper 10.1

Rathgeb, E.P., Theimer, T.H. and Huber, M.N. (1988). Buffering concepts for ATM switching networks. In *Proc. GLOBECOM'88*, Hollywood, 1277–81

Rathgeb, E.P., Theimer, T.H. and Huber, M.N. (1989). ATM switches – basic architectures and their performance. *International Journal of Digital and Analog Cabled Systems*, **2**(4), October, 227–36

Renger, T. and Briem, U. Leistungsuntersuchung von ATM Adaptation Layer Protokollen für Signalisierung, *private communication*.

Rose, M.T. (1993) *The Simple Book – An Introduction to Internet Management.* Englewood Cliffs, NJ: Prentice-Hall

Rothermel, K. and Seeger, D. (1988). Traffic studies of switching networks for asynchronous transfer mode (ATM). In *Proc. 12th Int. Teletraffic Congress*, Torino, paper 1.3A.5

Sato, K., Okamato, S. and Hadama, H. (1993). Optical path layer technologies to enhance B-ISDN performance. In *Proc. Int. Conf. on Communications*, Geneva, 1300–7

Schaffer, B. (1990). ATM switching in the developing telecommunication network. In *Proc. XIII Int. Switching Sym.*, Stockholm, vol. I, 105–10

Schoute, F.C. (1988). Simple decision rules for acceptance of mixed traffic streams. In *Proc. 12th Int. Teletraffic Congress*, Torino, paper 4.2A.5

Siebenhaar, R. and Bauschert, T. (1993). Vergleich von Algorithmen zur Verbindungsannahme in ATM-Netzen. *Tagungsband zur ITG/GI-Fachtagung Kommunikation in Verteilten Systemen*, Munich, pp. 114–28

Siemens (1989). The Intelligent Integrated Broadband Network – Telecommunications in the 1990s. Siemens, Munich

Sunshine, C.A. (1990). Network interconnection and gateways. *IEEE Journal on Selected Areas in Communications*, **8**(1), December, 4–11

Theimer, T.H. (1991). Performance comparison of routing strategies in ATM switches. In *Proc. XIII Int. Teletraffic Congress*, Copenhagen, 923–8

Tolmie, D. and Renwick, J. (1993). HIPPI: simplicity yields success. *IEEE Network Magazine*, **7**(1), January, 28–33

Turner, J.S. (1987). Design of a broadcast packet switching network. *IEEE Transactions on Communications*, **36**(6), June, 734–43

Vorstermans, J.P. and De Vleeschouwer, A.P. (1988). Layered ATM systems and architectural concepts for subscribers' premises networks. *IEEE Journal on Selected Areas in Communications*, **6**(9), December, 1545–55

Watson, G., Ooi, S., Skellern, D. and Cunningham, D. (1992). HANGMAN Gb/s network. *IEEE Network*, **6**(4), July, 10–18

Wu, C.L. and Feng, T.Y. (1980). On a class of multistage interconnection networks. *IEEE Transactions on Computers*, **29**(8), August, 694–702

Zitzen, W. (1990). Metropolitan area networks: taking LANs into the public network. *Telecommunications*, June, 53–60

Index

The index covers Chapters 1 to 15 and Appendix A. Index entries are to page numbers, those in italic referring to Figures and Tables. Alphabetical arrangement is word-by-word, where a group of letters followed by a space is filed before the same group of letters followed by a letter, for example, 'address resolution' will appear before 'addressing'.